Space Vehicle Design

Michael D. Griffin
James R. French

EDUCATION SERIES
J. S. PRZEMIENIECKI
Series Editor-in-Chief
Air Force Institute of Technology
Wright-Patterson Air Force Base, Ohio

Published by
American Institute of Aeronautics and Astronautics, Inc.
370 L'Enfant Promenade, SW, Washington, DC 20024-2518

American Institute of Aeronautics and Astronautics, Inc., Washington, DC

Library of Congress Cataloging in Publication Data

Griffin, Michael D. (Michael Douglas). 1949-
  Space vehicle design/Michael D. Griffin, James R. French.

p. cm. — (AIAA education series)
1. Space vehicles—Design and construction. I. French, James R.
II. Title. III. Series.
TL875.G68   1991   629.47′1—dc20   90-23705
ISBN 0-930403-90-8

Fourth Printing

FOREWORD

The publication of *Space Vehicle Design* by Michael D. Griffin and James R. French satisfies an urgent need for a comprehensive text on space systems engineering. This new text provides both suitable material for senior-level courses in aerospace engineering and a useful reference for the practicing aerospace engineer. The text incorporates several different engineering disciplines that must be considered concurrently as a part of the integrated design process and optimization. It also gives an excellent description of the design process and its accompanying tradeoffs for subsystems such as propulsion, power sources, guidance and control, and communications.

The text starts with an overall description of the basic mission considerations for spacecraft design, including space environment, astrodynamics, and atmospheric re-entry. Then the various subsystems are discussed, and in each case both the theoretical background and the current engineering practice are fully explained. Thus the reader is exposed to the overall systems-engineering process, with its attendant conflicting requirements of individual subsystems.

Space Vehicle Design reflects the authors' long experience with the spacecraft design process. It embodies a wealth of information for designers and research engineers alike. But most importantly, it provides the fundamental knowledge for the space systems engineer to evaluate the overall impact of candidate design concepts on the various component subsystems *and* the integrated system leading to the final design selection.

With the national commitment to space exploration, as evidenced by the continuing support of the Space Station and the National Aero-Space Plane programs, this new text on space system engineering will prove a timely service in support of future space activities.

J. S. PRZEMIENIECKI
Editor-in-Chief
AIAA Education Series

Texts Published in the AIAA Education Series

Re-Entry Vehicle Dynamics
  Frank J. Regan, 1984
Aerothermodynamics of Gas Turbine and Rocket Propulsion
  Gordon C. Oates, 1984
Aerothermodynamics of Aircraft Engine Components
  Gordon C. Oates, Editor, 1985
Fundamentals of Aircraft Combat Survivability Analysis and Design
  Robert E. Ball, 1985
Intake Aerodynamics
  J. Seddon and E. L. Goldsmith, 1985
Composite Materials for Aircraft Structures
  Brian C. Hoskins and Alan A. Baker, Editors, 1986
Gasdynamics: Theory and Applications
  George Emanuel, 1986
Aircraft Engine Design
  Jack D. Mattingly, William Heiser, and Daniel H. Daley, 1987
An Introduction to the Mathematics and Methods of Astrodynamics
  Richard H. Battin, 1987
Radar Electronic Warfare
  August Golden Jr., 1988
Advanced Classical Thermodynamics
  George Emanuel, 1988
Aerothermodynamics of Gas Turbine and Rocket Propulsion,
  Revised and Enlarged
  Gordon C. Oates, 1988
Re-Entry Aerodynamics
  Wilbur L. Hankey, 1988
Mechanical Reliability: Theory, Models and Applications
  B. S. Dhillon, 1988
Aircraft Landing Gear Design: Principles and Practices
  Norman S. Currey, 1988
Gust Loads on Aircraft: Concepts and Applications
  Frederic M. Hoblit, 1988
Aircraft Design: A Conceptual Approach
  Daniel P. Raymer, 1989
Boundary Layers
  A. D. Young, 1989
Aircraft Propulsion Systems Technology and Design
  Gordon C. Oates, Editor, 1989
Basic Helicopter Aerodynamics
  J. Seddon, 1990
Introduction to Mathematical Methods in Defense Analyses
  J. S. Przemieniecki, 1990
Space Vehicle Design
  Michael D. Griffin and James R. French, 1991

American Institute of Aeronautics and Astronautics, Inc., Washington, DC

TABLE OF CONTENTS

ix   Preface

1   **Chapter 1.   Introduction**
     1.1   Scope of Book
     1.2   Systems Engineering Process
     1.3   Requirements and Tradeoffs

15   **Chapter 2.   Mission Design**
     2.1   Basic Mission Classes

41   **Chapter 3.   Environment**
     3.1   Earth Environment
     3.2   Launch Environment
     3.3   Space Environment

85   **Chapter 4.   Astrodynamics**
     4.1   Fundamentals of Orbital Mechanics
     4.2   Non-Keplerian Motion
     4.3   Basic Orbital Maneuvers
     4.4   Interplanetary Transfer
     4.5   Perturbation Methods
     4.6   Orbital Rendezvous

163 **Chapter 5.   Propulsion**
     5.1   Rocket Propulsion Fundamentals
     5.2   Ascent Flight Mechanics
     5.3   Launch Vehicle Selection

231 **Chapter 6.   Atmospheric Entry**
     6.1   Fundamentals of Entry Flight Mechanics
     6.2   Fundamentals of Entry Heating
     6.3   Entry Vehicle Designs
     6.4   Aeroassisted Orbit Transfer

275 **Chapter 7.   Attitude Determination and Control**
     7.1   Introduction
     7.2   Basic Concepts
     7.3   Review of Rotational Dynamics
     7.4   Rigid Body Dynamics
     7.5   Space Vehicle Disturbance Torques
     7.6   Passive Attitude Control

7.7   Active Control
7.8   Attitude Determination
7.9   System Design Considerations

325   Chapter 8.   Configuration and Structural Design
8.1   Configuration Design
8.2   Mass Properties
8.3   Structural Loads
8.4   Large Structures
8.5   Materials

371   Chapter 9.   Thermal Control
9.1   Introduction
9.2   Conductive Heat Transfer
9.3   Convective Heat Transfer
9.4   Radiative Heat Transfer
9.5   Radiant Heat Transfer Between Surfaces
9.6   Spacecraft Thermal Modeling and Analysis
9.7   Thermal Environment
9.8   Spacecraft Energy Balance

395   Chapter 10.   Power
10.1   Power Subsystem Functions
10.2   Power Subsystem Evolution
10.3   Power Subsystem Design Drivers
10.4   Power Subsystem Elements
10.5   Design Practice
10.6   Batteries
10.7   Primary Power Source
10.8   Solar Arrays
10.9   Preliminary Solar Array Sizing
10.10   Radioisotope Thermoelectric Generators
10.11   Fuel Cells
10.12   Power Processor
10.13   Future Concepts

421   Chapter 11.   Telecommunications
11.1   Introduction
11.2   Command Subsystem
11.3   Hardware Redundancy
11.4   Autonomy
11.5   Command Subsystem Elements

459   Subject Index

ABOUT THE AUTHORS

Michael D. Griffin has participated in many different space missions while employed at Computer Science Corporation, the NASA Jet Propulsion Laboratory, the Applied Physics Laboratory of The Johns Hopkins University, the American Rocket Company, and the Strategic Defense Initiative Organization. As an adjunct professor at the University of Maryland, The Johns Hopkins University, and George Washington University, he has taught a variety of courses in aerospace engineering and applied mathematics. He is an AIAA Associate Fellow and recipient of its Space Systems Award, and a registered professional engineer. He holds six degrees in the fields of aerospace engineering, electrical engineering, physics, and business administration.

James R. French, after receiving a BSME from MIT in 1958, worked at the Rocketdyne Division of Rockwell International on the development and testing of the H-1, F-1 and J-2 engines for the Apollo/Saturn launch vehicles, and at TRW on the Apollo Lunar Module descent engine. While with the NASA Jet Propulsion Laboratory from 1967 to 1986, he participated in the Mariner, Viking, Voyager, and SP-100 programs and many advanced-mission studies. During 1986 and 1987 he was VP-Engineering of the American Rocket Company. Since 1987 he has been in private practice as a consultant in space systems engineering . An AIAA Associate Fellow, he has been a member of several AIAA Technical Committees, chaired the AIAA Space Systems Technical Committee, and has received the Shuttle Flag Award.

PREFACE

The idea for this text originated in the early 1980s with a senior-level aerospace engineering course in Spacecraft Design, taught by one of us at the University of Maryland. It was then a very frustrating exercise to provide appropriate reference materials for the students. Space vehicle design being an extraordinarily diverse field, no one text—in fact, no small group of texts—was available to unify the many disciplines of spacecraft systems engineering. As a consequence, in 1983 we decided to collaborate on a unifying text. The structure and academic level of the book followed from our development of a professional seminar series in spacecraft design. To meet the needs of engineers and others attending the seminars, the original academic course notes were radically revised and greatly expanded; when complete, the notes formed the outline for the present textbook.

The book meets, we believe, the needs of an upper-level undergraduate or Master's-level graduate course in aerospace vehicle design, and should likewise prove useful at the professional level. In this regard, our text represents somewhat of a departure from the more conventional academic style; it generally omits first-principle derivations in favor of integrating results from many specialized technical fields as they pertain to vehicle design and engineering tradeoffs at the system level.

It has been a long and torturous path to publication. Writing the manuscript was the easy part; publication was much more difficult. In the mid-1980s various publishers (not AIAA) showed discomfort with a perceived low-volume, "niche" product and backed away from the commitment we wanted. Job changes and the authors' busy schedules forced additional delays. And despite all the time it has taken to obtain the finished product, we both see many changes and improvements we would have liked to have made—but that would doubtless be true no matter how long we had worked.

In any event, the job is done for now. To all who have begun conversations with us in the last several years with, "When is the book coming out?," here it is. We hope you find it worth the wait.

Michael D. Griffin
James R. French
November, 1990

1
INTRODUCTION

1.1 SCOPE OF BOOK

In this book we attempt to treat the major engineering specialty areas involved in space vehicle and mission design from the viewpoint of the systems engineer. To attain this breadth, the depth of coverage in each area is necessarily limited. This is not a book for the specialist in attitude control, propulsion, astrodynamics, heat transfer, structures, etc., who seeks to enhance his knowledge of his own area. It is a book for those who wish to see how their own specialty is incorporated into a final spacecraft design and for those who wish to add to their knowledge of other disciplines.

To this end we have subordinated our desires to include involved analyses, detailed discussions of design and fabrication methods, etc. Equations are rarely derived, and never when they would interfere with the flow of the text; however, we take pains to state the assumptions behind any equations used. We believe that the detailed developments appropriate to each specialty area are well covered in other texts or in the archive journals. We refer the reader to these works where appropriate. Our goal in this work is to show how the knowledge and constraints from various fields are synthesized at the overall system level to obtain a completed design.

We intend this book to be suitable as a text for use in a senior-level design course in a typical aerospace engineering curriculum. Very few students emerge from four years of schooling in engineering or physical science feeling comfortable with the larger arena in which they will practice their specialty. This is rarely their fault; academic work by its nature tends to concentrate on that which is known and done and to educate the student in such techniques. This it does very well, subject of course to the cooperation of the student. What is not taught is how to function in the face of the unknown, the uncertain, and the not-yet-done. This is where the practicing engineer or scientist must learn to synthesize his knowledge, to combine the specialized concepts he has learned in order to obtain a new and useful result. This does not seem to be a quality that is taught in school.

It is also our intention that this book be useful as a reference tool for the working engineer. With this in mind, we have included as much state-of-the-art data as was practicable in the various areas that we treat. Thus, although we discuss the methods by which, say, rocket vehicle performance is analyzed, we are under no illusion that analytical methods produce the final answers in all cases of interest. We therefore include much more data in tabular and graphic form on the actual performance and construction of various rocket vehicles. We follow the same philosophy for attitude control, guidance, power, telecommunications, and of the other specialty areas and systems discussed here. However, this is not a "cookbook" or a compendium of standard results that can be applied to every problem. No book nor course of instruction can serve as a solution manual to all engineering problems. In fact, we take as an article of faith that, in any interesting engineering work, one is paid to solve previously unsolved problems. The most that any text can do is to provide a guide to the fundamentals. This we have tried to do by providing both data and analytical results, with a chain of references leading to appropriate sources.

1.2 SYSTEMS ENGINEERING PROCESS

What Is Systems Engineering?

The responses to the question that is the title of this section are many and varied. To some who claim to practice systems engineering, the activity seems to mean maintaining detailed lists of vehicle components mass and the name, number, and pedigree of each conductor that crosses the boundary between any two subsystems. To others it means computer architecture and software with little or no attention to hardware. To still others it means sophisticated computer programs for management and decision making, and so on.

In the opinion of the authors, definitions such as these are too restrictive and limited. As with the fabled blind men describing the elephant, each perceives some element of fact, but none fully describes the beast. As an aid to understanding the purpose of this book, we offer the following definition:

Space systems engineering is the art and science of developing an operable system capable of meeting mission requirements within imposed constraints including (but not restricted to) mass, cost, and schedule.

Clearly, all of the concepts mentioned earlier plus many more play a part in such an activity. Some may feel that the definition is too broad. That, however, is precisely the point. Systems engineering, properly done, is perhaps the broadest of engineering disciplines. The space systems engineer has the responsibility of defining a system based on requirements and constraints and overseeing its creation from a variety of technologies and subsystems.

In such a complex environment, conflict is the order of the day. The effective resolution of such conflict in an effective and productive manner is the goal of systems engineering. For all our high technology and sophisticated analytical capability, the solution is not always clear. This, plus the fact that one is dealing with people as much as with hardware or software, accounts for the inclusion of the word "art" in our definition. There will come a time in any system development when educated human judgment and understanding will be worth more than any amount of computer output. This in no way demeans the importance of detailed analysis and the specialists who do it, but, applied without judgment or conducted in an atmosphere of preconception and prejudice, such an analysis can be a road to failure. This truth has been demonstrated more than once, unfortunately, in the history of both military and civilian technical developments. It is the task of the systems engineer to avoid these pitfalls and make the technical decisions that best serve the achievement of the goal outlined in our definition.

Systems Engineering Requirements

To perform the task, there are certain characteristics that, if not mandatory, are at least desirable in the systems engineer. These are presented and amplified in this section.

The systems engineer must have an understanding of the goals of the project. These may be scientific, military, or commercial. Whatever the case, it is not possible to do the best job without an understanding of where the effort is headed. Decisions made without full knowledge of the context are subject to error that would otherwise be avoided. Not only must the systems engineer understand the goals, but it is incumbent upon him to pass this information on to his team so that they too understand the purpose of the effort. Communication is a two-way path.

A broad comprehension of the technical issues is mandatory. It is not expected that the systems engineer be an expert in all disciplines. No one human can aspire to the full breadth and depth of knowledge required in all fields, nor is it expected. That is why there is a team of specialists available. However, to make proper use of the resources afforded by such a team, the systems engineer must be sufficiently conversant with each of the technical areas involved to comprehend the issues and information available and to make the proper decisions. It is imperative that any technical issue must be evaluated in the context of the full system, not just the one(s) most obviously involved. This can only be done if there is sufficient comprehension of all technologies. Ideally the systems engineer should be able to carry out first-cut analyses in most technical areas to verify the information provided and to provide a preliminary assessment of the effect of proposed or edicted changes. (This, as much as any single factor, is the reason for this book.)

Because of the natural bent of our university system to create specialists rather than generalists, especially in advanced degree programs, individuals with substantial educational credentials often exhibit great depth but lack breadth. A combination of success in a specialized area plus the academic credentials often results in such a person being assigned to higher positions that require a systems-oriented viewpoint. If the individual recognizes this and develops the proper broad perspective, this can work very well. If, however, this person maintains the narrow view, becoming a "specialist in systems engineering clothing," problems may be anticipated because of excessive concentration in some areas to the neglect of others. This is not to say that the job will not get done, but it will probably not be done as well or proceed as smoothly as it might otherwise.

Given that the systems engineer cannot do it all himself and requires the assistance of a team, it follows that an important characteristic of the systems engineer is the ability to make maximum use of the capabilities of others. Part of this involves the somewhat undefinable characteristic called "leadership." However one might define it, the manifestation of leadership in which we are interested involves obtaining the maximum productivity from the team. Again, this is a matter of degree. A team of capable people will usually produce a product even with poor leadership. However, the same team, properly led, is vastly more effective and producing a product is generally an enjoyable experience for all participants. This aspect of the systems engineering task is discussed in more detail later in this chapter.

The essence of the previous paragraph is the hackneyed word "teamwork." It is truly appropriate in this instance in that, if the design team does not function as a fully integrated team rather than as a group of individuals, then effectiveness is diminished. The systems engineer has as one of his duties fostering the team spirit.

In any complex system, there is normally more than one way to do a job. Often various requirements will conflict and capabilities will not match perfectly. Success requires compromise. Indeed it often seems that the essence of the task is engineering a series of compromises that get the task done. To some who feel that technical activities should be pure, free of compromise, and always have a clear answer, the real world of engineering, especially systems engineering, will come as a considerable revelation. Willingness to compromise (within reasonable limits) is a vital characteristic of the systems engineer.

The key ingredient in systems design and in engineering the compromises discussed earlier is sound engineering judgment. Not everything can be analyzed, sometimes because the data or tools are not available and sometimes because of time or funding limitations. Very often results are ambiguous or can only be understood in context. The judgment of the team and ultimately of the systems engineer must be the final decision mechanism in such cases.

In some degree, judgment is a characteristic with which one is born. However, to be meaningul its use must be grounded in both education and experience.

The Team

We have referred repeatedly to the team and its importance. This cannot be overemphasized. A multidiscipline team of competent people is the most powerful tool at the disposal of the systems engineer. The quality of the product is a direct reflection of the capability of the team and the quality of its leadership. Computers and other analytical tools can enhance the productivity of the team and make the task easier but cannot substitute for human judgment and knowledge.

The reason for having a team is simple: No single person can have the detailed knowledge in all discipline areas that is required to carry out a complex engineering task. The protean "mad scientist" of fiction who understands all technology to the point that he can carry out a complex project (e.g., a rocket to the moon) unaided is in fact pure fiction. This fact does not preclude people trying. The authors can point to any number of projects, nameless in this volume to protect us from the wrath of the guilty, that were in fact done as a "one-man show" to the extent that one individual tried, single-handedly, to integrate the inputs of the specialists rather than leading the team in a coordinated integration effort. Uniformly, the output is a system of excess complexity and cost and/or lesser capability than it might have been.

A properly run team is an example of synergy in that it is greater than the sum of its parts. If all of the same people were used but kept closeted, contacting only the systems engineer, they would obviously be no less intelligent than they are when they are part of a team. Yet experience shows that the well-run team outperforms a nonintegrated array of specialists. The authors attribute this to the vigorous interaction between team members and to the sharing of knowledge, viewpoints, and concerns that often cause a solution to surface that no individual would have conceived of working alone. Often this is serendipitous; the discussion of one problem may suggest a solution for some other concern that is apparently not related. Again this can only happen in a closely knit team with frequent interaction.

How frequently should the team meet? Reasonably frequent meetings are necessary to promote the concept and sense of a team. There is no cut-and-dried rule. Meetings should be sufficiently frequent to maintain momentum and reinforce a habit of attendance. On the other hand, they should not be so frequent as to become a bore or to fritter away valuable time. Except in rare instances, formal, full-team meetings should not be more frequent than once a week. Intervals greater than two weeks are generally not desirable because of the loss of momentum and habit. Of course there will be multiple individual and subgroup interactions on specific topics once the team is accustomed to working together.

As the leader of the team, there are certain responsibilities that the systems engineer carries. He must ensure that all members contribute. The difference in human personality can result in the meeting being dominated by extroverts to the exclusion of introverts who have as much to say but lack the aggressiveness to assert themselves. The systems engineer must ensure that every individual has his fair share of time and contributes. This may require some leading questions or pushing for an answer but is as much a part of the job as is suppressing the excess verbosity of another individual.

A phenomenon that often plagues team meetings is digression. With any reasonably large group of bright people, many topics will arise that are not germane to the topic at hand, and the group can easily be seduced into following the new tack and ignoring the real topic. It is the duty of the team leader to prevent excessive deviation from the topic and thus maintain focus. Of course, in a long, intense meeting, an occasional digression can be refreshing and ease the tension. This must be allowed with judgment and caution to prevent valuable time from being wasted.

Equally distressing is the tendency of some individuals to ramble at great length, repeating themselves and heaping on unnecessary detail. The leader must cut this off (with due concern for feelings and sensitivity) when, in his judgment, the point of useful return is past. In a similar vein, two or three individuals working fine details of a problem below the level of interest to the team should be directed to table the discussion for that time and set up a separate meeting. Again, it is a judgment call as to when the point of productivity is past.

1.3 REQUIREMENTS AND TRADEOFFS

As noted earlier, the goal of the process led by the systems engineer is to develop a system to meet the requirements of the project. How are these requirements derived? Rarely if ever are even top-level requirements edicted in a complete, detailed, and unequivocal form. They are actually the result of a complex, interactive process invoking a variety of factors that may not be obvious to the uninitiated. The following sections will discuss in general terms the derivation of requirements.

Top-Level Requirements

The basic goals and constraints of a given space mission will generally be defined by the user or customer for the resulting system. Such goals will usually be in terms of target and activity, e.g., "Orbit Mars and observe atmospheric phenomena with particular emphasis on . . ." or "Develop a geosynchronous communications satellite capable of carrying 24 transponders operating in . . . ." At the same time, various constraints may be levied such as project start date, launch date, total cost, first year cost, etc. The top-level system requirements will then be derived from these goals and constraints.

Inputs for development of top-level requirements may come from a variety of sources. For example, science-oriented missions will typically have associated a science working group (SWG) composed of specialists in the field. (Usually these individuals will not be potential investigators on the actual mission to prevent any possible conflict of interest.) This group will provide detailed definition of the science goals of the mission in terms of specific observations to be made, types of instruments, and sensors that might be used, etc. The SWG requirements and desires must be iterated against the constraints and capabilities that otherwise define the mission. This will often be the systems engineer's most difficult interface. The requirements of the various investigations are often in conflict with one another or with the realities of practical engineering. Scientific investigators in single-minded pursuit of a goal often tolerate compromise poorly. Development of an innovative mission and system design that satisfies as many requirements and desires as possible while simultaneously achieving a compromise of the conflicting ones is a major test for both engineering and diplomatic skills. Furthermore, the dragon will not remain dead but continues to revive as the mission and system design and the science payload become better defined.

Nonscientific missions usually have a similar source of inputs. This group may go by various names but can generically be referred to as a user working group. This group represents the user community in terms of needs and desires. As with science requirements, some of these requirements may be in conflict, and resolution and compromise will be required. In many cases, spacecraft may be single-purpose devices, e.g., a communications relay satellite. In such a case, the problems with resolution of conflicting requirements are greatly reduced.

The study team itself has the primary responsibility for the development of top-level requirements for the system by turning the mission goals and desires into engineering requirements on the spacecraft that can later be converted into specific numerical requirements. As always, this is a process involving compromise and iteration in order to come up with a realistic set of requirements. Interaction between various subsystem and technology areas is essential for understanding the impact of requirements on the complete system and minimizing expensive surprises later.

In some cases, particularly when the mission dictates operating at the limits of available technology, various expert technology advisory groups may contribute to the process. Such groups may provide current data or projections of probable direction and degree of development during the course of the project.

Functional Requirements

Once the top-level requirements are defined, the next step is to derive from them the functional requirements that define what the system and the subsystems of which it is composed must accomplish in order to carry out the mission. Functional requirements are derived by converting the top-level requirements into specific engineering requirements such as velocity change, orbit parameters, fields of view, pointing direction, pointing accuracy, and a variety of other considerations.

The derivation of the functional requirements must be done within the context of technical capability and the constraints of cost and schedule. This is a critical juncture in the project. Unthinking acceptance of unrealistic requirements on one subsystem or arbitrary assumption of technology availability can lead to major problems with schedule and/or cost. It is very easy to accept a number for, as an example, pointing accuracy without critical assessment of what the requirement may mean in terms of attitude control sensors and effectors, structural accuracy and rigidity, etc. Excessively tight requirements can drive costs up or delay a tight schedule. First the requirements should be evaluated as to necessity. Is this accuracy essential to the mission, or did someone simply pick a number that sounded good? Sometimes a tight requirement will be pushed in a deliberate effort to justify use of an exciting new technology. If the goal of the mission is to advance technology, this may be fine; if the goal is to obtain observational data for the lowest cost in minimum time, it is essential to reduce the requirement to an acceptable level.

The preliminary version of the functional requirements is based on the top-level requirements and preliminary assessment of capability. At this stage, detail will be limited in most areas, and a great many of the more detailed requirements will be determined (TBD).

The TBDs will be filled in with real numbers early in the design phase. As the design progresses, the functional requirements evolve and mature. It is important for the design team to drive toward early completion of the requirements and filling in the TBDs. Of course, some of the requirements will change as the design matures; however, the act of pushing for early determination helps to accelerate the achievement of stability in requirements.

An early freeze of functional requirements is desirable, some would even say vital, to program stability and cost control. Probably no single factor has been more to blame for cost and schedule overruns than changing requirements. This may happen at the top level or at the functional requirements level. In the former case, the systems engineer has little control, although it is his duty to point out to his management and sponsors the impact of changes. At the functional requirements level, the systems engineer has substantial control and should exercise it. Absolute inflexibility is, of course, highly undesirable, since circumstances change and some modifications to functional requirements are inevitable. On the other hand, a relaxed attitude allowing easy and casual change without adequate coordination and review is an invitation to disaster.

Functional Block Diagram

The functional block diagram (FBD) is a tool that many people equate with the practice of systems engineering. Indeed the FBD is a highly useful tool for visualizing relationships between elements of the system. It is applicable at all levels.

The FBD may be used to demonstrate the relationships of major mission/system elements; for example, such elements as spacecraft, ground tracking system, mission operations, facility, user, etc. At the next level, it might be used to indicate the interaction of major subsystems within a system. An example is a diagram showing the relationship between major spacecraft subsystems that comprise a spacecraft.

The basic concept can be carried to as low a level as desired. An FBD showing the relationships between the major assemblies within a subsystem, e.g., solar arrays, batteries, and conditioning and control electronics within the power subsystem, can be most useful. One must be careful not to push it too far, however. Although in principle the FBD could be carried to the point of showing relationships between individual components, this really is not useful. More to the point, it can be actively harmful. Remember that once the decision is made to create such documentation, it must be kept up to date. If not up to date, any piece of documentation becomes a source of misinformation that can lead to costly and possibly dangerous errors. Maintaining documentation accuracy can be a major task.

It is too easy in the modern era to be seduced into creating overly complex and unncessary paper systems. (The computer is particularly dangerous in this case with the multitude of software systems to "help the manager manage.") Once created, these paper systems seem to take on a life of their own and expand and propagate. Even with computerized concepts, huge amounts of time and money can be wasted in excessive documentation. The systems engineer should think through the documentation requirements for his activity and implement a plan to meet the requirements. Avoid unnecessary "bells and whistles" that sound great but do not contribute. They will exact their price later.

Tradeoff Analysis

Tradeoff analysis is the essence of mission and system design. The combination of requirements, desires, and capabilities that go into defining a mission and the system that accomplishes it rarely fit together smoothly. The goal of the system designer is to obtain the best compromise among these factors, to meet the requirements as thoroughly as possible, to accommodate various desires, and to do so within the technical, financial, and schedule resources available.

Much has been said and written about how to do tradeoffs at the system and subsystem level. At one time it was admittedly a "judgment call." Decisions were made through the application of some person's experience and intuition applied to the desires and requirements, the analysis, and the available test data. More recently, what has become virtually a new industry has arisen to "systemize" (some would say "legitimize") the process. Elaborate mathematical decision-theoretic analyses and the use of computers to implement them are commonplace. It is debatable whether better results are achieved in this fashion; without doubt, it has led to greater diffusion of responsibility for decisions. This can hardly be a virtue, since any engineer worthy of the name must be willing to stand behind his work. In the case of the systems engineer, his work consists of the decisions he makes.

What is sometimes overlooked is the fact that, even with the use of computer analyses, engineering decisions are still, at bottom, based on the judgment of individuals or groups who determine the weighting factors, figures of merit, and algorithms that go into the models. Although technical specialists in various subsystems provide the expertise in their particular areas, it is the responsibility of the systems engineer to ensure that all pertinent factors are included and properly weighted. This should not be construed as an argument against the use of computers or any other labor-saving device that allows a more detailed analysis to be done or a wider range of options to be explored. It is rather to point out that such means are only useful with the proper inputs and in the hands of one with the knowledge and understanding to evaluate the output intelligently.

It may be instructive to consider some examples of tradeoffs in which a systems engineer might become involved. Note that we do not give the answers per se, merely the problems and some of the considerations involved in solving them. As we have indicated, there is rarely only one right answer. The answer, a completed system design, will be specific to the circumstances.

Spacecraft propulsion trades. On-board spacecraft propulsion requirements vary widely, ranging from trajectory correction maneuvers of 100 or 200 m/s to orbit insertion burns requiring a change in velocity (ΔV) on the order of 1000–2000 m/s. Options for meeting these requirements may include solid propulsion, liquid monopropellant or bipropellant, or some form of electric propulsion. Some missions may employ a combination.

Solid motors have the virtue of being simple and reliable. The specific impulse (see Chapter 5) is not as high as for most bipropellant systems, but the mass ratio (preburn to postburn mass; again see Chapter 5) is usually better. If the mission requires a single large impulse, a solid may be the choice. On the other hand, relatively high acceleration is typical with such motors, which may not be acceptable for a delicate structure in a deployed configuration.

A requirement for multiple impulses usually dictates the use of a liquid-propulsion system. The choice of a monopropellant or bipropellant is not necessarily obvious, however. The specific impulse of monopropellants tends to be one-half to two-thirds that of bipropellants; on the other hand, a monopropellant system has half the number of valves and tanks and operates with a cooler thrust chamber. For a given total impulse the mass of monopropellant carried must be greater, but if launch mass capability allows this, the greater simplicity of a monopropellant system may win out even for large ΔV requirements. Often a solid rocket will provide the major velocity change, whereas a low thrust mono- or bipropellant system will provide thrust vector control during the solid burn as well as subsequent orbit maintenance and correction maneuvers.

Electric propulsion offers very low thrust and very high specific impulse. Obviously it is most attractive on vehicles that have considerable electric power available. Applications requiring continuous low thrust for long periods, very high impulse resolution (small "impulse bits"), or minimum propellant consumption may favor these systems. Some examples that have been identified are communications satellites in geosynchronous orbit (see Chapter 2), where long-period, low-impulse stationkeeping requirements exist, and comet rendezvous missions, where the total impulse needed exceeds that available with chemical propulsion systems.

Communications system trades. Telecommunications requirements are driven by the amount of information to be transmitted, the time available to do so, and the distance over which it must be sent. This is true regardless of whether the signal is a human voice or a stream of digital science data. Given the required data rate, the tradeoff devolves to one between antenna gain (which, if it is a parabolic dish, translates directly to size) and broadcast power. In our discussion, we assume that the antenna is a parabolic dish. For a given data rate and a specified maximum bit error rate with known range and power, the required antenna size is defined as a function of operating frequency.

Antenna size can easily become a problem, since packaging for launch may be difficult or impossible. Antennas that fold for launch and are deployed for operation in space may avoid the packaging difficulty but introduce cost and reliability problems. Also, such antennas are of necessity usually rather flexible, which, for large sizes, may result in rather poor figure control. Without good figure control, the potential gain of a large antenna cannot be realized. Larger antennas have other problems as well. Increased gain (with any antenna) implies a reduced beamwidth that results in a requirement for more accurate antenna and/or spacecraft pointing knowledge and stability. This can reverberate through the system, often causing overall spacecraft cost and complexity to increase. Orientation accuracy for many spacecraft is driven by the requirements of the communications system.

Higher broadcast power is another alternative, but it will naturally have a significant effect on the power subsystem, driving up mass and/or solar array size. If flight-qualified amplifiers of adequate power do not exist, expensive development and qualification of new systems must be initiated.

Use of higher frequencies (e.g., X band as opposed to S band) allows increased data rates for a given antenna size and power, but, since the effective gain of the dish is higher at higher frequencies, again results in a requirement for increased pointing accuracy. Also, if communication with ground stations must be guaranteed, the use of X band can become a problem, since these frequencies may be effectively unusable when it is raining at the tracking station.

In the final analysis, the solution may not lie within the communications system at all. More sophisticated on-board processing or data encoding can reduce the amount of data that needs to be transmitted to achieve the same information transfer (or reduce the bit error rate) to a point compatible with constraints on power, mass, antenna size, and frequency. Of course, this alternative is not free either. More computational capability will be required, and careful (e.g., expensive) prelaunch analysis must be done to ensure that the data are not unacceptably degraded in the process. The cost of developing and qualifying the software for on-board processing is also a factor.

Power system trades. Spacecraft power sources to date have been limited to choices between solar photovoltaic, isotope-heated thermoelectric, and chemical (batteries or fuel cells) sources. Generally speaking, batteries or fuel cells are acceptable as sole power sources only for short-duration missions, measured in terms of days or at most a few weeks. Batteries in particular are restricted to the shorter end of the scale because of limited efficiency and unfavorable power-to-mass ratio. Fuel cells are much more efficient but are more complex. They have the advantage of producing potable water, which can be an advantage for manned missions.

Solar photovoltaic arrays have powered the majority of spacecraft to date. The simplicity and reliability of these devices make them most attractive. They can be used as close to the sun as the orbit of Mercury, although careful attention to thermal control is required. New technology in materials and fabrication will allow use even closer than Mercury orbit. Such arrays can provide power as far out as the inner regions of the asteroid belt. With concentrators, they may be useful as far from the sun as the orbit of Jupiter, although the complexity of deployable concentrators has limited interest in these devices so far. In the future, man-tended assembly or deployment in space may render such concepts more attractive. Batteries are usually required as auxiliary sources when solar arrays are used to provide overload power or power during maneuvers and eclipse periods.

For long missions far from the sun, or for missions requiring full operation during the night on a planetary surface, radioisotope thermoelectric generators (RTGs) have been the choice (as with Voyagers or the Viking landers). These units are long-lived and produce steady power in sunlight or darkness. They tend to be heavy, and the radiation produced can be a problem for electronics and science instruments, especially gamma ray spectrometers.

All of the sources mentioned earlier have difficulty when high power is desired. Deployable solar arrays in the 10–20 kW range have been built. Larger arrays have been proposed and are probably possible, but present a variety of problems in terms of drag, maneuverability, articulation control, interaction with spacecraft attitude control, etc. Solar thermal cycles using Rankine, Brayton, or Stirling cycles have been proposed in order to take advantage of the higher efficiency of these cycles as compared to that of solar cells; however, none has yet been flown. As mentioned, all solar power systems suffer from operational constraints due to eclipse periods and distance from the sun.

Nuclear power plants based on critical assemblies (reactors) offer great promise for the future, with their combination of high power at moderate weight for long periods. As will be discussed later, such units introduce substantial additional complexity into both mission and spacecraft design. In the final analysis, the spacecraft designer must trade off the characteristics and requirements of all systems to choose the best power source or combination of sources for his mission.

The preceding examples of tradeoff considerations are by no means all that will be encountered in the design of a spacecraft system. They are simply examples of high-level trades on major engineering subsystems. The process becomes more complex and convoluted as the system develops and occurs at every level in the design. Every technologist in every subsystem area will have his favored approach to sell, often with little regard to its system value. The task of the systems engineer is to evaluate the overall impact of these concepts on all of the other subsystems and the integrated system before making a selection.

Technology tradeoffs. A difficult area for decision making is that of new vs existing technology. The systems engineer is often caught between opposing forces. On one side is program and project management, who, in general, are primarily interested in getting the job done on schedule, within budget, and with minimum uncertainty. To this end, management tends to apply pressure to "do what you did last time." In other words, minimize the introduction of new concepts or technology with the attendant risk and uncertainty.

On the other side is the host of technical specialists associated with the various spacecraft subsystems. These people are more likely to be interested in the leading edge of technology in their field and to have very little interest in flying the "same old thing" again, particularly if several years have elapsed with attendant technical development.

The dichotomy here is real and the decision may be of profound significance. To maximize capability, remain competitive, encourage new development, etc., it is clearly desirable to bring new technology into play as early as possible. Yet one must avoid being seduced by promise or potential that is not yet real. It is almost axiomatic that any project that pushes too far beyond current technology in too many areas will, even if ultimately successful, come in behind schedule and/or over budget. Examples are legion, with the Space Shuttle as one prominent instance.

In a properly managed program it will be the lot of the systems engineer to, if not make the technology decision, at least make recommendations to management so that they can decide properly.

Many concerns must be considered in making technology decisions. Some of these will be discussed in the remainder of this section.

One of the first questions to be addressed is the basic one: "Will the existing technology do the job?" If well-known technology as embodied in existing designs and hardware will do everything required by a comfortable margin, then there is little incentive for management to do something new just because it is the latest thing. On the other hand, if the task mandates the use of new technology if it is to be accomplished at all, the decision is made for you. It then becomes the task of the systems engineer to work with management to accurately define the cost/schedule impacts and risks that may be involved as well as the impact on the total system.

Cost impacts of incorporation of new technology are highly variable. Savings may be realized because of higher efficiency, lower mass, lower volume, or all of the above. These effects can propagate through the entire system, reducing structure mass, power demands, etc. Note that changes like this only reduce cost if the entire system is being designed. If the spacecraft in question is one more in a long series and other subsystems are already designed (or even already built), then full realization of these potential advantages would require redesign of most of the other subsystems, becoming in effect a new system design and actually increasing overall costs.

This example points to a major risk involved with the introduction of new technology and emphasizes the need for the systems engineer to constantly keep in view the complete system and how changes in one subsystem may propagate in unforeseen ways. A subsystems engineer might appear and propose introduction of a new technology item in his subsystem after design is well advanced. The advantages cited might be higher efficiency, greater capability, or just the fact that it is the latest technology. It will probably be argued that the cost increase within the subsystem will be small or nonexistent. The subsystems engineer's interest (and the depth of his argument) will usually end at that point. The systems engineer must look beyond this and address other questions that include but are not necessarily limited to the following examples.

If ground support and test equipment already exist, will they be compatible with the new change, or will extensive modifications be required? Will new special test and handling requirements be levied (e.g., static electricity precautions, inert gas purge, etc)? Probably the most important question relates to impact on other subsystems. Is this change truly transparent, or will new requirements (e.g., new noise limits, special power requirements or restrictions, etc.) be imposed? Will the new item impact mission planning due to greater radiation sensitivity (or require shielding mass, which negates some of the purported advantages)? Failure to assess these issues and to coordinate with the other subsystems during the initial decision process can lead to very costly surprises later on after major commitments have been made.

Another area of concern is that of actual availability of components based on the new technology. Demonstrations in the laboratory, even fabrication of test components, do not correspond to actual availability. Even if commercial parts are available, the space qualified units required for most projects may not be. Thus, commitment to the new item could imply that your project has to pay the cost of setting up a production line and/or a space qualification program. This may not only be costly but incompatible with the schedule.

The availability question has two sides. It may be equally difficult to obtain older technology components if several years have passed since the previous buy. This is especially true in the rapidly evolving electronics component field. A case in point is the case of the Voyager spacecraft, in which it was desired to duplicate many electronic subsystems from the Viking Orbiter. To the dismay of project management, it was found that the manufacturer was terminating production of certain critical integrated circuits and was not interested in keeping the line open in order to produce the relatively small volume of parts needed. Because the redesign needed to incorporate new components would have been expensive and unacceptable in terms of schedule, the project ended up paying to maintain the production line for the required parts. Similarly, increasingly restrictive environmental rules or world politics may restrict availability of structural alloys readily available a few years earlier.

The listing of caveats could easily be construed as being anti-new technology unless there is no choice. This is by no means the case. All things being equal, one would almost always go with the new technology. Unfortunately, new technology is usually promoted in glowing terms with little consideration of possible negative aspects. Someone must accurately assess all sides of the issue in order to make the proper decision. That person is the systems engineer, with proper support of technical experts. It is noteworthy that excessive concern with the problems mentioned earlier can place management in a very "bearish" mood regarding new technology. This can result in adherence to old approaches long after newer, safer, more effective capabilities have become available and well proved. It is just as much the duty of the systems engineer to try to break this logjam as it is to keep from falling into the new technology traps already discussed. The challenge is to know the difference.

2
MISSION DESIGN

2.1 BASIC MISSION CLASSES

Space vehicle design requirements do not, except in very basic terms, have an existence that is independent of the mission to be performed. In fact, it is almost trivial to note that the type of mission to be flown and the performance requirements that are imposed define the spacecraft design that results. Just as a wide variety of aircraft exist to satisfy different broad classes of tasks, so may most space missions be categorized as belonging to one or the other general types of flight. Missions to near Earth orbit, for example, will impose fundamentally different design requirements than planetary exploration missions, no matter what the end goal in each case. In this chapter we examine a variety of different mission classes, with a view to the high-level considerations that are thus imposed on the vehicle design process.

Low Earth Orbit

Low Earth orbit (LEO) can be loosely defined as any orbit that is at less than geosynchronous altitude. By far the majority of missions flown to date have been to LEO, and it is probable that this trend will continue. A great variety of missions have been flown. These include flight tests, Earth observations for scientific, military, and various utilitarian purposes, and observations of local or deep space phenomena. Future missions can be expected to have similar goals plus the addition of new classes for purely commercial purposes. Examples in this latter category may include space-processing tasks and Earth resources surveys.

Flight tests. In the early days of orbital flight, every mission was in some sense a flight test, regardless of its primary goals, simply because of the uncertainty in technology and procedures. With increasing technical and operational maturity, however, many missions have become essentially routine. In such cases, flight tests are conducted only for qualification of new vehicles, systems, or techniques.

Flight tests in general are characterized by extensive instrumentation packages devoted to checking vehicle or system performance. Mission profiles are often more complex than an operational mission because of the desire to verify as many modes of operation as possible. There is a close analogy with aircraft flight testing, where no real payload is carried and the performance envelope is explored to extremes that are not

15

expected to be encountered under ordinary conditions. An important difference arises in that aircraft testing will involve many hours of operation over many flights, probably with a number of test units. Space systems, on the other hand, are usually restricted to one or very few test units and one flight per operational unit. The Space Shuttle provides the first instance of multiple flight tests of the same unit. Even in this case, the number of such flights was very low by aircraft standards. It is interesting to recall that Apollo 11, the first lunar landing mission, was only the fifth manned flight using the command module, the third to use a lunar module, and in fact only the 21st U.S. manned mission. One can hardly imagine, for example, Lindbergh having flown the Atlantic on the basis of such limited experience.

Because of the limited number of flight tests usually allowed for space systems, it is essential that a maximum value be obtained from each one. Not only must the mission profile be designed for the fullest possible exercise of the system, but the instrumentation package must provide the maximum return. LEO offers an excellent environment for test missions. The time to reach orbit is short, the energy expenditure is as low as possible for a space mission, communication is nearly instantaneous, and many hours of flight operation may be accumulated by a single launch to orbit.

As indicated earlier, the Apollo manned lunar program is an excellent example of this type of testing. The various vehicles and procedures were put through a series of unmanned and manned exercises in LEO prior to lunar orbit testing and the lunar landing. Even the unmanned first flight of the Saturn 5/Apollo Command Service Module (CSM) illustrates the philosophy of striving for maximum return on each flight. This flight featured an "all-up" test of the three Saturn 5 stages, plus restart of the third stage in Earth orbit, as required for a lunar mission, followed by a re-entry test of the Apollo command module. Viewed as a daring (and spectacularly successful) gamble at the time, it is seen in retrospect that little if any additional program risk was incurred. If the first stage had failed, nothing would have been learned about the second and higher stages—exactly the situation if dummy upper stages had been used until a first stage of proven reliability had been obtained. And a failure in any higher stage would still have resulted in obtaining more information than would have been the case with dummy upper stages. Of course, the cost of all-up testing can be much higher if repeated failures are incurred. But even here, equipment costs must be traded off against manpower costs incurred when extra flights are included. Even if equipment costs alone are considered, one must note that, when testing upper stages, many perfectly good lower stages must be used to provide the correct flight environment.

The systematic flight-test program for Apollo, leading to a lunar landing after a series of manned and unmanned flights, is apparent in Table 2.1. This table is not a complete summary of all Apollo flight tests. Between 1961 and 1966 some 10 Saturn 1 flights were conducted, of which 3 were used to launch the Pegasus series of scientific missions. Also, two pad-abort and four high-altitude tests of the Apollo launch escape system were conducted during this period. However, only "boilerplate" versions of the Apollo spacecraft were used for these missions, and only the first stage of the Saturn 1 was ever employed for a manned flight, and even then its use was not crucial to the program. Table 2.1 summarizes the tests conducted involving major use of flight hardware.

As may be seen in Table 2.1, one class of flight test that does not actually require injection into orbit is entry vehicle testing. There is seldom any advantage to long-term orbital flight for such tests. The entry must be flown in some approximation of real time, and an instrumented range is often desired. Therefore, such tests are usually suborbital ballistic lobs with the goal of placing the entry vehicle on some desired trajectory. Propulsion may be applied on the descending leg in order to achieve high entry velocity on a relatively short flight. This was, in fact, done on the previously mentioned unmanned Apollo test flights in order to simulate lunar return conditions. Note that such flight tests may not be required to match precisely the geometry and velocity of a "real-life" mission. If the main parameter of interest is, for example, heat flux into the shield, this

Table 2.1 Summary of Apollo test missions

Date	Mission	Comments
Feb. 26, 1966	AS-201	Saturn 1B first flight. Suborbital mission testing command module entry systems at Earth orbital speeds. Partial success due to loss of data.
Aug. 25, 1966	AS-202	Successful repeat of AS-201.
July 5, 1966	AS-203	Orbital checkout of S-4B stage. No payload.
Nov. 9, 1967	AS-501 (Apollo 4)	Saturn 5 first flight. Test of Apollo SPS restart capability and re-entry performance at lunar return speeds.
Jan. 22, 1968	AS-204 (Apollo 5)	Earth orbit test of lunar module descent and ascent engines.
April 4, 1968	AS-502 (Apollo 6)	Repeat of Apollo 4. Third stage failed to restart. SPS engines used for high-speed re-entry tests.
Oct. 11, 1968	AS-205 (Apollo 7)	First manned Apollo flight. Eleven-day checkout of CSM systems.
Dec. 21, 1968	AS-503 (Apollo 8)	First manned lunar orbital flight. Third flight of Saturn 5.
March 3, 1969	AS-504 (Apollo 9)	Earth orbital checkout of lunar module and CSM/LM rendezvous procedures.
May 18, 1969	AS-505 (Apollo 10)	Lunar landing rehearsal; test of all systems and procedures except landing.
July 16, 1969	AS-506 (Apollo 11)	First manned lunar landing. Sixth Saturn 5 flight, fifth manned Apollo flight, third use of lunar module.

may be achieved at lower velocity by flying a lower-altitude profile than would be the case for the actual mission.

Entry flight tests are often performed in the Earth's atmosphere for the purpose of simulating a planetary entry. Typically, it is impossible to simulate the complete entry profile because of atmospheric and other differences; however, critical segments may be simulated by careful selection of parameters. The Viking Mars entry system and the Galileo probe entry system were both tested in this way. The former used a rocket-boosted ballistic shot launched from a balloon, whereas the latter involved a parachute drop from a balloon to study parachute deployment dynamics.

Launch vehicle tests usually involve flying the mission profile while carrying a dummy payload. In some cases it is possible to minimize range and operational costs by flying a lofted trajectory that does not go full range or into orbit. For example, propulsion performance, staging, and guidance and control for an orbital vehicle can be demonstrated on a suborbital, high-angle, ICBM-like flight.

Earth observation.   Earth observation missions cover the full gamut from purely scientific to completely utilitarian. Both extremes may be concerned with observations of the surface, the atmosphere, the magnetosphere, or the interior of the planet, and of the interactions of these entities among themselves or with their solar system environment.

Missions concerned with direct observation of the surface and atmosphere are generally placed in low circular orbits in order to minimize the observation distance. Selecting an orbit altitude is generally a compromise between field of view, ground track spacing, observational swath width, and the need to maintain orbit stability against atmospheric drag without overly frequent propulsive corrections or premature mission termination. In some cases orbital period may be a factor because of the need for synchronization with a station or event on the surface. In other cases the orbital period may be required to be such that an integral number of orbits occur in a day. This is particularly the case with navigation satellites.

Orbital inclination is usually driven by a desire to cover specific latitudes, sometimes compromised by launch vehicle and launch site azimuth constraints. For full global coverage, polar or near-polar orbits are required. Military observation satellites make frequent use of such orbits, often in conjunction with orbit altitudes chosen to produce a period that is a convenient fraction of the day or week, thus producing very regular coverage of the globe. In many cases it is desired to make all observations or photographs at the same local sun angle or time (e.g., under conditions that obtain locally at, say, 1500 h). Orbital precession effects due to the perturbing influence of Earth's equatorial bulge may be utilized to provide this capability. A near-polar, slightly retrograde orbit with the proper altitude will precess at the same angular rate as the Earth revolves about the sun, thus maintaining constant sun angle throughout the year (see Chapter 4).

The LEO missions having the most impact on everyday life are the weather satellites. Low-altitude satellites provide close-up observations, which, in conjunction with global coverage by spacecraft in high orbit, provide the basis for our modern weather forecasting and reporting system. Such spacecraft are placed in the previously mentioned sun-synchronous orbits of sufficient altitude for long-term stability. The RCA-built Television and Infrared Observation Satellite (Tiros) series has dominated this field for years, undergoing very substantial technical evolution in that time. These satellites are operated by the National Oceanic and Atmospheric Administration (NOAA). The Department of Defense operates similar satellites under a program called the Defense Meteorological Satellite Program.

Ocean survey satellites, of which Seasat was an early example, have requirements similar to those of the weather satellites. All of these vehicles aim most of their instruments toward the region directly beneath the spacecraft or near its ground track. Such spacecraft are often referred to as "nadir-pointed."

Many military missions flown for observational purposes are similar in general requirements and characteristics to those discussed earlier. Specific requirements may be quite different, being driven by particular payload and target considerations.

Missions dedicated to observation of the magnetic field, radiation belts, etc., will usually tend to be in elliptical orbits because of the desire to map the given phenomena in terms of distance from the Earth as well as latitude. For this reason, substantial orbital eccentricity and a variety of orbital inclinations may be desired. Requirements by the payload range from simple sensor operation without regard to direction to tracking particular points or to scanning various regions.

Many satellites require elliptic orbits for other reasons. It may be desired to operate at very low altitudes either to sample the upper atmosphere (as with the Atmospheric Explorer series) or to get as close as possible to a particular point on the Earth for high resolution. In such cases, higher ellipticity is required to obtain orbit stability, since a circular orbit at the desired periapsis altitude might last only a few hours.

Space observation.   Space observation has come into full flower with our ability to place advanced scientific payloads in orbit. Gone are the days when the astronomer was restricted essentially to the visible spectrum. From Earth orbit we can examine space and the bodies contained therein across the full spectral range and with resolution no longer severely limited by the atmosphere. (The Mount Palomar telescope has a diffraction-limited resolving power some 20 times that realized in practice because of atmospheric turbulence.) This type of observation took its first steps with balloons and sounding rockets, but came to full maturity with orbital vehicles.

Predictably, our sun was one of the first objects to be studied with space-based instruments, and interest in the subject continues unabated. Spacecraft have ranged from the Orbiting Solar Observatory built by Ball Aerospace to the impressive array of solar observation equipment that was carried on the manned Skylab mission. Orbits are generally characterized by the desire that they be high enough that drag and atmospheric effects can be ignored. Inclination is generally not critical, although in some cases it may be desired to orbit in the ecliptic plane. If features on the sun itself are to be studied, fairly accurate pointing requirements are necessary, since the solar disk subsumes only 0.5 deg as seen from Earth.

Many space observation satellites are concerned with mapping the sky in various wavelengths, looking for specific sources, and/or the universal background. Satellites have been flown to study spectral regimes from gamma radiation down to infrared wavelengths so low that the detectors are cooled to near absolute zero to allow them to function. An excellent example here is the highly successful International Infrared Astronomical Telescope (IRAS) spacecraft, with liquid helium at 4.2 K used for cooling. In the x-ray band, the High Energy Astronomical Observatories (HEAO-2) spacecraft succeeded in producing the first high-resolution (comparable to ground-based optical telescopes) pictures of the sky and various sources at these wavelengths. Although most such work has concentrated on stellar and galactic sources, there has recently been some interest in applying such observations to bodies in our solar system, e.g., ultraviolet observations of Jupiter or infrared observations of the asteroids.

The Hubble Space Telescope represents the first space analog of a full-fledged Earth-based observatory. This device, a sizeable optical system even by ground-based standards, will have the capability for deep space and planetary observations of various types. Periodic servicing by the Shuttle and occasional retrieval for modifications and overhaul on the ground make this the closest thing yet to a permanent observatory in space. Observations by the Space Telescope will extend man's reach to previously unknown depths of space; however, it operates chiefly in the visible band, and so smaller, more specialized observatories will still be needed for coverage of gamma, x-ray, and infrared wavelengths.

Radio astronomers also suffer from the attenuating effects of the atmosphere in certain bands, as well as limits on resolution due to the impracticality of large, ground-based dish antennas. Although so far unrealized, there is great potential for radio astronomy observations from space. Antennas can be larger, lighter, and more readily steerable. Moreover, the use of extremely high precision atomic clocks allows signals from many different antennas to be combined coherently, resulting in the possibility of space-based antenna apertures of almost unlimited size. Radio observations with such antennas could eventually be made to a precision exceeding even the best optical measurements.

Space observatories are precision instruments featuring severe constraints on structural rigidity and stability, internally generated noise and disturbances, pointing accuracy and stability, etc. Operation is usually complicated by the need to avoid directly looking at the sun or even the Earth. Orbit requirements are not generally severe but may be constrained by the need for Shuttle accessibility while at the same time avoiding unacceptable atmospheric effects.

Space-processing payloads. As discussed in Chapter 3, the space environment offers certain unique features that are impossible or difficult, and thus extremely expensive, to reproduce on the surface of a planet. Chief among these are weightlessness (not the same as absence of gravity; tidal forces will still exist) and nearly unlimited access to hard vacuum. These factors offer the possibility of manufacturing in space many items that cannot easily be produced on the ground. Examples that have been considered include large, essentially perfect crystals for the semiconductor industry, various types of pharmaceuticals, and alloys of metals, which, because of their different densities, are essentially immiscible on Earth.

Space-processing payloads to date have been small and experimental in nature. Such payloads have flown on several Soviet missions and on U.S. missions on sounding rockets, Skylab, and the Shuttle. The advent of the Shuttle, with its more routine access to LEO, has resulted in substantial increases in the number of experiments being planned and flown. The Shuttle environment has made it possible for such experiments to be substantially less constrained by spacecraft design considerations than in the past. Furthermore, it is now possible for a "payload specialist" from the sponsoring organization to fly as a Shuttle crew member with only minimal training. It seems quite probable that, in the not too distant future, much of the traffic to and from LEO will consist of launching raw materials and returning products.

Processing stations will evolve into Shuttle-deployed free flyers in order to achieve the efficiency of continuous operation and tighter control over the environment (important for many manufacturing processes) than would be possible in the multiuser Shuttle environment. Such stations would require periodic replenishment of feedstock and removal of the products. This might be accomplished with the Shuttle or other vehicles as dictated by economics and the current state of the art. In any case, it introduces a concept previously seldom considered in spacecraft design: the transport and handling of bulk cargo.

Autonomy, low recurrent cost, and reliability will probably be the hallmarks of such delivery systems. The Soviet "Progress" series of resupply vehicles used in the Salyut and Mir space station programs may be viewed as early attempts in the design of vehicles of this type. However, the Progress vehicles still depend on the station crew to effect most of the cargo transfer (liquid fuel is piped essentially without crew involvement). It may be desirable for economic reasons to have future resupply operations of this nature carried out by unmanned vehicles. This will add some interesting challenges to the design of spacecraft systems. It seems certain that there will be a strong and growing need for robotics technology and manufacturing methods in astronautics.

In the longer term, the high-energy aspects of the space environment may be as significant as the availability of hard vacuum and $0\,g$. The sun produces about $1400\,\text{W/m}^2$ at Earth, and this power is essentially uninterrupted for many orbits of possible future interest. The advance of solar energy collection and storage technology cannot fail to have an impact on the economic feasibility of orbital manufacturing operations. In this same vein, it is also clear that the requirement to supply raw material from Earth for space manufacturing processes is a tremendous economic burden on the viability of the total system. Again, it seems certain that, in the long term, development of unmanned freighter vehicles capable of returning lunar and/or asteroid materials to Earth orbit will be undertaken. With the advent of this technology, and the use of solar energy, the economic advantage in many manufacturing operations could fall to products manufactured in geosynchronous or other high Earth orbits.

Geosynchronous Earth Orbit

Geosynchronous Earth orbit (GEO), and particularly the specific geosynchronous orbit known as "geostationary," is some of the most valuable "property" in space. The brilliance of Arthur Clarke's foresight in suggesting the use of communications satellites in GEO has been amply demonstrated. However, in addition to comsats, weather and scientific spacecraft occupy many slots in GEO.

As the name implies, a spacecraft in GEO is moving in synchrony with the Earth; i.e., the orbit period is that of Earth's day, 24 h. This does not imply that the satellite appears in a fixed position in the sky from the ground, however. Only in the special case of a 24-h circular equatorial orbit will the satellite appear to hover in one spot over the Earth. Other synchronous orbits will produce ground tracks with average locations that remain over a fixed point; however, there may be considerable variation from this average during the 24-h period. The special case of the 24-h circular equatorial orbit is properly referred to as "geostationary."

A 24-h circular orbit with nonzero inclination will appear from the ground to describe a nodding motion in the sky; that is, it will travel north and south each day along the same line of longitude, crossing the equator every 12 h. The latitude excursion will, of course, be equal to the orbital inclination. If the orbit is equatorial and has a 24-h period but is not exactly circular, it will appear to oscillate along the equator, crossing back and forth through lines of longitude. If the orbit is both noncircular and of nonzero inclination (the usual case, to a slight extent, due to various injection and stationkeeping errors), the spacecraft will appear to describe a figure eight in the sky, oscillating through both latitude and longitude about its average point on the equator. If the orbit is highly inclined or highly elliptical, then the figure eight will become badly distorted. In all cases, however, a true 24-h orbit will appear over the same point on Earth at the same time each day. An orbit with a slightly different period will have a slow, permanent drift across the sky as seen from the ground. Such slightly nonsynchronous orbits are used to move spacecraft from one point in GEO to another by means of minor trajectory corrections.

It is also interesting to consider very high orbits which are not synchronous but which have periods that are simply related to a 24-h day. Examples are the 12-h and 48-h orbits. Of interest are the orbits used by the Soviet Molniya spacecraft for communications relay. Much of the Soviet Union lies at very high latitudes, areas that are poorly served by geostationary comsats. The Molniya spacecraft use highly inclined, highly elliptical orbits with 12-h periods that place them, at the high point of their arc, over the Soviet Union twice each day for long periods. Minimum time is spent over the unused southern latitudes. While in view, communications coverage is good, and these orbits are easily reached from the high-latitude launch sites possessed by the Soviet Union. The disadvantage, of course, is that some form of antenna tracking control is required.

The utility of the geostationary or very nearly geostationary orbit is of course that a communications satellite in such an orbit is always over the same point on the ground, thus greatly simplifying antenna tracking and ground-space-ground relay procedures. Nonetheless, as long as the spacecraft drift is not so severe as to take it out of sight of a desired relay point, antenna tracking control is reasonably simple and is not a severe operational constraint, so that near-geostationary orbits are also quite valuable. The same feature is also important with weather satellites; it is generally desired that a given satellite be able to have essentially continuous coverage of a given area on the ground, and it is equally desirable that ground antennas be readily able to find the satellite in the sky.

The economic value of such orbits was abundantly emphasized during the 1979 World Administrative Radio Conference (WARC-79), when large groups of underdeveloped nations, having little immediate prospect of using geostationary orbital slots, nonetheless successfully prosecuted their claims for reservations of these slots for future use. Of concern was the possibility that, by the time these nations were ready to use the appropriate technology, the geostationary orbit would be too crowded to admit further spacecraft. With present-day technology and political realities, this concern is somewhat valid. There are limits on the proximity within which individual satellites may be placed.

The first limitation is antenna beamwidth. With reasonably sized ground antennas, at frequencies now in use (mostly C band; see Chapter 12), the antenna beamwidth is about 3 deg. To prevent inadvertent commanding of the wrong satellite, international agreements limit geostationary satellite spacing to 3 deg. Competition for desirable spots among nations lying in similar longitude belts is becoming severe. A trend to higher frequencies and other improvements (receiver selectivity and the ability to reject signals not of one's own modulation method are factors here) is allowing a reduction to 2-deg spacing, which will alleviate but not eliminate the problem. Political problems also appear, in that each country wants its own autonomous satellite, rather than to be part of a communal platform, a step that could eliminate the problem of inadvertent commands by using a central controller.

There is also beginning to be a physical hazard. Older satellites have worn out, and without active stationkeeping will drift in orbit, posing a hazard to other spacecraft. Also, jettisoned launch stages and other hardware are in near-GEO orbits. All of this drifting hardware constitutes a hazard to operating systems, which is increasing due to the increasing size of newer systems. There is some evidence that one or two collisions have already occurred. Mission designers are sensitive to the problem, and procedures are often implemented, upon retiring a satellite from active use, to lift it out of geostationary orbit prior to shutdown.

Communications satellites. Of all the facets of space technology, the one that has most obviously affected the everyday life of the average citizen is the communications satellite, so much so that it is taken for granted. In the early 1950s a tightly scheduled plan involving helicopters and transatlantic aircraft was devised to transport films of the coronation of Queen Elizabeth II so that it could be seen on U.S. television the next day. In contrast, the 1981 wedding of Prince Charles was telecast live all over the world without so much as a comment on the fact of its possibility. Less spectacular but of even greater impact is the current ease and reliability of long distance business and private communication by satellite. Gone are the days of "putting in" a transcontinental or transoceanic phone call and waiting for the operator to call back hours later. Today, direct dialing to most developed countries is routine, and we are upset only when the echo-canceling does not work properly.

The communications satellites that have brought about this revolution are to the spacecraft designer quite paradoxical, in the sense that in many ways they are quite simple (we exclude, of course, the communications gear itself, which is increasingly capable of feats of signal handling and processing that are truly remarkable). Since, by definition, a communications satellite is always in communication with the ground, such vehicles have required very little in the way of autonomous operational capability. Problems can often be detected early and dealt with by direct ground command. Orbit placement and correction maneuvers can if desired be done in an essentially real-time, "fly-by-wire" mode. Most of the complexity (and much of the mass) is in the communications equipment, which is the raison d'être for these vehicles. Given the cost of placing a satellite in orbit and the immense commercial value of every channel, the tendency is to cram the absolute maximum of communications capacity into every vehicle. Lifespan and reliability are also important, and reliability is usually enhanced by simple designs.

The value of and demand for communications channels, together with the spacing problems discussed earlier, are driving vehicle design in the direction of larger, more complex multipurpose communications platforms. Indeed, economic reality is pushing us toward the very large stations originally envisioned by Clarke for the role, but with capabilities far exceeding anything imagined in those days of vacuum tubes, discrete circuit components, and point-to-point wiring. Also noteworthy is that comsats thus far have been unmanned. This trend will probably continue, although there may be some tendency, once very large GEO stations are built, to allow for temporary manned occupancy for maintenance or other purposes. Pioneering concepts assumed an essential role for man in a communications satellite; as Clarke has said, it was viewed as inconceivable (if it was considered at all) that large, complex circuits and systems could operate reliably and autonomously for years at a time.

A high degree of specialization is already developing in comsat systems, especially in carefully designed antenna patterns that service specific and often irregularly shaped regions on Earth. This trend can be expected to continue in the future. The large communications platforms discussed earlier will essentially (in terms of size, not complexity) be elaborate antenna farms with a variety of specialized antennas operating at different frequencies and aimed at a variety of areas on the Earth and at other satellites.

It will be no surprise that the military services operate comsat systems as well. In a number of cases, such as the later TRW Fltsatcom models, these vehicles have become quite elaborate with multiple functions and frequencies. Reliability and backup capability are especially important in these applications, as well as provision for secure communications. Of interest to the spacecraft design engineer is the growing trend toward "hardening" of these spacecraft. In the event of war, nuclear or conventional, preservation of communications capability becomes essential. Spacecraft generally are rather vulnerable to intense radiation pulses, whether from nuclear blasts in space (generating electromagnetic pulses as well) or laser radiation from the ground. The use of well-shielded electrical circuits and, where possible, fiber-optic circuits can be expected. There is, in fact, some evidence of Soviet "blinding" of U.S. observation satellites using ground-based lasers. Designers can also expect to see requirements for hardening spacecraft against blast and shrapnel from potential "killer" satellites.

Weather satellites.   Weather satellites in GEO are the perfect complement to the LEO vehicles discussed earlier. High-altitude observations can show cloud, thermal, and moisture patterns over roughly one-third of the globe at a glance. This provides the large-scale context for interpretation of the data from low-altitude satellites, aircraft, and surface observations.

Obviously, it is not necessary for a satellite to be in a geostationary or even a geosynchronous orbit to obtain a wide-area view. But, as discussed, it is still considered very convenient, and operationally desirable, for the spacecraft to stand still in the sky for purposes of continued observation, command, and control. Crowding of weather satellites does not present the problems associated with comsats, however, since entirely different frequency bands can be used for command and control purposes. The only real concern in this case is collision avoidance.

The Geostationary Operational Environmental Satellites (GOES) system designed by Hughes Aircraft Company is an excellent example of this type of satellite. Even though the purpose is different, many of the requirements of weather and communications satellites are similar, and the idea of combined functions, especially on larger platforms, may well become attractive in the Shuttle era.

Space observation.   To date, there has been relatively little deep space observation from GEO. Generally speaking, there has been little reason to go to this energetically expensive orbit for observations from deep space. There are some exceptions; the International Ultraviolet Explorer (IUE) observatory satellite used an elliptic geosynchronous orbit with a 24,300-km perigee altitude and a 47,300-km apogee altitude. This orbit allowed more viewing time of celestial objects with less interference from Earth's radiation belts than would have been the case for a circular orbit, while still allowing the spacecraft to be in continuous view of the Goddard Space Flight Center tracking stations. Also, at higher altitudes the Earth subtends a smaller arc and more of the sky is visible. This can be important for sensitive optical instruments, which often cannot be pointed within many degrees of bright objects like the sun, moon, or Earth, because of the degradation of observations resulting from leakage of stray light into the optics. As more sensitive observatories for different spectral bands proliferate, there may be a desire to place them as far as possible from the radio, thermal, and visible light noise emanating from Earth. This will certainly be important with the advent of large antenna arrays for radio astronomy. Gravity-gradient and atmospheric disturbances will need to be minimized in this case, and this will imply high orbits.

In this connection, an interesting possibility for the future is the Orbiting Deep Space Relay Satellite (ODSRS), which has been studied on various occasions under different names. This spacecraft, which would inherently be quite large, would serve as a replacement or supplement for the existing ground-based Deep Space Network (DSN). The DSN currently consists of large dish-antenna facilities in California, Australia, and Spain, with the placement chosen so as to enable continuous observation and tracking of interplanetary spacecraft irrespective of Earth's rotation. The ODSRS concept has several advantages. Long-term, continuous tracking of a spacecraft would be possible and would not be limited by Earth's rotation. Usage of higher frequencies would be possible, thus enhancing data rates and narrowing beamwidths. This in turn would allow spacecraft transmitters to use lower power. The atmosphere poses a significant problem to the use of extremely high frequencies from Earth-based antennas. Attenuation in some bands is quite high, and rain can obliterate a signal (X-band signals are attenuated by some 40 dB in the presence of rain). Furthermore, a space-borne receiver can be easily cooled to much lower temperatures than is possible on Earth, improving its signal-to-noise ratio. The ODSRS would receive incoming signals from deep space and relay them to ground at frequencies compatible with atmospheric passage. Between tracking assignments, it could have some utility as a radio telescope.

Spacecraft performing surveys of the atmosphere, radiation belts, magnetic field, etc., around the Earth may be in synchronous, subsynchronous, or supersynchronous orbits that may or may not be circular. This might be done so as to synchronize the spacecraft with some phenomena related to Earth's rotation, or simply to bring it over the same ground station each day for data transmission or command and control.

As our sophistication in orbit design grows and experimental or other requirements pose new challenges, more complex and subtle orbits involving various types of synchrony as well as perturbations and other phenomena will be seen. We have only scratched the surface in this fascinating area.

Lunar and Deep Space

Missions to the moon and beyond are often very similar to Earth orbital missions in terms of basic goals and methods. However, because of the higher energy requirements, long flight times, and infrequent launch opportunities for current propulsion systems, evolution of these missions from the basic to the more detailed and utilitarian type has been arrested compared to Earth orbital missions. In general, deep space missions fall into one of three categories: inner solar system targets, outer solar system targets, and solar orbital.

Inner planetary missions. The target bodies included in this category are those from Mercury to the inner reaches of the asteroid belt. The energy required to reach these extremes from Earth is roughly the same, a vis-viva energy of $30-40$ km^2/s^2 (see Chapter 4). Even though the region encompasses a variation in solar radiative and gravitational intensity of about 60, it can be said to be dominated by the sun. Within this range, it is feasible to design solar-powered spacecraft and to use solar orientation as a factor in thermal control. Flight times to the various targets are measured in months, rather than years, for most trajectory designs of interest.

As would be expected, our first efforts to explore another planet were directed toward the nearby moon. Indeed, the first crude efforts by both the U.S. and the USSR to fly by or even orbit the moon came only months after the first Earth orbiters. Needless to say, there were at first more failures than successes. The first U.S. Pioneer spacecraft were plagued with various problems and were only partly successful. Probably the scientific highlight of this period was the return of the first crude images of the unknown lunar farside by the Soviet Luna 3 spacecraft. The lunar program then settled into what might be considered the classic sequence of events in the exploration of a planetary body. The early Pioneer flybys were followed by the Ranger family, designed to use close-approach photography of a single site followed by destruction on impact. Reconnaisance orbiters, the Lunar Orbiter series, came next, followed by the Surveyor program of soft landers. Finally, manned exploration followed with the Apollo program.

Although omitting the hard landers, the Soviet program followed a similar path and was clearly building toward manned missions until a combination of technical problems and the spectacular Apollo successes terminated the effort. A number of notable successes were achieved, however. Luna 9 made a "soft" (actually a controlled crash, with cameras encased in a balsa wood sphere for survival) landing on the moon some months prior to Surveyor 1. The propaganda impact of this achievement was somewhat lessened by the early decoding and release of the returned pictures from Jodrell Bank Observatory, in England. Also, the Lunokhod series of unmanned craft managed to achieve both surface mobility and sample return, though not before the Apollo landings.

Exploration of the other inner planets, so far as it has gone, has followed essentially the scenario previously outlined. Both the U.S. and USSR have sent flyby and orbital missions to Venus and Mars. The Soviet Union has landed a series of Venera spacecraft on Venus (where the survival problems dwarf anything so far found outside the sun or Jupiter), and the U.S. has achieved two spectacularly successful Viking landings (also orbiters) on Mars. The asteroid belts have not so far been a target, although many mission concepts have been advanced.

The innermost planet, Mercury, has so far been the subject only of flybys and by only one spacecraft, Mariner 10. The use of a Venus gravity assist (see Chapter 4) to reach Mercury, plus the selection of a resonant solar orbit, allowed Mariner 10 to make three passes of the planet. This mission was one of the most astrodynamically complex missions yet flown, and one of the most successful. Mariner 10 provided our first good look at this small, dense, heavily cratered member of the solar system.

Table 2.2 summarizes the missions to date to the moon and the inner planets.

Outer planetary missions.   The outer planets, of which all but Pluto have been visited, have been the subject only of flyby missions. Pioneers 10 and 11 led the way, with Pioneer 10 flying by Jupiter and Pioneer 11 visiting both Jupiter and Saturn. These missions were followed by Voyagers 1 and 2, both of which have flown by both Jupiter and Saturn, surveying both the planets and many of their moons. The rings of Jupiter and several new satellites of Saturn were discovered. All four vehicles acquired suffi-

Table 2.2   Summary of lunar and inner planet missions

Date	Mission	Comments
Late 1950s	Luna	Early Soviet missions. First pictures of far side of Moon.
Late 1950s	Pioneer	Early U.S. missions to lunar vicinity.
Early 1960s	Luna	Continued Soviet missions. First unmanned lunar landing.
Early 1960s	Ranger	U.S. lunar impact missions. Detailed photos of surface.
1966–1968	Surveyor	U.S. lunar soft lander. Five successful landings.
1966–1968	Lunar Orbiter	U.S. photographic survey of Moon.
1968–1972	Apollo	U.S. manned lunar orbiters & landings. First manned landing.
1968	Zond	Soviet unmanned tests of a manned lunar swingby mission.
Late 1960s	Luna	Soviet unmanned lunar sample return.
Early 1970s	Lunakhod	Soviet unmanned teleoperated lunar rover.
1962 & 1965	Mariner 2 & 5	U.S. Venus flyby missions. Mariner 2 first planetary flyby.
1964 & 1969	Mariner 4, 6, 7	U.S. Mars flyby missions.
1971	Mariner 9	U.S. Mars orbiter. First planetary orbiter.
1973	Mariner 10	U.S. Venus/Mercury flyby.
1975	Viking 1 & 2	U.S. Mars orbiter/lander missions.
1990	Magellan	U.S. Venus radar mapper.
1960s, 1970s	Mars	Series of Soviet Mars orbiter/lander missions.
1970s, 1980s	Venera	Long running series of Soviet Venus featuring orbiters and landers.
1989	Phobos	Soviet Mars orbiter missions.

cient energy from the flybys to exceed solar escape velocity, becoming, in effect, mankind's first emissaries to the stars. The two Pioneers and Voyager 1 will not pass another solid body in the foreseeable future (barring the possibility of an unknown 10th planet or a "brown dwarf" star), but Voyager 2 carried out a Uranus encounter in 1986 and a Neptune flyby in 1989. Achievement of these goals is remarkable since the spacecraft has far exceeded its 4-yr design lifetime. Even though the instrumentation designed for Jupiter and Saturn is not optimal at the greater distances of Uranus and Neptune, excellent results were achieved.

It is interesting to note that the scientific value of the Pioneers and Voyager 1 did not end with their last encounter operation. Long distance tracking data on these spacecraft are being used to obtain information on the possibility, and potential location, of the suspected 10th planet of the solar system. Such expectations are due to the inability to reconcile the orbits of the outer planets, particularly Neptune, with the theoretical predictions including all known perturbations. Both Neptune and Pluto (somewhat fortuitously, it now seems) were discovered as a result of such observations. Tracking data from the Pioneers and Voyagers can return more data, and more accurate data, in a few years than in several centuries of planetary observations. Moreover, since these spacecraft are departing the solar system at an angle to the ecliptic, they provide data otherwise totally unobtainable.

By the logical sequence outlined previously, Jupiter should be the next target for an orbiter and an atmospheric probe. Such is in fact the case. The Galileo program is designed to implement these goals, as well as to achieve many successive flybys of the Jovian moons from its Jupiter orbit. Although delayed by many factors, most recently the Challenger accident, Galileo was launched in 1989 on a circuitous path involving a Venus flyby and two Earth flybys on route to Jupiter. This complexity is a result of cancellation of the high energy Centaur stage and substitution of a lower energy IUS.

Spacecraft that visit the outer planets cannot depend on solar energy for electrical power and heating. Use of concentrators might extend the range of useful solar power out to Jupiter, but at the cost of considerable complexity. The spacecraft that have flown to these regions, as well as those that are planned, depend on power obtained by radioactive decay processes. These power units, called in general radioisotope thermoelectric generators (RTGs) use banks of thermoelectric elements to convert the heat generated by radioisotope decay into electric power. The sun is no longer a factor at this point, and all heat required, for example, to keep propellants warm, must be supplied by electricity. On the positive side, surfaces designed to radiate heat at modest temperatures, such as electronics boxes, can do so in full sunlight, a convenience for the configuration designer that is not available inside the orbit of Mars.

The small bodies. Comets and asteroids, the small bodies of the solar system, tended to be ignored during the early phases of space exploration, although various mission possibilities were discussed. Although most of the scientific interest (and public attention) focuses on comets, the asteroids present a subject of great interest also. Not only are they of scientific interest, but, as we have discussed, some may offer great promise as sources of important raw materials for space fabrication and colonization projects.

The main belt asteroids are sufficiently distant from the sun that they are relatively difficult to reach in terms of energy and flight time. Except for the inner regions of the belt, solar power is not really practical. For example, an asteroid at a typical 2.8 AU distance from the sun suffers a decrease in solar energy by a factor of 8.84 compared with that available at the orbit of Earth. RTGs or, in the future, possibly full-scale nuclear reactors are required.

However, many asteroids have orbits that stray significantly from the main belt, some passing inside the orbit of Earth. These asteroids are generally in elliptic orbits, many of which are significantly inclined to the ecliptic plane. Orbits having high eccentricity and/or large inclinations are quite difficult, in terms of energy, to reach from Earth. However, a few of these bodies are in near-ecliptic orbits with low eccentricity and are the easiest extraterrestrial bodies to reach after the moon. In fact, if one includes the energy expenditure required for landing, some of these asteroids are easier to reach than the lunar surface. Clearly, these bodies have potential for future exploration and exploitation. Relatively few of these Earth-approaching asteroids are known as yet, but analysis indicates that there should be large numbers of them. More are being discovered all the time.

Comets in general are in quite eccentric orbits, often with very high inclination. Some orbits are so eccentric that it is debatable whether they are in fact closed orbits at all. In any case, the orbital periods, if the term is meaningful, are very large for such comets. Some comets are in much shorter but still highly eccentric orbits; the comet Halley, with a period of 76 yr, lies at the upper end of this short-period class. The shortest known cometary period is that of Encke, at 3.6 yr.

As stated, most comets are in high-inclination orbits, of which Halley's Comet is an extreme example, with an inclination of 160 deg. This means that it circles the sun in a retrograde direction at an angle of 20 deg to the ecliptic. With few exceptions, comet rendezvous (as distinct from intercept) is not possible using chemical propulsion. High-energy solar or nuclear-powered electric propulsion or solar sailing can, with reasonable technological advances, allow rendezvous with most comets.

Orbit design considerations. Although we will consider this topic in more detail in Chapter 4, the field of orbit and trajectory design for planetary missions is so rich in variety that an overview is appropriate at this point. Transfer trajectories to other planets are determined at the most basic level by the phasing of the launch and target planets. Simply put, both must be in the proper place at the proper time. This is not nearly as constraining as it may sound, particularly with modern computer-aided techniques. A wide variety of transfer orbits can often be found to match launch dates that are proper from other points of view, such as the availability of hardware and funding.

The conventional transfer trajectory is a solar orbit designed around an inferior conjunction (for inner planets) or opposition (for outer planets). Such orbits, although they do not possess the flexibility described earlier, are usually the best compromise of minimum energy and minimum flight time. These orbits typically travel an arc of somewhat less than 180 deg (type 1 transfer) or somewhat more than 180 deg (type 2 transfer) about the sun. A special case here is the classical two-impulse, minimum-energy Hohmann transfer. This trajectory is completely specified by noting that it involves a 180-deg arc between the launch and target planets and is tangent to both the departure and arrival orbits. However, the Hohmann orbit assumes coplanar circular orbits for the two planets, a condition that is in practice never met exactly. Since the final trajectory is rather sensitive to these assumptions, a true Hohmann transfer is not used. Furthermore, flight times using such a transfer would be unreasonably long for any planetary target outside the orbit of Mars. Ingenuity in orbit design or added booster power or both must be used to obtain acceptable mission durations for flights to the outer planets.

The expenditure of additional launch energy is the obvious approach to reducing flight times. This involves placing the apsis of the transfer orbit well beyond the target orbit, thus causing the vehicle to complete its transfer to the desired planet much more quickly. In the limit, this section can be made to appear nearly a straight line, but at great energy cost at both departure and arrival. A planetary transfer such as this is beyond our present capabilities.

The other extreme is to accept longer flight times to obtain minimum energy expenditure. In its simplest form this involves an orbit of 540 deg of arc. The vehicle flies to the target orbit (the target is elsewhere), back to the launch orbit (the launch planet is elsewhere), then finally back to the target. Such an evolution sometimes saves energy relative to shorter trajectories through more favorable nodal positioning or other factors. This gain must be traded off against other factors such as increased operations cost, budgeting of on-board consumables, failure risk, and utility of the science data.

A more complicated but more commonly used option involves application of a velocity change sometime during the solar orbit phase. This can be done propulsively or by a suitable target flyby (increasingly the method of choice) of a third body, or by some combination of these. The propulsive ΔV approach is simplest. A substantial impulse applied in deep space may, for example, allow an efficient change in orbital plane, thus reducing total energy requirements. A more exacting technique is to fly past another body in route and use the swingby to gain or lose energy (relative to the sun, not the planet providing the gravity assist). Mariner 10 used this technique at Venus to reach Mercury, and Pioneer 11 and the two Voyagers used it at Jupiter to reach Saturn. Voyager 2, of course, used a second gravity assist at Saturn to continue to Uranus. The Venus and Earth swingbys mentioned in conjunction with the Galileo mission supply both plane change and added energy.

The gravity-assist technique, now well established, was first used with Mariner 10. In fact, the only means of reaching Mercury with current launchers and a mass sufficient to allow injection into Mercury orbit with chemical propulsion is via a multirevolution transfer orbit with one or more Venus flybys to reduce the energy of the orbit at Mercury arrival to manageable levels. Of course, in planetary exploration, the additional time spent in doing swingbys is hardly a penalty; we have not yet reached the point where so much is known about any planet that an additional swingby is considered a waste of time. As noted, this is now a mature technique. It will be exploited to the fullest during the Galileo mission to Jupiter, where repeated pumping of the spacecraft orbit through gravity assists from its moons will be used to raise and lower the orbit and change its inclination. The orbit in fact will never be the same twice. These "tours" allow the maximum data collection about the planet and its satellites while permitting a thorough survey of the magnetic field and the space environment.

The final class of methods whereby difficult targets can be reached without excessive propulsive capability involves the use of the launch planet for gravity assist maneuvers. The spacecraft is initially launched into a solar orbit synchronized in order to intercept the launch planet again, usually after one full revolution of the planet, unless a midcourse V is applied. The subsequent flyby can be used to change the energy or inclination of the transfer orbit, or both. It is also possible to apply a propulsive ΔV during the flyby. Such mission profiles have been frequently studied as options for outer planetary missions, and, as discussed, are being applied to Galileo.

The orbits into which spacecraft are placed about a target planet are driven by substantially the same criteria as for spacecraft in Earth orbit. For instance, the Viking orbiters were placed in highly elliptical 24.6-h orbits (one Martian day) so that they would arrive over their respective lander vehicles at the same time each day to relay data. Future Mars geoscience mappers will utilize polar sun-synchronous orbits like those used by similar vehicles at Earth. A possibility for planetary orbiters is that, rather than being synchronized with anything at the target planet, they would be in an orbit with a period synchronized with Earth. For example, the spacecraft might be at periapsis each time a particular tracking station was in view.

Low-thrust planetary trajectories that are required for electric propulsion and solar sailing are quite different from the ballistic types described thus far, since the thrust is applied constantly over very long arcs in the trajectory. Such trajectories also may make use of planetary flybys to conserve energy or reduce mission duration. The most notable difference is at the departure and target planets. At the former, unless boosted by chemical rockets to escape velocity, the vehicle must spend months spiraling out of the planetary gravity field. In some cases this phase may be as long as the interplanetary flight time. At the target, the reverse occurs.

This situation results from the very low thrust-to-mass ratio of such systems. In one instance where solar-electric propulsion was proposed for a Mars Sample Return mission, it was found that the solar-electric vehicle did not have time to spiral down to an altitude compatible with the use of a chemically propelled sample carrier from the surface. To return to Earth, it had to start spiraling back out before reaching a reasonable rendezvous altitude. Higher thrust-to-mass ratios such as those offered by nuclear-electric propulsion or advanced solar sails would overcome this problem. Solar-electric propulsion and less capable solar sails are most satisfactory for missions not encountering a deep gravity well. Comet and asteroid missions and close or out-of-ecliptic solar missions are examples.

Advanced Mission Concepts

Thus we have dealt with mission design criteria and characteristics primarily for space missions that have flown or are planned for flight in the near future. In a sense, design tasks at all levels for these missions are known quantities. Though space flight has still not progressed to the level of routine airline-type operations, nonetheless, much experience has been accumulated since Sputnik 1, to the point where spacecraft design for many types of tasks can be very prosaic. In many areas, there is a well-established way to do things, and designs evolve only within narrow limits.

This is not true of missions that are very advanced by today's standards. Such missions include the development of large structures for solar power satellites or antenna farms, construction of permanent space stations, lunar and asteroid mining, propellant manufacture on other planets, and many other activities that can be envisioned only dimly at present. For these advanced concepts, the designer's imagination is still somewhat free to roam, limited only by established principles of sound engineering practice. In this section, we examine some of the possibilities for future space missions that have been advocated in recent years, with attention given to the mission and spacecraft design requirements they will pose.

Large space structures. Many of the advanced mission concepts that have surfaced have in common the element of requiring the deployment in Earth orbit of what are, by present standards, extremely large structures. Examples of such systems include solar power satellites, first conceived by Dr. Peter Glaser, and the large, centralized antenna platforms alluded to previously in connection with communications satellites. These structures will have one outstanding difference from Earth-based structures of similar size, viz., their extremely low mass. If erected in a 0-g environment, these platforms will never need to cope with the stresses of Earth's field and need to be designed only to offer sufficient rigidity for the task at hand. This fact alone will offer many opportunities for both success and failure in exploiting the capabilities of large space platforms.

Orbit selection for large space structures will in principle be guided by much the same criteria as for smaller systems; that is, the orbit design will be defined by the mission to be performed. However, the potentially extreme size of the vehicles involved will offer some new criteria for optimization. Thus, systems of large area and low mass will be highly susceptible to aerodynamic drag and will generally need to be in very high orbits in order to avoid excessive requirements for drag compensation propulsion. For such platforms, solar pressure can become the dominant orbital perturbation. Similarly, systems with very large mass will tend toward low orbits to minimize the expense of construction with materials ferried up from Earth. When the time comes that many large platforms are deployed in high Earth orbit, it may be expected that the use of lunar and asteroid materials for construction will become economically attractive. In terms of energy requirements, the moon is closer to geosychronous orbit than is the surface of the Earth. The consequences of this fact have been explored in a number of studies.

Other characteristics of expected large space systems have also received considerable analytical attention. As mentioned, structures such as very large antennas or solar power satellites will be of quite low mass for their size by Earth standards. Yet these structures, particularly antennas, require quite precise shape control to achieve their basic goals. On Earth, this requirement is basically met through the use of sufficient mass to provide the needed rigidity, a requirement that is not usually inconsistent with that for sufficient strength to allow the structure to support itself in Earth's gravitational field. As mentioned, in a 0-g environment this will not be the case. Very large structures of low mass will have very low characteristic frequencies of vibration and quite possibly very little damping at these frequencies. Thus, it has been expected that some form of active shape control will often be required, and much effort has been expended in defining the nature of such control schemes.

Translation control requires similar care. For example, it will hardly be sufficient to attach a single engine to the middle of a solar power satellite some tens of square kilometers in size and ignite it. Not much of the structure will remain with the engine. It may be expected that electric or other low-thrust propulsion systems will come into their own with the development of large space platforms.

Space stations. Concepts for manned space stations have existed since the earliest days of astronautics. Von Braun's 1952 study, published in *Collier's*, remains a classic in this field. The first-generation space stations, the Soviet Salyut and American Skylab vehicles, fall far short of von Braun's ambitious concepts. This from some viewpoints is quite surprising; early work seems often to have assumed that construction of large, permanent stations would be among the first priorities to be addressed once the necessary space transportation capability was developed. This has not turned out to be the case. Political factors, such as the "moon race," have influenced the course of events, but technical reality has also been recognized. Repeated studies have failed to show any single driving requirement for the deployment of a space station. The consensus that has instead emerged is that, if a permanent station or stations existed, many uses would be found for it that currently require a separate satellite, or are simply not done. Nonetheless, the first steps have been taken, and it seems inevitable that in the not too distant future, large, permanent, manned orbital stations will be deployed to satisfy some of the scientific, military, and civil mission requirements now met through the use of many smaller satellites and also to provide capabilities that do not now exist.

Selection of space station orbits will be driven by the same factors as for smaller spacecraft, a tradeoff between operational requirements, energy to achieve orbit, and difficulty of maintaining the desired orbit. For small space stations such as the Soviet Salyut series, maneuvering is not especially difficult, and periodic orbit maintenance can be accomplished with thrusters. The large, flexible assemblies proposed for future stations may be more difficult to maneuver and for this reason may tend to favor higher orbits. As mentioned, some type of electric propulsion will probably be required for orbit maintenance in this case, both because of its reduced propellant requirements and its low thrust.

Space stations designed for observation, whether civil or otherwise, will have characteristics similar to their smaller unmanned brethren. They will generally be found in high-inclination low orbits, perhaps sun-synchronous, for close observation, or in high orbits where a more global view is required. On the other hand, stations of the Space Operations Center type, which are used as way stations en route to geosynchronous orbit or planetary missions as well as for scientific purposes, will probably be in fairly low orbits at inclinations compatible with launch site requirements.

Space stations of the von Braun rotary wheel type may never be realized because of the realization that artificial gravity is not necessary for crew flight times up to several months' duration. This has been demonstrated by both Soviet and American missions, wherein proper crew training and exercise have allowed the maintenance of satisfactory physical conditioning. By eliminating the need for artificial gravity, the need for a symmetric, rotating design is also eliminated. This greatly simplifies configuration and structural design, observational techniques, and operations, especially flight operations with resupply vehicles. For an interesting visual demonstration of the problems of docking with a rotating structure, the reader is urged to view Stanley Kubrick's film, *2001: A Space Odyssey*.

The problem of supplying electric power for space station operations is substantial. Skylab, Salyut, and Mir have used solar panel arrays with batteries for eclipse periods. This will probably remain the best choice for stations with power requirements measured in a few tens of kilowatts. As power requirements become large, which history indicates is inevitable, the choice becomes less clear. The large areas of high-power solar arrays pose a major drag and gravity-gradient stabilization problem in LEO, and their intrinsic flimsiness poses severe attitude control problems even in high orbit. The use of dynamic conversion of solar heat to electricity is promising in reducing the collection area but has other problems.

The only presently viable alternative to solar power for a permanent station is a nuclear system, and here we are generally talking about nuclear reactors rather than the RTGs discussed earlier. RTGs do not have a sufficiently high power-to-weight ratio to be acceptable when high power levels are required. Chemical energy systems such as fuel cells are not practical for permanent orbiting stations when the reactants must be brought from Earth. This conclusion could change in the short term if a practical means of recovering unused launch vehicle propellant could be devised, and in the long term if use of extraterrestrial materials becomes common. But in the meantime, nuclear power offers the only compact, long-lived source of power in the kilowatt to megawatt range.

Nuclear power also raises substantial problems. The high-temperature reactor and thermal radiators, the high level of ionizing radiation, and the difficulty of systems integration caused by these factors present substantial engineering problems. No less serious is public concern with possible environmental effects due to the uncontrolled re-entry of a reactor. This first happened with the Soviet Cosmos 954 vehicle, which fortunately crashed in a remote region of Canada. The cleanup operations involved were not trivial.

Of similar importance is the environmental control system of the station. The more independent of resupply from the ground it can be, the more economical the permanent operation of the station will become. The ultimate goal of a fully recycled, closed environmental system will be long in coming, but even a reasonably high percentage of water and oxygen recycling will be of significant help. The possibility of an ecological approach to oxygen recycling may allow production of fresh fruits, vegetables, and decorative plants. The latter may be of only small significance to the resupply problem but may do wonders for crew morale. Similar concern with environmental issues has gone into U.S. Navy nuclear submarines, which spend long periods submerged. It will be of interest in space station design to examine the methods by which crew morale is maintained in such situations. That the issue is not trivial is shown by records of more than one U.S. space flight, where flight crew boredom and overwork have on occasion led to some acrimonious exchanges with ground control.

Space colonies.   Long-term-habitability space stations can be expected to provide the initial basis for the design of space colonies or colonies on other planets or asteroids. The borderline between space stations or research or work stations on other planets and true colonies is necessarily somewhat blurred, but the use of the term "colonies" is generally taken to imply self-sufficient habitats with residents of all types who expect to live out their lives in the colony. Trade with Earth is presumed, as a colony with no economic basis for its existence probably will not have one. On the other hand, it seems reasonable that "research stations" or "lunar mining bases" could grow into colonies, given the right circumstances.

O'Neill and his co-workers have been the most ardent recent proponents of the utility and viability of space colonies. In the O'Neill concept, the colonies will have as their economic justification the construction of solar power satellites for Earth, using raw materials derived from lunar or asteroid bases. It would seem that other uses for such habitats could be found as well; as mentioned previously, it may well be that eventually much of Earth's heavy manufacturing is relocated to sites in space in order to take advantage of the availability of energy and raw materials. In any case, O'Neill envisions truly extensive space habitats, tens of kilometers in dimension, and featuring literally all of the comforts of home, including grass, trees, and houses in picturesque rural settings.

Whether or not these developments ever come to pass (and the authors do not wish to say that they cannot; well-reasoned economic arguments for developing such colonies have been advanced), such concepts would seem to be the near-ultimate in spacecraft design. In every way, construction of such habitats would pose problems that, without doubt, are presently unforeseen. The engineering of space colonies and colonies on other planets will demand the use of every specialty known on Earth today, from agriculture to zoology, and these specialists will have to learn to transfer their knowledge to extraterrestrial conditions. The history of the efforts of Western Europeans simply to colonize other regions of Earth in the sixteenth and seventeenth centuries suggests both that it will be done and that it will not be done easily.

Use of lunar and asteroid materials. Even our limited exploration of the moon has indicated considerable potential for supplying useful material. We have not in our preliminary forays observed rich beds of ore such as can be found on Earth. Some geologists have speculated that such concentrations may not exist on the moon, and it certainly seems reasonable to suppose that they do not exist near the surface, which is a regolith composed of material pulverized and dispersed in countless meteoric impacts. However, the common material of the lunar crust offers a variety of useful materials, most prominently aluminum and oxygen. And surprisingly, titanium is in relatively large supply in the lunar samples so far seen. A more useful metal for space manufacturing would be hard to find. The metals exist as oxides or in more complex compounds. A variety of processes have been suggested for the production of useful metals and oxygen; which material is the product and which is the by-product depends on the prejudices of the reader.

Because of the cost of refining the material on the moon and transporting it to Earth, it is improbable that such materials would be economically competitive with materials produced here on Earth. An exception would be special alloys made in 0-*g* or other substances uniquely depending on the space environment for their creation. However, extraterrestrial materials may well compete with materials ferried up from Earth for construction in orbit or on the moon itself. This is the primary justification for lunar and asteroid mining, and it seems so strong that it must eventually come to pass, when the necessary base of capital equipment exists in space.

It may well be that products (as opposed to raw materials) manufactured in space will compete successfully with comparable products manufactured on Earth. Early candidates will be goods whose price is high for the mass they possess and whose manufacture is energy-intensive, hampered by gravity and/or atmospheric contaminants, and highly suitable for automated production. Semiconductors and integrated circuits, pharmaceuticals, and certain alloys have been identified in this category. Other activities may follow; one can imagine good and sufficient reasons for locating genetic engineering research and development efforts in an isolated space-based laboratory.

With the accumulation in orbit of sufficient capital equipment to allow large-scale use of lunar or other extraterrestrial materials, and the development of effective solar energy collection methods, the growth of heavy manufacturing must follow. As noted, the surface of the moon is much closer to either GEO or LEO in terms of energy expenditure than is the surface of Earth. Any really large projects will probably be more economical with lunar material, even considering the necessary investment in lunar mining bases. And some resources are more readily used than others; even relatively modest traffic from LEO to GEO, the moon, or deep space will probably benefit from oxygen generated on the moon and sent down to Earth orbit.

The as yet unproven theory that water ice is trapped in permanently dark, very cold regions near the lunar poles is of great interest. Water is not only vital for life-support functions (though with closed systems, humans generate water as a by-product of other activities, thus reducing the life-support problem to that of food alone), it is useful in a variety of chemical processes, and especially in the production of hydrogen. Thus far it appears that no economically viable supply of hydrogen exists on the moon except in these hypothetical ice reservoirs. Hydrogen is useful as a propellant and in a variety of chemical reactions. If it does not exist on the moon, it will have to be imported from Earth, at least in the short term. Although its low mass makes importation of hydrogen tolerable, the desirability of finding it on the moon is obvious.

The use of asteroid materials has equally fascinating potential. Taken as a class, asteroids offer an even more interesting spectrum of materials than has so far been identified on the moon. The metallic bodies consist mostly of nickel-iron, which should be a reasonably good structural material as found and would be refinable into a variety of others. The carbonaceous chondrite types seem to contain water, carbon, and organic materials as well as silicates. These would have the obvious advantage of being water and hydrogen sources; indeed, some models of the Martian climate have postulated that such asteroids are the source of what Martian water exists. The most common, and probably least useful, asteroids are composed mostly of silicate materials; essentially, they are indistinguishable from common Earth dirt.

Although, as we have mentioned, most asteroids lie in the main belt between Mars and Jupiter, a modest number lie in orbits near to or crossing that of Earth. Some of these are energetically quite easy to reach, but with the problem that the low round-trip energy requirement is achieved at the cost of travel times on the order of 3 yr or more. Launch windows are restricted to a few weeks every 2 or 3 yr. Thus, although it is true that some asteroids are easier to reach than the surface of the moon, this must be balanced against lunar round-trip time of a few days, together with the ability to make the trip nearly any time. Thus, although asteroid materials of either the Earth-approaching or main-belt variety will probably become of substantial importance eventually, it seems likely that lunar materials will do so first, if only because of convenience.

Propellant manufacturing.   Propellant manufacturing is a special case involving the use of resources naturally occurring on the various bodies of the solar system. It was mentioned in passing under the more general subject of lunar and asteroid resources, but it is by no means restricted to these bodies. In the inner solar system, Mars seems to offer the most promise for application of in situ propellant manufacturing technology.

As noted previously, for the manufacture of a full set of propellants (both fuel and oxidizer), water is both necessary and sufficient. However, carbon, which is also in short supply on the moon, is also important. The atmosphere of Mars provides carbon dioxide in abundance, and water is known to exist in the polar ice caps and most probably in the form of permafrost over much of the planet. Propellant manufacturing has been studied both for unmanned sample return missions and for manned missions. The advantages are comparable to those that accrue by refueling airliners at each end of a flight, rather than designing them to carry fuel for a coast-to-coast round-trip.

Because of the difficulty of mining permafrost or low-temperature ice, it has been suggested that the first propellant manufacturing effort might use the atmosphere exclusively. Carbon dioxide can be taken in by compression and then, in a cell using thermal decomposition and an oxygen permeable membrane, split into carbon monoxide and oxygen. The oxygen can then be liquified and burned with a fuel brought from Earth. Methane is the preferred choice, since it has high performance, a high oxidizer-to-fuel ratio (to minimize the mass brought from Earth), and is a good refrigerant. The latter quality contributes to the process of liquifying the oxygen and keeping both propellants liquid until enough oxidizer is accumulated and the launch window opens.

It should be noted that the combination of carbon monoxide and oxygen is a potential propellant combination. The theoretical performance is not especially good and, so far as is known, it has never been tested. The performance might be adequate for short-range vehicles supporting a manned base on Mars, however, and would certainly be convenient. It is even suitable for orbital vehicles although propellant mass is large. A final advantage is that, since the exhaust product is carbon dioxide, there would be no net effect on the Martian atmosphere.

Making use of Martian water broadens the potential options considerably. Besides the obvious hydrogen/oxygen combination, use of both water and carbon dioxide allows the synthesis of other chemicals such as methane. Methane is an excellent fuel and is more easily storable than hydrogen. Methanol can also be created, either as a fuel or for use in other chemical processes. From this brief glimpse, it can be seen that water and carbon or carbon dioxide form the basis for propellant manufacturing as well as other chemical processes.

Because carbonaceous chondrites presumably contain both water and carbon compounds, it is probable that these bodies have potential for various types of chemical synthesis as well. The satellites of the outer planets contain considerable water; indeed, some are mostly water. Whether useful carbon-containing compounds are available is less certain, but at least the hydrogen/oxygen propellant combination will be available.

In all propellant manufacturing processes, the key is power. Regardless of the availability of raw materials, substantial energy is required to decompose the water or carbon dioxide. Compression and liquefaction of the products also require energy. The possible sources of energy are solar arrays, nuclear systems using radioisotopic decay, and critical assemblies (reactors). Solar energy is only practical in the inner solar system, and then probably only for small production rates.

Nuclear waste disposal. Disposal of long-lived highly radioactive waste in space has been discussed for a number of years. The attraction is obvious; it is the one disposal mode that, properly implemented, has no chance of contaminating the biosphere of Earth because of leakage or natural disaster.

The least demanding technique would be to place the waste into an orbit of Earth that is at sufficient altitude that no conceivable combination of atmospheric drag or orbital perturbations would cause the orbit to decay. Even though this is workable, it is not considered satisfactory by some, since the material is still within the Earth's sphere of influence and thus might somehow come down. A more practical objection is that as use of near-Earth space increases, it might be undesirable to have one region rendered unsafe.

Another suggestion is to place all the material on the moon, say, in a particular crater. This generally avoids the orbit stability problem but has the disadvantage of rendering one area of the moon quite unhealthy. Energy cost would be high as well, since the material would need to be soft-landed to avoid scattering on impact.

From an emotional viewpoint at least, interplanetary space seems the most desirable arena for disposal, preferably in an orbit far from that of Earth. One approach would steal a page from the Mariner 10 mission. For a total energy expenditure less than that for a landing on the moon, the material could be sent on a trajectory to flyby Venus. This could move the perihelion of the orbit to a point between Venus and Mercury. A relatively minor velocity change at the perihelion of the orbit would then lower aphelion inside the orbit of Venus. The package would then be in a stable, predictable orbit that would never again come close to Earth.

The major problem with the space disposal of nuclear waste is the emotional fear of a launch failure spreading the material widely over the surface of the Earth. Although a number of concepts could be applied to minimize the risk, it seems doubtful that this concept will become acceptable to the public in the near future.

3
ENVIRONMENT

In the broadest sense, the spacecraft environment includes everything to which the spacecraft is exposed from its beginning as raw material to the end of its operating life. This includes the fabrication, assembly, and test environment on Earth, transportation from point to point on Earth, launch, the space environment, and possibly a re-entry or destination environment at another planet. In some cases, especially where the rigorous safety standards applied to manned flight are concerned, even the origin of the materials used and the details of the processes by which they are fashioned into spacecraft components may be important.

Both natural and man-made environments are imposed upon the spacecraft. Contrary to the popular view, the rigors of launch and the space environment itself are often not the greatest hazards to the spacecraft. The spacecraft is designed to be launched and to fly in space. If the design is properly done, these environments are not a problem; a spacecraft sometimes seems at greatest risk on Earth in the hands of its creators. Spacecraft are often designed with only the briefest consideration of the need for ground handling, transportation, and test. As a result, these operations and the compromises and accommodations necessary to carry them out may in fact represent a more substantial risk than anything that happens in a normal flight.

However, the preceding comments imply that the spacecraft is designed for proper functioning in flight. To do this it is necessary to know the range of conditions encountered. This includes not only the flight environment but also the qualification test conditions that must be met in order to demonstrate that the design is correct. To provide confidence that the design will be robust in the face of unexpectedly severe conditions, these tests are typically more stringent than the expected actual environment.

In this chapter we shall discuss the Earth, launch, and space environments, but in somewhat different terms. The launch and flight environments are usually quite well defined for specific launch vehicles and missions. These conditions, and the qualification test levels that are derived from them, will be treated as the actual environment for which the vehicle must be designed. The Earth environment is assumed to be controllable, within limits, to meet the requirements of a spacecraft, subsystem, or component. Also, the variety of Earth environments, modes of handling and transport, etc., is so great as to preclude a detailed quantitative discussion in this volume. Accordingly, the discussion will be of a more general nature when addressing Earth environments.

3.1 EARTH ENVIRONMENT

Throughout its tenure on Earth, the spacecraft and its components are subjected to a variety of potentially degrading environments. The atmosphere itself is a primary source of problems. Containing both water and oxygen, the Earth's atmosphere is quite corrosive to a variety of materials, including many of those used in spacecraft, such as lightweight structural alloys. Corrosion of structural materials can cause stress concentration or embrittlement, possibly leading to a failure during launch. Corrosion of pins in electrical connectors may lead to excessive circuit resistance and thus unsatisfactory performance. Because of these effects it is desirable to control the relative humidity and in extreme cases to exclude oxygen and moisture entirely by use of a dry nitrogen or helium purge. This is normally required only for individual subsystems such as scientific instruments; in general, the spacecraft can tolerate exposure to the atmosphere if humidity is not excessive. However, too low a relative humidity is also poor practice both from consideration of worker comfort and from a desire to minimize buildup of static electric charge (discussed later in more detail). A relative humidity in the 40–50% range is normally a good compromise.

Another environmental problem arising from the atmosphere is airborne particulate contamination, or dust. Even in a clean normal environment, dust will accumulate on horizontal surfaces fairly rapidly. For some spacecraft a burden of dust particles is not significant; however, in many cases it can have undesirable effects. Dust can cause wear in delicate mechanisms and can plug small orifices. Dislodged dust particles drifting in space, illuminated by the sun, can look very much like stars to a star sensor or tracker on the spacecraft. This confusion can and has caused loss of attitude reference accuracy in operating spacecraft. Finally, dust typically hosts a population of viruses and bacteria that are unacceptable on a spacecraft destined for a visit to a planet on which Earth life might be viable.

Because of the concern for preventing dust contamination, spacecraft and their subsystems are normally assembled and tested in "clean room" environments. Details of how such environments are obtained are not of primary interest here. In general, clean rooms (see Fig. 3.1) require careful control of surfaces in the room to minimize dust generation and supply of conditioned air through high-efficiency particulate filters. In more stringent cases a unidirectional flow of air is maintained, entering at the ceiling or one wall and exiting at the opposite surface.

The most advanced type of facility is the so-called laminar flow clean room, in which the air is introduced uniformly over the entire surface of a porous ceiling or wall and withdrawn uniformly through the opposing surface or allowed to exit as from a tunnel. Actual laminarity of flow is unlikely, especially in a large facility, but the very uniform flow of clean air does minimize particulate collection. Small component work is done at "clean benches," workbench type facilities where the clean environment is essentially restricted to the benchtop. The airflow exhausts toward the worker seated at the bench, as in Fig. 3.2.

Clean room workers usually must wear special clothing that minimizes

Fig. 3.1   Clean room. (Courtesy of Astrotech Space Operations.)

particulate production from regular clothing or the body. Clean room garb typically involves gloves, smocks or "bunnysuits," head covering, and foot covering. All this must be lint-free. In some cases masks are required as well. Because of the constant airflow and blower noise and the restrictive nature of the clothing, clean room work is often tiring even though it does not involve heavy labor.

Clean facilities are given class ratings such as Class 100,000, Class 1000, or Class 100 facilities. The rating refers to the particulate content of a cubic foot of air for particles between specified upper and lower size limits; thus, lower numbers represent cleaner facilities. Class 100 is the cleanest rating normally discussed and is extremely difficult to maintain in a large facility, especially when any work is in progress. Even Class 1000 is difficult in a facility big enough for a large spacecraft and one in which several persons might be working. A Class 10,000 facility is the best that might normally be achievable under such conditions and represents a typical standard for spacecraft work. Fresh country air would typically yield a rating of approximately Class 300,000. Clean rooms are usually provided with anterooms for dressing and airlocks for entry. Airshowers and sticky floormats or shoe scrubbers provide final cleanup.

A major hazard to many spacecraft components is static electricity. The triboelectric effect may yield very substantial voltages on human skin, plastics, and other surfaces. Some electronic components, in particular, integrated circuits or other components using metal-oxide semiconductor (MOS) technology, are extremely sensitive to high voltage and can easily be damaged by a discharge such as might occur off the fingertip of a technician. To prevent such occurrences, clean room workers must be grounded when handling hardware. This is usually done using conductive flooring and conductive shoes or ankle ground straps. For especially sensitive cases a ground strap on the wrist may be worn.

Since low relative humidity contributes to static charge accumulation, it is desirable that air in spacecraft work areas not be excessively dry. The compromise with the corrosion problem discussed earlier usually results in a chosen relative humidity of about 40–50%. Plastic cases and covers and tightly woven synthetic garments, all favored for low particle generation, tend to build up very high voltages unless treated to prevent it. Special conductive plastics are available, as are fabric treatment techniques. However, the conductive character can be lost so that clean room articles must be constantly monitored.

In theory, with all electronic components mounted and all electrical connections mated, the spacecraft should be safe from static discharge. In practice, however, the precautions discussed earlier are generally observed by anyone touching or handling the spacecraft. The primary risk arises from contact with the circuit proper that occurs when pins are touched in an unmated connector. Unnecessary contact of this type should be avoided.

Transporting the spacecraft from point to point on Earth may well subject it to more damaging vibration and shock than launch. Road vibration and shock during ground transportation can be higher than those imposed by launch and the duration is much longer, usually hours or days compared with the few minutes involved in launch. For short trips, as from building to building within a facility, the problem can best be handled by moving the spacecraft very slowly over a carefully selected and/or prepared route. For longer trips where higher speed is required, special vehicles employing air cushion suspension are usually required. These vehicles may be specially built for the purpose, or may simply be commercial vans specialized for delicate cargo. Truck or trailer suspensions can deteriorate in service, and it is usually desirable to subject them to instrumented road tests before committing expensive and delicate hardware to a long haul.

Flying is generally preferable to ground transportation for long trips. Jets are preferred to propeller-driven aircraft because of the lower vibration and acoustic levels. High g loads can occur at landing or as a result of turbulence and the spacecraft must be properly supported to provide protection. The depressurization/pressurization cycle involved in climb and descent can also be a problem. For example, a closed vessel, although designed for several atmospheres of internal pressure, can easily collapse if it bleeds down to an internal pressure equivalent to several thousand feet altitude during flight and then is quickly returned to sea level. This is particularly a problem when transporting propulsion stages having large tanks with relatively thin walls.

When deciding between flying or road transportation, it should be recalled that in general it will be necessary to transport by road to the airport, load on the plane, and then reverse the procedure at the other end. For trips of moderate length, a decision should be made as to whether flying, with all the additional handling involved, is in fact better than completing the entire trip on the ground.

In all cases, whether transporting by ground or air, it is essential that the spacecraft be properly secured to the carrier vehicle structure. This requires careful design of the handling and support equipment. Furthermore, all delicate structures that could be damaged by continued vibration should be well secured or supported.

For some very large structures, the only practical means of long-range transportation is via water. Barges were used for the lower stages of the Saturn 5 launch vehicle and continue to be used to transport the Shuttle external tank from Michoud, Louisiana, to Cape Canaveral.

The cleanliness, humidity, and other environmental constraints discussed earlier usually must continue in force during transportation. In many cases, as with the shipment by boat of the Hubble Space Telescope from the Sunnyvale, California, fabrication site to Cape Canaveral, this presents a significant logistical challenge.

3.2 LAUNCH ENVIRONMENT

Launch imposes a highly stressful environment on the spacecraft for a relatively brief period. During the few minutes of launch the spacecraft is subjected to significant axial loads by the accelerating launch vehicle plus lateral loads from steering and wind gusts. There will be substantial mechanical vibration and severe acoustic energy input. The latter is especially pronounced just after liftoff as the rocket engine noise reflects from the ground. Aerodynamic noise also contributes, especially in the vicinity of Mach 1. During the initial phase of launch, atmospheric pressure will drop from essentially sea level values to space vacuum. Aerodynamic heating of the nose fairing during low-altitude ascent and directly by residual atmosphere after fairing jettison may impose thermal loads that drive some aspects of the spacecraft design. Stage shutdown and separation as well as fairing jettison will produce shock transients.

To ensure that the spacecraft is delivered to its desired orbit or trajectory in condition to carry out the mission, it must be designed for and qualified to the expected stress levels, with a margin of safety. To facilitate preliminary design, launch vehicle user handbooks specify pertinent parameters such as acoustic, vibration, and shock levels. For vehicles with a well-established flight history, the data are based on actual in-flight measurements. Vehicles in the developmental phase provide estimated or calculated data based on modeling and comparison with similar vehicles.

Environmental data of the type presented in user handbooks are suitable for preliminary analysis in the early phases of spacecraft design and are useful in establishing initial structural design requirements. Since the spacecraft and launch vehicle interact, however, the actual environment will vary somewhat from one spacecraft payload to another, and the combination of launch vehicle and spacecraft must be analyzed as a coupled system.[1] As a result, the actual environment anticipated for the spacecraft changes with its maturing design and the resulting changes in the total system. Since this in turn affects the spacecraft design, it is clear that an iterative process is required.

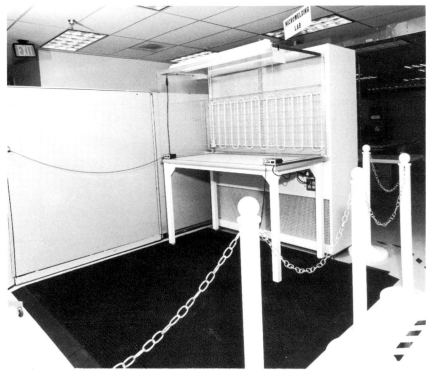

Fig. 3.2   Clean bench. (Courtesy of Ball Aerospace Systems Division.)

The degree of analytical fidelity required in this process is a function of mass margins, fiscal resources, and schedule constraints. For example, structural modeling of the Viking Mars Orbiter/Lander was detailed and thorough since mass margins were tight. On the other hand, the Solar Mesosphere Explorer, a low-budget Earth orbiter that had a very large launch vehicle margin, was subjected to limited analysis. Many structures were made from heavy plate or other material that was so overdesigned that it limited the need for detailed analysis. When schedule is critical, extra mass may well be allocated to the structural design to limit the need for detailed analysis and testing.

Acoustic loads are pervasive within the nose fairing or payload bay, with peaks sometimes occurring at certain locations. Vibration spectra are usually defined at the base of the attach fitting or adapter. Shock inputs are usually defined at the location of the generating device, typically an explosively actuated or mechanically released device.

In many cases the various inputs actually vary somewhat from point to point, especially in the case of shock spectra. For convenience in preliminary design, this is often represented by a single curve that envelops all the individual cases. Examples of this may be seen among the curves presented in this chapter. In general, use of such curves will lead to a conservative design that, at the cost of some extra mass, is well able to withstand the actual flight environment.

Launch Vehicle Data

In this section we present data drawn from user handbooks for the various major launch vehicles discussed in Chapter 5. The figures are grouped according to type of data rather than by launch vehicle. Figures 3.3–3.20 present vibration data for the various launch vehicles. Random vibration data are presented as curves of spectral density in g^2/Hz, essentially a measure of energy vs frequency of vibration. For the Shuttle, data are presented at the main longeron and keel fittings, whereas for the expendable vehicles it is at the attachment plane. The first two curves for the Shuttle (see Figs. 3.3 and 3.4) represent predicted data, and the third (Fig. 3.5) presents flight data for longeron vibration based on Shuttle

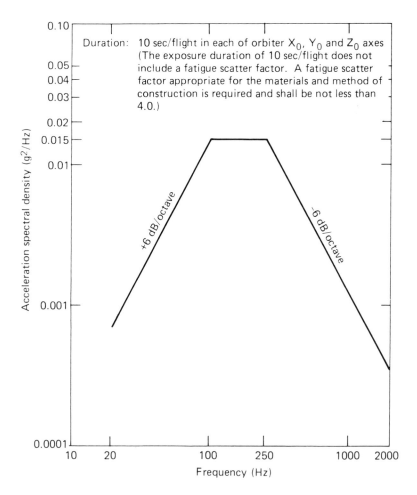

Fig. 3.3   **Shuttle vibration environment: unloaded main longeron trunion-fitting vibration.**

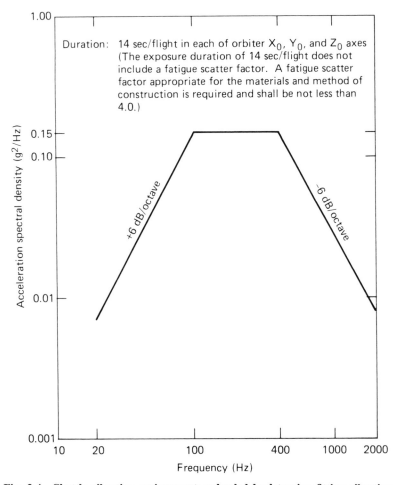

Fig. 3.4   Shuttle vibration environment: unloaded keel trunion fitting vibration.

flights STS 1-4. It is instructive to compare Figs. 3.3 and 3.5 and note that the flight data yield higher frequency vibration and higher Y-axis levels than predicted. This does not appear to be a serious problem, since trunion fitting slippage tends to isolate much of this vibration from the payload. Flight data for the keel fitting (not shown) are very close to the predicted curve (Fig. 3.4).

Provisions for mounting payloads in the Shuttle bay are discussed in Chapter 5. These mountings allow for limited motion in certain directions. This helps decouple payloads from orbiter structural vibrations. Furthermore, the presence of the payload mass itself tends to damp the vibration. These effects lead to a vibration attenuation factor CV. This is presented in

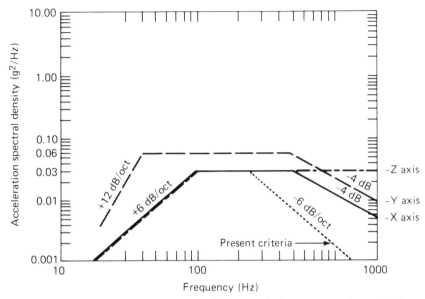

Fig. 3.5 Shuttle vibration environment: Orbiter main longeron random vibration criteria derived from flight data.

Fig. 3.6. It is applied as

$$ASD_{\text{payload}} = CV \times ASD_{\text{unloaded orbiter structure}} \tag{3.1}$$

where *ASD* is the acceleration spectral density.

Longitudinal vibration is generally caused by thrust buildup and tailoff of the various stages plus such phenomena as the "pogo" effect, which sometimes plagues liquid-propellant propulsion systems. This is manifested by thrust oscillations generally in the 5- to 50-Hz range. The phenomenon results from coupling of structural and flow system oscillations and can usually be controlled by a suitably designed gas-loaded damper in the propellant feed lines.

Lateral vibrations usually result from wind gust and steering loads as well as thrust buildup and tailoff.

Expendable vehicle data, presented as longitudinal and lateral sinusoidal vibration data, random vibration, and acoustic and shock spectra, are presented in Figs. 3.7–3.20 and Tables 3.1 and 3.2.

3.3 SPACE ENVIRONMENT

The space environment is characterized by a very hard (but not total) vacuum, very low (but not zero) gravitational acceleration, possibly intermittent or impulsive nongravitational accelerations, ionizing radiation, extremes of thermal radiation source and sink temperatures, severe thermal gradients, micrometeoroids, and orbital debris. Some or all of these features may drive various aspects of spacecraft design.

Vacuum

Hard vacuum is of course one of the first properties of interest in designing for the space environment. Many key spacecraft design characteristics and techniques are due to the effects of vacuum on electrical, mechanical, and thermal systems. Material selection is crucially affected by

Table 3.1 Envelope of maximum estimated noise levels internal to payload fairing for Titan 34D/IUS launch and flight

Duration of estimated noise levels: total of 60 s from liftoff through maximum dynamic pressure (max Q). Zero dB reference 0.0002 dyne/cm^2.

$\frac{1}{3}$-Octave band center frequency, Hz	$\frac{1}{3}$-Octave band sound pressure level, dB
25	145
31.5	121
40	122.5
50	124
63	125.5
80	127
100	129
125	130.5
160	131.5
200	132.5
250	133.5
315	134
400	134.5
500	134.5
630	134
800	133.5
1000	133
1250	131
1600	129.5
2000	128.5
2500	126.5
3150	125
4000	123
5000	121.5
6300	120
8000	118
10,000	116

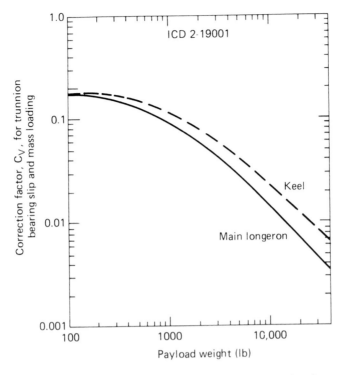

Fig. 3.6   Shuttle vibration environment: vibration attenuation factor.

Table 3.2   Center-of gravity load factors

Event/Axis	Steady-state acceleration	Dynamic acceleration
Liftoff		
Axial	+1.5 g	±1.5 g
Lateral		±5.0 g
Torsional		±0.05 g/in.
Maximum airloads		
Axial	+2.0 g	±1.0 g
Lateral		±2.5 g
Torsional		±0.05 g/in.
Stage I shutdown		
Axial	0 to +4.0 g	±4.0 g
Lateral		±3.0 g
Stage II shutdown		
Axial	0 to +2.5 g	±7.5 g
Lateral		±2.0 g

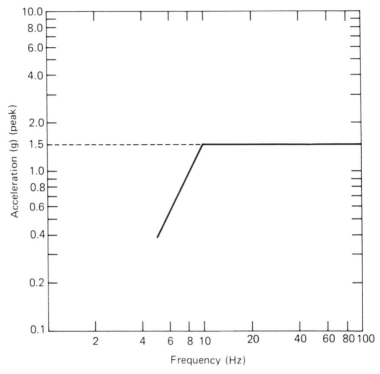

Fig. 3.7   Ariane vibration environment: longitudinal sinusoidal vibrations.

its vacuum behavior. Many materials that see routine engineering use for stressful ground engineering applications are inappropriate even for relatively benign spacecraft applications.

Most materials will outgas to at least some extent in a vacuum environment. Metals will usually have an outer layer into which gases have been adsorbed during their tenure on Earth, and which is easily released once in orbit. Polymers and other materials composed of volatile compounds may outgas extensively in vacuum, losing substantial fractions of their initial mass. Some basically nonvolatile materials, such as graphite-epoxy and other composites, are hygroscopic and absorb considerable water from the air. This water will be released, over a period of months, once the spacecraft is in orbit. Some plating materials will, when warm, migrate in vacuum to colder areas of the spacecraft when they recondense; cadmium is notorious in this regard.

Outgassing materials can be a problem for several reasons. In polymeric or other volatile materials, the nature and extent of the outgassing can lead to serious changes in the basic material properties. Even where this does not occur, as in water outgassing material from graphite-epoxy, structural distortion will result. Such composites are often selected because of their high stiffness-to-weight ratio and low coefficient of thermal expansion, for

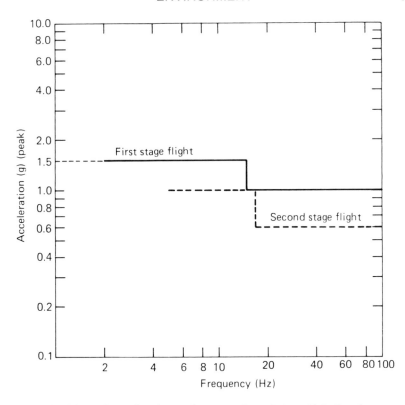

Fig. 3.8   Ariane vibration environment: lateral sinusoidal vibrations.

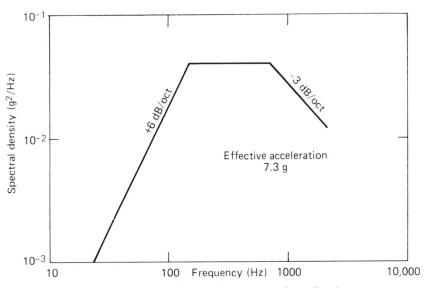

Fig. 3.9   Ariane vibration environment: random vibrations.

applications where structural alignment is critical. Obviously, it is desirable to preserve on orbit the same structure as was fabricated on the ground. Outgassing is also a problem in that the vapor can recondense on optical or other surfaces where such material depositions would degrade the device performance. Even if the vapor does not condense, it can interfere with certain delicate measurements. For example, ultraviolet astronomy is effectively impossible in the presence of even trace amounts of water vapor.

Outgassing is usually dealt with by selecting, in advance, those materials where it is less likely to be a problem. In cases where the material is needed because of other desirable properties, it will be "baked out" during a lengthy thermal vacuum session and then wrapped with tape or given some other coating to prevent reabsorption of water and other volatiles. Obviously, other spacecraft instruments and subsystems may need to be protected while the bake-out procedure is in progress.

Removal of the adsorbed O_2 layer in metals that do not form an oxide layer, such as stainless steel, can result in severe galling, pitting, and cold welding between moving parts where two pieces of metal come into contact. Such problems are usually avoided by not selecting these materials for dynamic applications in the space environment.

Partial Vacuum

Although the vacuum in low Earth orbit, for example at 200 km, is better than anything obtainable on the ground, it is still by no means total. At Shuttle operating altitudes, enough residual atmosphere remains to interact in a significant fashion with a spacecraft. Drag and orbit decay due to the residual atmosphere are discussed in Chapter 4; it may be necessary to include propulsion for drag compensation in order to prevent premature re-entry and destruction of the spacecraft. Of greater interest here, however, are possible chemical interactions between the upper atmosphere atomic and molecular species and the spacecraft materials.

It was noted during early Shuttle missions that a pronounced blue glow appeared on various external surfaces while in the Earth's shadow. This was ascribed to recombination of atomic oxygen into molecular oxygen on contact with the Shuttle skin. Although it presented no problems to the Shuttle itself, the glow is a significant problem for certain scientific observations. It was also noted from samples returned following the 1984 rendezvous and on-orbit repair of the Solar Maximum Mission spacecraft that the Kapton thermal blanketing material had been severely eroded by the action of atomic oxygen.

The combined effects of thermal extremes and the near-vacuum environment may alter the reflective and emissive characteristics of the external spacecraft surfaces. When these surfaces are tailored for a particular energy balance, as is often the case, degradation of the spacecraft thermal control system performance can result.

A particularly annoying partial vacuum property is the fact that very-low-density gases are easily ionized, providing excellent but unintended conductive paths between points in electronic hardware that are at moderate to high potential differences. This tendency is aggravated by the fact

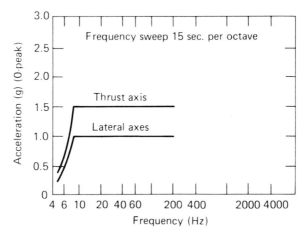

(a) Recommended sinusoidal vibration environment

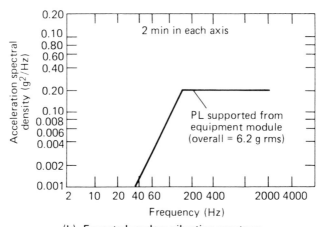

(b) Expected random vibration spectrum

Fig. 3.10   Atlas-Centaur environment.

that, at high altitudes, the residual molecular and atomic species are already partly ionized by solar ultraviolet light and various collision processes. The design of electronic equipment intended for use in launch vehicles is of course strongly affected by this fact, as is the design of spacecraft that are intended for operation in very low orbits. A key point is that, even though a spacecraft system (such as a command receiver or inertial navigation system) is intended for use only when in orbit, it may be turned on during ascent. If this is so, then care needs to be exercised to prevent electrical arcing during certain phases of flight. To this end, spacecraft equipment that may be on during the ascent phase should be operated during the evacuation phase of thermal vacuum chamber testing.

Table 3.3   1976 U.S. Standard atmosphere

Altitude, km	Temperature, K	Pressure, N/m^2	Density, kg/m^3
0	288.150	1.01325 E + 5	1.2250 E + 0
1	281.651	8.9876 E + 4	1.1117 E + 0
2	275.154	7.9501 E + 4	1.0066 E + 0
3	268.659	7.0121 E + 4	9.0925 E − 1
4	262.166	6.1660 E + 4	8.1935 E − 1
5	255.676	5.4048 E + 4	7.3643 E − 1
6	249.187	4.7217 E + 4	6.6011 E − 1
7	242.700	4.1105 E + 4	5.9002 E − 1
8	236.215	3.5651 E + 4	5.2579 E − 1
9	229.733	3.0800 E + 4	4.6706 E − 1
10	223.252	2.6499 E + 4	4.1351 E − 1
11	216.774	2.2699 E + 4	3.6480 E − 1
12	216.650	1.9399 E + 4	3.1194 E − 1
13	216.650	1.6579 E + 4	2.6660 E − 1
14	216.650	1.4170 E + 4	2.2786 E − 1
15	216.650	1.2111 E + 4	1.9476 E − 1
16	216.650	1.0352 E + 4	1.6647 E − 1
17	216.650	8.8497 E + 3	1.4230 E − 1
18	216.650	7.5652 E + 3	1.2165 E − 1
19	216.650	6.4674 E + 3	1.0400 E − 1
20	216.650	5.5293 E + 3	8.8910 E − 2
21	217.581	4.7289 E + 3	7.5715 E − 2
22	218.574	4.0475 E + 3	6.4510 E − 2
23	219.567	3.4668 E + 3	5.5006 E − 2
24	220.560	2.9717 E + 3	4.6938 E − 2
25	221.552	2.5492 E + 3	4.0084 E − 2
26	222.544	2.1883 E + 3	3.4257 E − 2
27	223.536	1.8799 E + 3	2.9298 E − 2
28	224.527	1.6161 E + 3	2.5076 E − 2
29	225.518	1.3904 E + 3	2.1478 E − 2
30	226.509	1.1970 E + 3	1.8410 E − 2
31	227.500	1.0312 E + 3	1.5792 E − 2
32	228.490	8.8906 E + 2	1.3555 E − 2
33	230.973	7.6730 E + 2	1.1573 E − 2
34	233.743	6.6341 E + 2	9.8874 E − 3
35	236.513	5.7459 E + 2	8.4634 E − 3
36	239.282	4.9852 E + 2	7.2579 E − 3
37	242.050	4.3324 E + 2	6.2355 E − 3
38	244.818	3.7713 E + 2	5.3666 E − 3
39	247.584	3.2882 E + 2	4.6268 E − 3
40	250.350	2.8714 E + 2	3.9957 E − 3
41	253.114	2.5113 E + 2	3.4564 E − 3
42	255.878	2.1996 E + 2	2.9948 E − 3
43	258.641	1.9295 E + 2	2.5989 E − 3

(Table continued on next page.)

Table 3.3 (cont.) 1976 U.S. Standard atmosphere

44	261.403	1.6949 E + 2	2.2589 E − 3
45	264.164	1.4910 E + 2	1.9663 E − 3
46	266.925	1.3134 E + 2	1.7142 E − 3
47	269.684	1.1585 E + 2	1.4965 E − 3
48	270.650	1.0229 E + 2	1.3167 E − 3
49	270.650	9.0336 E + 1	1.1628 E − 3
50	270.650	7.9779 E + 1	1.0269 E − 3
51	270.650	7.0458 E + 1	9.0690 E − 4
52	269.031	6.2214 E + 1	8.0562 E − 4
53	266.277	5.4873 E + 1	7.1791 E − 4
54	263.524	4.8337 E + 1	6.3901 E − 4
55	260.771	4.2525 E + 1	5.6810 E − 4
56	258.019	3.7362 E + 1	5.0445 E − 4
57	255.268	3.2782 E + 1	4.4738 E − 4
58	252.518	2.8723 E + 1	3.9627 E − 4
59	249.769	2.5132 E + 1	3.5054 E − 4
60	247.021	2.1958 E + 1	3.0968 E − 4
61	244.274	1.9157 E + 1	2.7321 E − 4
62	241.527	1.6688 E + 1	2.4071 E − 4
63	238.781	1.4515 E + 1	2.1178 E − 4
64	236.036	1.2605 E + 1	1.8605 E − 4
65	233.292	1.0929 E + 1	1.6321 E − 4
66	230.549	9.4609 E + 0	1.4296 E − 4
67	227.807	8.1757 E + 0	1.2503 E − 4
68	225.065	7.0529 E + 0	1.0917 E − 4
69	222.325	6.0736 E + 0	9.5171 E − 5
70	219.585	5.2209 E + 0	8.2829 E − 5
71	216.846	4.4795 E + 0	7.1966 E − 5
72	214.263	3.8362 E + 0	6.2374 E − 5
73	212.308	3.2802 E + 0	5.3824 E − 5
74	210.353	2.8008 E + 0	4.6386 E − 5
75	208.399	2.3881 E + 0	3.9921 E − 5
76	206.446	2.0333 E + 0	3.4311 E − 5
77	204.493	1.7286 E + 0	2.9448 E − 5
78	202.541	1.4673 E + 0	2.5239 E − 5
79	200.590	1.2437 E + 0	2.1600 E − 5
80	198.639	1.0524 E + 0	1.8458 E − 5
81	196.688	8.8923 E − 1	1.5750 E − 5
82	194.739	7.5009 E − 1	1.3418 E − 5
83	192.790	6.3167 E − 1	1.1414 E − 5
84	190.841	5.3105 E − 1	9.6940 E − 6
85	188.893	4.4568 E − 1	8.2196 E − 6
86	186.87	3.7338 E − 1	6.958 E − 6
87	186.87	3.1259 E − 1	5.824 E − 6
88	186.87	2.6173 E − 1	4.875 E − 6
89	186.87	2.1919 E − 1	4.081 E − 6

(Table continued on next page.)

Table 3.3 (cont.) 1976 U.S. Standard atmosphere

90	186.87	$1.8359\text{ E}-1$	$3.416\text{ E}-6$
91	186.87	$1.5381\text{ E}-1$	$2.860\text{ E}-6$
92	186.96	$1.2887\text{ E}-1$	$2.393\text{ E}-6$
93	187.25	$1.0801\text{ E}-1$	$2.000\text{ E}-6$
94	187.74	$9.0560\text{ E}-2$	$1.670\text{ E}-6$
95	188.42	$7.5966\text{ E}-2$	$1.393\text{ E}-6$
96	189.31	$6.3765\text{ E}-2$	$1.162\text{ E}-6$
97	190.40	$5.3571\text{ E}-2$	$9.685\text{ E}-7$
98	191.72	$4.5057\text{ E}-2$	$8.071\text{ E}-7$
99	193.28	$3.7948\text{ E}-2$	$6.725\text{ E}-7$
100	195.08	$3.2011\text{ E}-2$	$5.604\text{ E}-7$
110	240.00	$7.1042\text{ E}-3$	$9.708\text{ E}-8$
120	360.00	$2.5382\text{ E}-3$	$2.222\text{ E}-8$
130	469.27	$1.2505\text{ E}-3$	$8.152\text{ E}-9$
140	559.63	$7.2028\text{ E}-4$	$3.831\text{ E}-9$
150	634.39	$4.5422\text{ E}-4$	$2.076\text{ E}-9$
160	696.29	$3.0395\text{ E}-4$	$1.233\text{ E}-9$
170	747.57	$2.1210\text{ E}-4$	$7.815\text{ E}-10$
180	790.07	$1.5271\text{ E}-4$	$5.194\text{ E}-10$
190	825.31	$1.1266\text{ E}-4$	$3.581\text{ E}-10$
200	854.56	$8.4736\text{ E}-5$	$2.541\text{ E}-10$
210	878.84	$6.4756\text{ E}-5$	$1.846\text{ E}-10$
220	899.01	$5.0149\text{ E}-5$	$1.367\text{ E}-10$
230	915.78	$3.9276\text{ E}-5$	$1.029\text{ E}-10$
240	929.73	$3.1059\text{ E}-5$	$7.858\text{ E}-11$
250	941.33	$2.4767\text{ E}-5$	$6.073\text{ E}-11$
260	950.99	$1.9894\text{ E}-5$	$4.742\text{ E}-11$
270	959.04	$1.6083\text{ E}-5$	$3.738\text{ E}-11$
280	965.75	$1.3076\text{ E}-5$	$2.971\text{ E}-11$
290	971.34	$1.0685\text{ E}-5$	$2.378\text{ E}-11$
300	976.01	$8.7704\text{ E}-6$	$1.916\text{ E}-11$
310	979.90	$7.2285\text{ E}-6$	$1.552\text{ E}-11$
320	983.16	$5.9796\text{ E}-6$	$1.264\text{ E}-11$
330	985.88	$4.9630\text{ E}-6$	$1.035\text{ E}-11$
340	988.15	$4.1320\text{ E}-6$	$8.503\text{ E}-12$
350	990.06	$3.4498\text{ E}-6$	$7.014\text{ E}-12$
360	991.65	$2.8878\text{ E}-6$	$5.805\text{ E}-12$
370	992.98	$2.4234\text{ E}-6$	$4.820\text{ E}-12$
380	994.10	$2.0384\text{ E}-6$	$4.013\text{ E}-12$
390	995.04	$1.7184\text{ E}-6$	$3.350\text{ E}-12$
400	995.83	$1.4518\text{ E}-6$	$2.803\text{ E}-12$

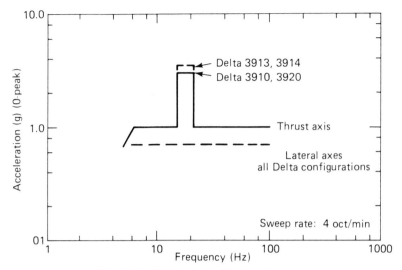

Fig. 3.11   Delta sinusoidal vibration.

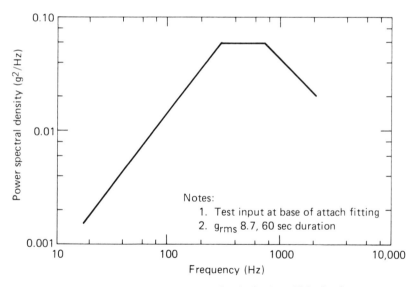

Fig. 3.12   Delta thrust and lateral axis random flight levels.

Table 3.3 and Fig. 3.21 presents the current U.S. Standard Atmosphere model,[2] and Fig. 3.22 shows the density of oxygen at low-orbit altitudes. It is seen that substantial variation of upper atmosphere properties with the 11-yr solar cycle exists. As will be discussed further in Chapters 4 and 8, this variation can be of great importance in both mission and spacecraft

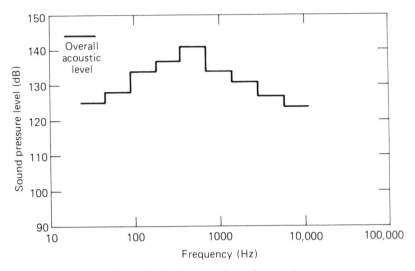

Fig. 3.13   Delta acoustic environment.

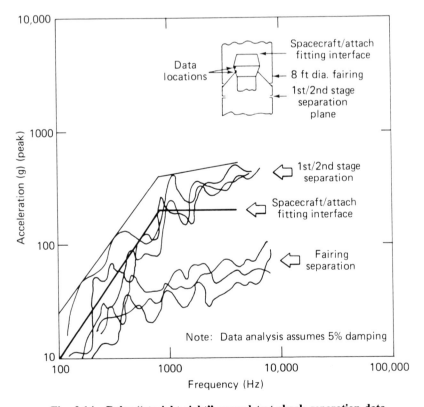

Fig. 3.14   Delta "straight-eight" ground test shock separation data.

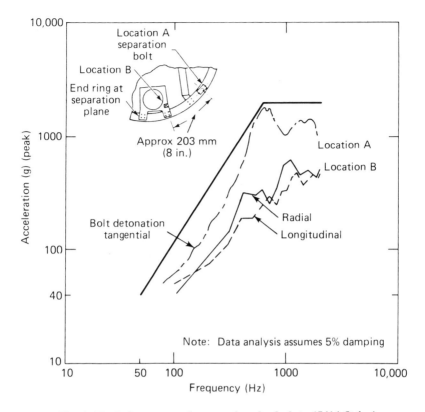

Fig. 3.15   Delta spacecraft separation shock data (5414 fitting).

design. Orbital operations during periods of greater solar activity, and consequently higher upper atmosphere density, produce both more rapid orbit decay and more severe aerodynamic torques on the spacecraft. This can in turn necessitate a greater mass budget for secondary propulsion requirements for drag makeup and similar compensations in the attitude control system design.

Weightlessness and Microgravity

It is commonly thought that orbital flight provides a weightless environment for a spacecraft and its contents. To some level of approximation this is true, but, as with most absolute statements, it is inexact. A variety of effects result in acceleration levels (i.e., "weight" per unit mass) between 10^{-3} and 10^{-11} g, where 1 g is the acceleration due to gravity at the Earth's surface, 9.81 m/s^2.

The acceleration experienced in a particular case will depend on the size of the spacecraft, its configuration, its orbital altitude if in orbit about a planet with an atmosphere, the solar cycle, and residual magnetic moment.

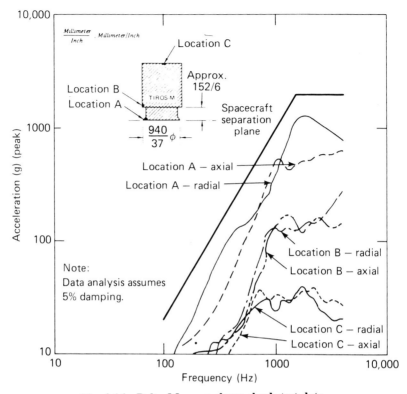

Fig. 3.16 Delta Marman clamp shock test data.

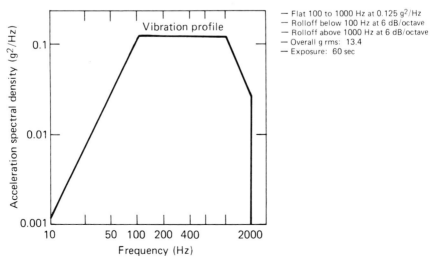

Fig. 3.17 Titan 34D vibration environment: IUS/spacecraft interface random vibration.

Note: 1. Vibration level converted from a 1600-lb spacecraft to a spacecraft weighing w lb is:

$$\text{Delta dB} = \frac{-20 \log w}{1600}$$

2. Based on 15-ft payload fairing

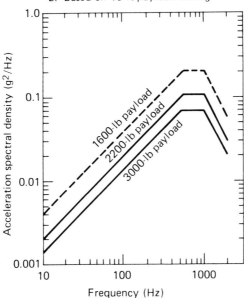

Fig. 3.18 Titan 3C vibration environment: transtage/spacecraft interface random vibration.

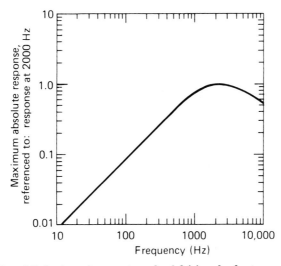

Fig. 3.19 Titan 3C shock environment: payload fairing shock at spacecraft interface, normalized at 2000 Hz.

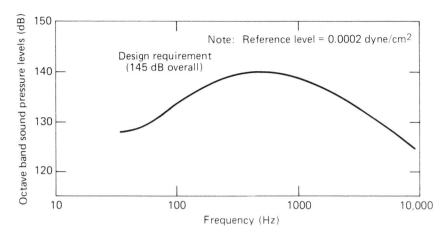

Fig. 3.20   Titan 3C payload fairing internal acoustic spectrum.

Additionally, the spacecraft will experience periodic impulsive disturbances resulting from attitude or translation control actuators, internal moving parts, or the activities of a human flight crew. If confined to the spacecraft interior, these disturbances may produce no net displacement of the spacecraft center of mass. However, for sensitive payloads such as optical instruments or materials-processing experiments that are fixed to the spacecraft, the result is the same.

The most obvious external sources of perturbing accelerations are environmental influences such as aerodynamic drag and solar radiation pressure, both discussed in Chapter 4. If necessary, these and other nongravitational effects can be removed, to a level of better than $10^{-11} g$, by a disturbance-compensation system to yield essentially drag-free motion. This concept is discussed in Chapter 4 and has seen use with navigation satellites, where the ability to remain on a gravitationally determined (thus highly predictable) trajectory is of value.

A perturbing acceleration that cannot be removed is the so-called gravity-gradient force. Discussed in more detail in Chapter 8, this force results from the fact that only the spacecraft center of mass is truly in a gravitationally determined orbit. Masses on the vehicle that are closer to the center of the Earth would, if in a free orbit, drift slowly ahead of those masses located farther away. Since the spacecraft is a more or less rigid structure, this does not happen; the internal elastic forces in the structure balance the orbital dynamic accelerations tending to separate masses orbiting at different altitudes.

Gravity-gradient effects are significant ($10^{-3} g$ or possibly more) over large vehicles such as the Shuttle or a space station. For most applications this may be unimportant; however, certain materials-processing operations are particularly demanding of low gravity conditions and thus may need to be conducted in free-flying modules, where they can be located near the center of mass. Higher altitude also diminishes the effect, which follows an inverse cube force law.

Although we have so far discussed only the departures from the idealized 0-g environment, it is nonetheless true that the most pronounced and obvious condition associated with space flight is weightlessness. As with other environmental factors, it has both positive and negative effects on space vehicle design and flight operations. The benefits of weightlessness in certain manufacturing and materials-processing applications are in fact a significant practical motivation for the development of a major space operations infrastructure. Here, however, we focus on the effects of 0 g on the spacecraft functional design.

The 0-g environment allows the use of relatively light spacecraft structures by comparison with earthbound designs. This is especially true where the structure is actually fabricated in orbit or packaged in such a way that it is not actually used or stressed until the transportation phase is complete. The U.S. space station will be an example in this regard, as was the lunar roving vehicle developed during the Apollo program. A possibly awkward side effect of large, low-mass structures is that they tend to have relatively low damping and hence are susceptible to substantial structural excitation. Readers who have seen the films of the Tacoma Narrows Bridge disaster, the classic case in this regard, will be aware of the potential for concern. Less dramatically, attitude stabilization and control of large space vehicles is considerably complicated by structural flexibility. This is discussed in more detail in Chapter 7.

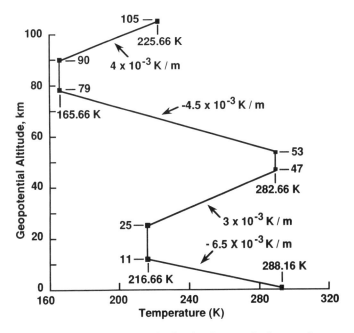

Fig. 3.21 Temperature distribution for standard atmosphere.

In some cases, the relatively light and fragile mechanical designs appropriate for use in space render ground testing difficult. Booms and other deployable mechanisms may not function properly, or at least the same way, in a 1-g field if designed for 0 or low g. Again, a case in point is the Apollo lunar rover. The actual lunar rover, built for one-sixth g, could not be used on Earth, and the lunar flight crews trained on a stronger version. In other cases, booms and articulating platforms may need to be tested by deploying them horizontally and supporting them during deployment in Earth's gravity field.

The calibration and mechanical alignment of structures and instruments intended for use in flight can be a problem in that the structure may relax to a different position in the strain-free 0-g environment. For this and similar reasons, spacecraft structural mass is often dictated by stiffness requirements rather than by concerns over vehicle strength. Critical instrument alignment and orientation procedures are often verified by the simple artifice of making the necessary measurements in a 1-g field, then inverting the device and repeating the measurements. If significant differences are not observed, the 0-g behavior is probably adequate.

Weightlessness complicates many fluid and gasdynamic processes, including thermal convection, compared with ground experience. The situation is particularly exacerbated when one is designing for human presence. Effective toilets, showers, and cooking facilities are much harder to develop for use in 0 g. When convection is required for thermal control or for breathing air circulation, it must be provided by fans or pumps. The same is true of liquids in tanks; if convection is required to maintain thermal or chemical

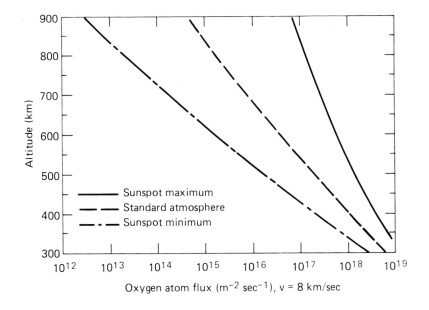

Fig. 3.22 Oxygen atom flux variation with altitude.

uniformity, it must be explicitly provided. Weightlessness is a further annoyance when liquids must be withdrawn from partially filled tanks, as when a rocket engine is ignited in orbit. Secondary propulsion systems will usually employ special tanks with pressurized bladders or wicking to ensure the presence of fuel in the combustion chamber. Larger engines are usually ignited following an ullage burn of a small thruster to force the propellant to settle in place over the intake lines to the engine.

As mentioned, a significant portion of the concern over spacecraft cleanliness during assembly is due to the desire to avoid problems from floating dust once in orbit. Similarly, careful control over assembly operations is necessary to prevent dropped or forgotten bolts, washers, electronic components, tools, and other paraphernalia from causing problems in flight. Again, this may be of particular concern for manned vehicles, where an inhaled foreign object could be deadly. It is for this reason that the Shuttle air circulation ports are screened; small objects tend to be drawn by air currents toward the intake screens, where they remain until removed by a crew member.

Weightlessness imposes other design constraints where manned operations are involved. Early attempts at extravehicular operations during the Gemini program of the mid-1960s showed that inordinate and unexpected effort was required to perform even simple tasks in 0 g. Gene Cernan on his Gemini 9 flight became so exhausted merely putting on his maneuvering backpack that he was unable to test the unit. Other astronauts experienced difficulty in handling their life-support tethers and in simply shutting the spacecraft hatch upon completion of the extravehicular activity (EVA).

These and other problems were in part caused by the bulkiness and limited freedom of movement possible in a spacesuit, but to a greater extent were due to the lack of body restraint normally provided by the combination of friction and the 1-g Earth environment. With careful attention to the placement of hand and foot restraints it proved possible to accomplish significant work during EVA without exhausting the astronaut. This was demonstrated by Edwin (Buzz) Aldrin during the flight of Gemini 12 and put into practice "for real" by the Skylab 2 crew of Conrad, Kerwin, and Weitz during the orbital repair of the Skylab workshop. Today, EVA is accepted as nearly routine, as shown during a number of successful retrieval, repair, and assembly operations from the Shuttle.

Radiation

Naturally occurring radiation may be a problem for many missions, primarily due to its effects on spacecraft electronic systems. Such radiation effects are of basically two kinds: degradation due to total dose and malfunctions induced by so-called single-event upsets. Fundamentally different mechanisms are involved in these two failure modes.

High-energy particulate radiation impacting a semiconductor device will locally alter the carefully tailored crystalline structure of the device. After a sufficient number of such events, the semiconductor is simply no longer the required type of material and ceases to function properly as an electronic device. Total dose effects can be aggravated by the radiation intensity; a solar flare can induce failures well below the levels normally tolerated by a given device, since at lower dose rates the device will anneal to some extent and "heal" itself.

The other physical effect that occurs upon impact of particulate radiation with other matter is local ionization as the incoming particle slows down and deposits energy in the material. In silicon, for example, one hole-electron pair is produced for each 3.6 eV of energy expended by the incoming particle. Thus, even a relatively low energy cosmic ray of some 10^7 eV will produce about 3×10^6 electrons, or 0.5 pC. This is not an insignificant charge level in modern integrated circuit devices and may result in a single-event upset.

The single-event upset phenomenon has come about as a result of successful efforts to increase speed and sensitivity and reduce power requirements of electronic components by packing more semiconductor devices into a given volume. This is done essentially by increasing the precision of integrated circuit manufacture so that smaller circuits and devices may be used. For example, the mid-1980s state of the art in integrated circuit manufacture enabled integrated circuit devices with characteristic feature sizes on the order of 1 μm to be fabricated. Small circuits and transistor junctions imply operation at lower current and charge levels; thus by the late 1970s device "critical charge" levels reached the 0.01- to 1.0-pC range, where a single ionizing particle could produce enough electrons to change a "0" state to a "1," or vice versa. This phenomenon was explained in a classic paper by May and Woods.[3] Its potential for harm if the change of state occurs in a critical memory location is obvious.

In practice, the damage potential of the single-event upset may exceed even that due to a serious software malfunction. If CMOS circuitry is used, the device can "latch up" into a state where it draws excessively high current, destroying itself. This is particularly unfortunate in that CMOS components require very little power for operation (lower than NMOS or PMOS alone and far lower than TTL) and are thus attractive to the spacecraft designer. Latch-up protection is possible, either in the form of external circuitry or built into the device itself. Built-in latch-up protection is characteristic of modern CMOS devices intended for use in high-radiation environments.

The most annoying property of single-event upsets is that, given a device that is susceptible to them, they are statistically guaranteed to occur (this is true even on the ground). One can argue about the rate of such events; however, as noted earlier, even one upset at the wrong time and place could be catastrophic. Protection from total dose effects can be essentially guaranteed with known and usually reasonable amounts of shielding, in combination with careful use of radiation hardened parts. However, there is no reasonable amount of shielding that offers protection against heavy nuclei galactic cosmic rays causing single-event upsets.[4,5]

Upset-resistant parts are available and should be used when analysis indicates the upset rate to be significant. (The level of significance is a debatable matter, with an error rate of 10^{-10}/day a typical standard. Note that, even with such a low rate, several upsets would be expected for a spacecraft with a megabit of memory and a projected 10-yr lifetime.) As pointed out, shielding will not provide full relief but can be used to advantage to screen out at least the lower energy particles, thus reducing the upset rate. However, in many applications even relatively low error rates cannot be tolerated, and other measures may be required. These basically fall into the category of error detection and correction. Such methods include the use of independent processors with "voting" logic and the addition of extra bits to the required computer word length to accommodate error detection and correction codes. Other approaches may also be useful in particular cases.

As mentioned, total dose effects are more tractable because of the more predictable dependence of the dose on the orbit and the mission lifetime. For low-orbit missions, radiation is typically not a major design consideration. For this purpose, low orbit may be defined as less than about 1000-km altitude. At these altitudes, the magnetic field of the Earth deflects most of the incoming solar and galactic charged particle radiation. Since the configuration of the magnetic field does channel some of the particles toward the magnetic poles (the cause of auroral displays), spacecraft in high inclination orbits will tend to receive somewhat greater exposure than those at lower inclinations. However, because periods are still relatively short and the levels moderate, the expected dosages are not typically a problem.

Figures 3.23 and 3.24 present the natural radiation environment vs altitude for spacecraft in Earth orbit. Figure 3.23 shows the radiation dose accumulated by electronic components over a 10-yr mission in circular, equatorial orbits. Since electronic components are normally not exposed directly to space but are contained in a structure, curves are presented for two thicknesses of aluminum structure to account for the shielding effect. The extremely high peaks, of course, correspond to the Van Allen radiation belts, toroidal belts of charged particle radiation trapped by the magnetic field, centered on the equator. Note that the shielding is more effective in the outer belt. This reflects the fact that the outer belt is predominantly electrons, whereas the much heavier protons dominate the inner belt. Figure 3.24 shows the radiation count vs energy level for selected Earth orbits.

Fortunately for the communications satellite industry, geostationary orbit at about six Earth radii is well beyond the worst of the outer belt and is in a region in which the shielding due to the spacecraft structure alone is quite effective. However, it may be seen that in a 10-yr mission a lightly shielded component could accumulate a total dose of 10^6 rad. To put this in perspective, Table 3.4 presents radiation resistance or "hardness" for various classes of electronic components. As this table shows, very few components can sustain this much radiation and survive. The situation becomes worse when one recognizes the need to apply a radiation design margin of the order of two in order to be certain that the components will complete the mission with unimpaired capability. For a dose of 1 Mrad and a design margin of two, all components must be capable of 2 Mrad. At this level the choices are few, thus mandating increased shielding in order to guarantee an adequate suite of components for design.

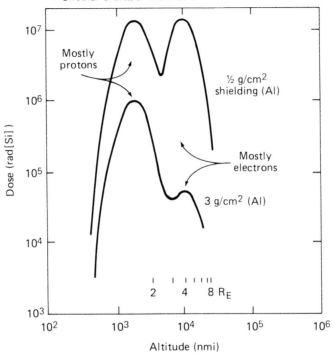

Fig. 3.23   Radiation environment for circular equatorial orbits.

**Table 3.4   Radiation hardness levels
for semiconductor devices**

Technology	Total Dose, rads (Si)
CMOS (soft)	10^3-10^4
CMOS (hardened)	$5 \times 10^4-10^6$
CMOS/SOS (soft)	10^3-10^4
CMOS/SOS (hardened)	$> 10^5$
ECL	10^7
I^{2}L	$10^5-4 \times 10^6$
Linear IC^{2}s	$5 \times 10^3-10^7$
MNOS	10^3-10^5
MNOS (hardened)	$5 \times 10^5-10^6$
NMOS	$7 \times 10^2-7 \times 10^3$
PMOS	$4 \times 10^3-10^5$
TTL/STTL	$> 10^6$

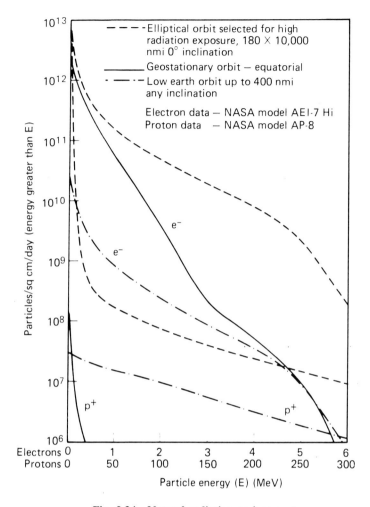

Fig. 3.24   Natural radiation environment.

The example discussed earlier is not unreasonable. Most commercial communications satellites are designed for on-orbit lifetime of 5–7 yr, and an extended lifetime of 10 yr is quite reasonable as a goal. In many cases these vehicles do not recoup the original investment and begin to turn a profit until several years of operation have elapsed.

If design requirements and operating environment do require shielding beyond that provided by the material thickness needed for structural requirements, it may still be possible to avoid increasing the structural thickness. Spot shielding is very effective for protecting individual sensitive components or circuits. Such shielding may be implemented as a box containing the hardware of interest. Another approach might be a potting compound loaded with shielding material. (Obviously, if the shielding substance is electrically conductive, care must be exercised to prevent any detrimental effect on the circuit.) An advantage offered by the nonstructural nature of spot shielding is that it allows for the possibility of using shielding materials, such as tantalum, that are more effective than the normal structural materials. This may allow some saving in mass.

Alterations in the spacecraft configuration may also be used advantageously when certain circuits or components are particularly sensitive to the dose anticipated for a given mission and orbit. Different portions of the spacecraft will receive different dosages according to the amount of self-shielding provided by the configuration. Thus, components placed near rectangular corners may receive as much as seven-fourths the dose of a component placed equally near the spacecraft skin, but in the middle of a large, thick panel. When some flexibility in the placement of internal electronics packages exists, these and other properties of the configuration may be exploited.

A spacecraft in orbit above the Van Allen belts or in interplanetary space is exposed to solar-generated radiation and galactic cosmic rays. The dose levels from these sources are often negligible, although solar flares can contribute several kilorads when they occur. Galactic cosmic rays, as discussed earlier, can produce severe single-event upset problems, since they consist of a greater proportion of high-speed, heavy nuclei against which it is impossible to shield.

Manned flight above the Van Allen belts is a case where solar flares may have a potentially catastrophic effect. The radiation belts provide highly effective shielding against such flares, and in any case a reasonably rapid return to Earth is usually possible for any such close orbit. (This assumption may need to be re-examined for the case of future space station crews.) Once outside the belts, however, the received intensity of solar flare radiation may make it impractical to provide adequate shielding against such an event. For example, although the average flare can be contained, for human physiological purposes, with $2-4$ g/cm^2 of shielding, infrequent major events can require up to 40 g/cm^2, an impractical amount unless a vehicle is large enough to have an enclosed, central area to act as a "storm cellar." It is worth noting that the Apollo command module, and certainly the lunar module, did not provide enough shielding to enable crew survival in the presence of a flare of such intensity as that which occurred in Aug. 1972.

Most of the bodies in the solar system do not have intense magnetic fields and thus have no radiation belts (by the same token, low-altitude orbits and the planetary surface are thus unprotected from solar and galactic radiation). This cannot be said of Jupiter, however. The largest of the planets has a very powerful magnetic field and intense radiation belts. Figure 3.25 indicates the intensity of the Jovian belts.

Natural radiation sources may not be the only problem for the spacecraft designer. Obviously, military spacecraft for which survival is intended (possibly "hoped" is the term) in the event of a nuclear exchange pose special challenges. Less pessimistically, future spacecraft employing nuclear reactors for power generation will require shielding methods not previously employed, at least on U.S. spacecraft. Even relatively low-powered ra-

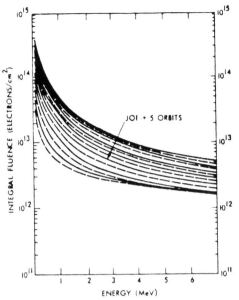

a) Integral electron fluences for
   the Galileo mission (JOI-
   Jupiter orbit insertion)

b) Electron dose vs aluminum
   shield thickness for the
   Galileo mission

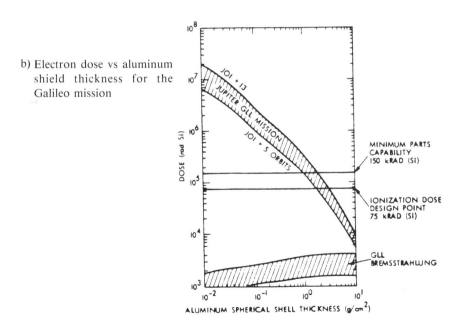

Fig. 3.25   Jupiter radiation environment.

dioisotope thermoelectric generators (RTGs), used primarily on planetary spacecraft, can cause significant design problems. These issues will be discussed in more detail in Chapter 11.

Finally, radiation may produce damaging effects on portions of the spacecraft other than its electronic systems. Polymers and other materials formed form organic compounds are known to be radiation-sensitive. Such materials, including Teflon and Delrin, are not used on external surfaces in high-radiation environments such as Jupiter orbit.[6] Other materials, such as Kevlar epoxy, which may be used in structural or load-bearing members, can suffer a 50–65% reduction in shear strength after exposure to large (3000 Mrad) doses such as those that may be encountered by a permanent space station.[7]

Micrometeoroids

Micrometeoroids are somewhat of a hazard to spacecraft, although substantially less than once imagined. NASA in 1969 published a model of the meteoroid environment[8] that was based on data from the Pegasus satellites flown in Earth orbit specifically for the purpose of obtaining micrometeoroid flux and penetration data, detectors flown on various lunar and interplanetary spacecraft, and optical and radar observation from Earth. This model still represents the best source of design information available today for near-Earth space. The model approximates near-Earth micrometeoroid flux vs particle mass by

$$\log_{10} N_t > m = -14.339 - 1.584 \log_{10} m - 0.063 (\log_{10} m)^2 \qquad (3.2)$$

when the particle mass m is in the range $10^{-12} \text{ g} < m < 10^{-6} \text{ g}$. For larger particles such that $10^{-6} \text{ g} < m < 1 \text{ g}$, the appropriate relation is

$$\log_{10} N_t > m = -14.37 - 1.213 \log_{10} m \qquad (3.3)$$

These relationships are presented graphically in Fig. 3.26. For specific orbital altitudes, gravitational focusing and the shielding effect of the planet must be considered in order to derive the specific meteoroid flux environment for the orbit in question.

Because of the gravitational attraction of the Earth, more meteoroids are found at low altitudes than farther out. A correction for this focusing effect must be applied when extrapolating near-Earth meteoroid flux data to high orbits or to deep space. Assuming an average meteoroid velocity in deep space of 20 km/s, Fig. 3.27 presents a curve of the defocusing factor by which the data from Fig. 3.26 are multiplied in order to obtain the expected flux at a given altitude.

The increase in particle flux for low-altitude planetary orbits tends to be offset by the shielding factor provided by the planet. The body shielding factor ζ is defined as the ratio of shielded to unshielded flux and is given by

$$\zeta = (1 + \cos\theta)/2 \qquad (3.4)$$

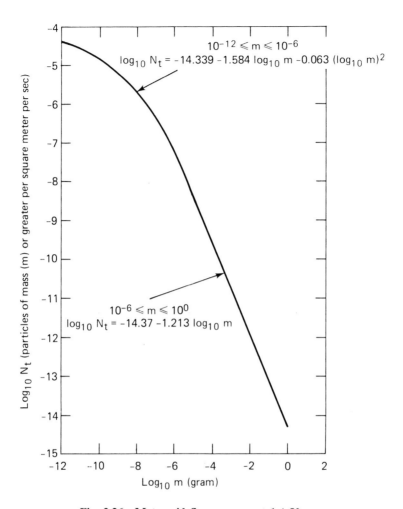

Fig. 3.26 Meteoroid flux vs mass at 1 A.U.

where

$$\sin\theta = R/(R + h) \tag{3.5}$$

where R is the shielding planet radius and h the spacecraft altitude.

Figure 3.28 shows the geometry for the body shielding factor. Although particles vary considerably in density and velocity, for most purposes a density of 0.5 g/cm^3 and a velocity of 20 km/s are used as average values.

It will be seen that most micrometeoroids are extremely small. To put the threat in perspective, a rule of thumb is that a particle of 1 μg will just penetrate a 0.5-mm-thick sheet of aluminum. For most applications, the spacecraft external structure, thermal blankets, etc., provide adequate protection against particles with any significant probability of impact. For longer missions or more severe environments, additional protection may be needed, as with the Viking Orbiter propulsion system. This presented a fairly large area over a relatively long mission. More significantly, however, micrometeoroid impact on the pressurized tanks was highly undesirable, since, although penetration was extremely unlikely, the stress concentrations caused by the crater could have caused an eventual failure. The problem was dealt with by making the outer layer of the thermal blankets out of Teflon-impregnated glass cloth.

The kinetic energy of micrometeoroids is typically so high that, upon impact, the impacting body and a similar mass of the impacted surface are vaporized. This leads to the concept of the "meteor bumper" proposed originally by Dr. Fred Whipple long before the first orbital flights. Although most spacecraft do not require protection of this magnitude, some very severe environments may dictate use of this concept. The concept involves placing a thin shield (material choice is not highly critical but preferably metal) to intercept the incoming particle a short distance from the main structure of the pressure vessel. The thickness of the shield is dictated by the anticipated size of the particles. Ideally the shield should be just thick enough to ensure vaporization of the largest particles that have significant probability of being encountered. The spacing between the shield and the main structure is designed to allow the jet of vaporized material, which still has substantial velocity, to spread over a larger area before striking the main structure. The result of such an event then is a hole in the shield and possibly a dent or depression in the inner structure. Without the shield, a particle of sufficient mass and kinetic energy to dictate this type of protection could cause major damage. Even if it did not penetrate, the impact could result in spalling of secondary particles, still

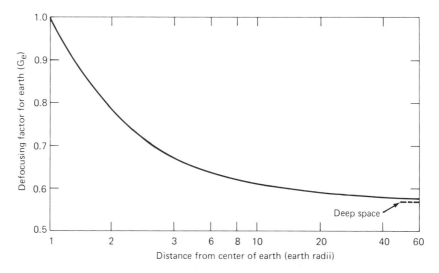

Fig. 3.27 Defocusing factor due to the Earth's gravity for an average meteoroid velocity of 20 km/s.

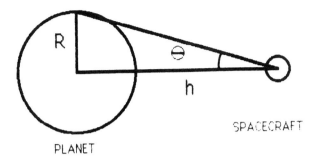

Fig. 3.28 Method for determining body shielding factor for randomly oriented spacecraft.

quite energetic, off the other side of the structure. Such particles could result in severe vehicle damage or crew injury.

The perceptive reader will see that this ability of an impacting particle to spall larger slower particles off the anti-impact side places a significant constraint on shield design. Any area that is made thicker than the optimum for vaporization, say, for attachment brackets, could become the source of secondary particles. These particles, being more massive than the original and possessing considerable kinetic energy, but not enough to vaporize them on impact, can be very damaging. It is clear from this brief discussion that design of such shields is an exacting task requiring both science and art. An actual flight application of this concept is the European Space Agency's Giotto probe, which flew through the dust cloud of Halley's Comet. In this instance the shield is only required on one side of the spacecraft. Relative velocity of the dust is 60–70 km/s. Figure 3.29 shows the Giotto configuration.

Cour-Palais[9] provides a very thorough discussion of mechanisms of meteoroid damage. Although a detailed knowledge of the phenomena involved is beyond the usual scope of systems engineering, a general understanding will be useful in assessing protection which may be required for a given spacecraft mission.

Orbital Debris

Naturally occurring particles are not the only, or, at some altitudes, the most severe impact hazard. In a quarter century of space operations, mankind has managed to create a major hazard in low Earth orbit. As of Jan. 1, 1985, the USAF Space Command was tracking 5408 space objects larger than 10 cm, and statistical samples obtained using ground telescopes show approximately 40,000 objects larger than 1 cm in diameter. Impact sensors on various spacecraft have demonstrated the presence of literally billions of small particles, consisting mostly of paint flecks and aluminum oxide, in the 0.01- to 0.5-mm range. In all cases cited by the USAF, the

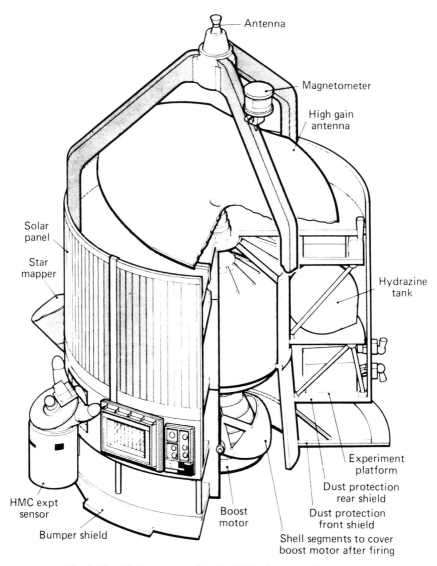

Fig. 3.29   Giotto spacecraft with Whipple meteor bumper.

debris level exceeds, and sometimes greatly exceeds, the natural meteoroid background.

This debris cloud has a variety of sources. Over 80 explosions or other breakups of spacecraft or rocket stages have occurred, contributing over 40,000 kg of debris. In some cases this has occurred deliberately, or at least no effort was made to prevent it, as with early Delta second stages, which usually overpressurized and exploded after jettison.

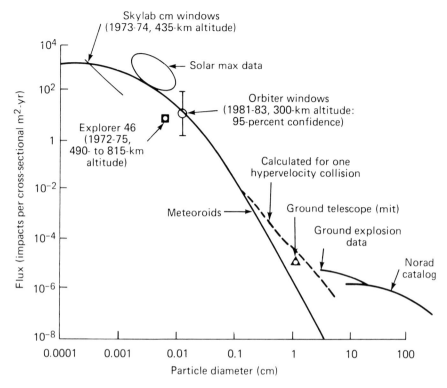

Fig. 3.30   Space-debris: observed flux corrected to 4-cm transmitting size.

The situation is unlikely to improve in the near future. Approximately 2×10^6 kg of spacecraft material resides at altitudes below 2000 km, most in the form of intact vehicles having characteristic dimensions on the order of 3 m. The varying orbital planes of these objects can produce high intersection angles and the potential for high collision velocities. As Kessler and Cour-Palais[10] have shown, such collisions are a statistical certainty and can be expected to contribute to an increasingly dense debris cloud. Routine space operations such as the firing of solid rocket motors, which generate extensive particulate debris, will also continue to add to the low-orbit hazard.

In the early days of space exploration, such considerations no doubt seemed unimportant, since space seemed to be "vast" and "limitless." Although these romantic descriptors are true in general, the volume occupied by moderate altitude, moderate inclination orbits around the Earth is by no means limitless and in fact becomes somewhat congested when populated by tens of thousands of particles moving at 8–10 km/s. The debris density is most severe at the medium altitudes. The debris flux appears to the worst in the altitude range of 600–1100 km. Below 200–

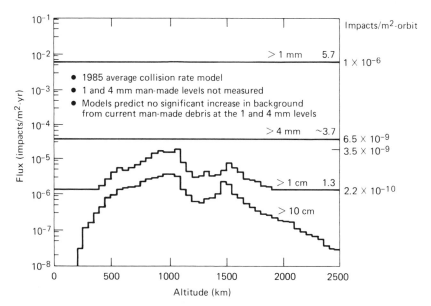

Fig. 3.31   Cumulative flux.

300 km, atmospheric drag causes the debris orbits to decay into the atmosphere. Above 1100 km, the flux tapers off because of the increasing volume of space and because operations have been limited in these orbits.

While geosynchronous orbit is becoming crowded, the debris problem has not reached the severity of the lower altitude environment. This is in part because the large, potentially explosive booster stages that have contributed substantially to the low-orbit debris cloud do not reach geosynchronous altitude. However, it is also true that the communications satellite community was among the first to recognize that measures to minimize orbital clutter should be routinely employed. To this end, it has become standard practice in the industry to lift outmoded or nonfunctional satellites out of the geostationary ring, with fuel for this purpose included in the satellite design budget.

Although no official debris model has yet been adopted, Figs. 3.30 and 3.31 present NASA results[11] widely regarded as the best currently available. Consisting primarily of particles of spacecraft and booster structure, the debris has a much higher density than the comet-derived meteoroid particles. For particles smaller than 1 cm, the density is taken to be 2.8 g/cm^3. For large particles, the particle density ρ is found to be approximately

$$\rho = 2.8/D^{0.074} \text{ g/cm}^3 \qquad (3.6)$$

where D is the average diameter in centimeters. The average relative velocity is assumed to be 10 km/s.

Because of the high flux of particles, the probability of a debris strike on a spacecraft in some orbits may be quite high. Worse still, there is a chance that the strike could involve a "large" particle of a few millimeters diameter. Such an impact could well be catastrophic. For example, NASA models of debris hazards for manned orbital operations assume fatal space suit damage from particles in the 0.3- to 0.5-mm range and catastrophic Shuttle damage from a 4-mm particle. Particles in the 1-mm range will cause a mandatory abort in some cases, such as impact with the radiators.

Although no such catastrophic debris strike is known to have occurred, impacts of lesser importance have been observed. For example, on the STS-7 mission the outer layer of a windshield on the Shuttle Challenger was cracked by what, upon postflight analysis, proved to be a fleck of paint. Also, many small impacts were observed in samples of thermal blanketing returned from the Solar Max spacecraft following its 1984 on-orbit repair. These and other experiences will undoubtedly lead to further efforts to update standard orbital debris models to reflect changing conditions.

Even a perfect debris model is not of much help to the spacecraft designer having the task of protecting his vehicle. Armoring the spacecraft may not be practical from a mass standpoint and in some cases may not even be possible. At this point, the only strategy is to avoid high-probability orbits. As mentioned, the problem will probably increase in severity for some years before greater awareness and increased use of reusable launch vehicles will reverse the trend. A spate of antisatellite (ASAT) vehicle tests of the type conducted by the USSR on several occasions and by the U.S. in Sept. 1985 could greatly aggravate the problem.

As an illustrative example of the effects of hypervelocity impact on orbital clutter, the Sept. 1985 ASAT test was estimated to have created approximately 10^6 fragments between 1-mm and 1-cm diam. This event alone thus produced, at an altitude sufficient to yield long-lived orbits, a debris environment in excess of the natural micrometeoroid background.

Thermal Environment

Space flight presents in many cases both a varied and extreme thermal environment to the space vehicle designer. Spacecraft thermal control is an important topic in its own right and will be treated in more detail in Chapter 10. However, it is appropriate in this section to survey some of the environmental conditions that must be addressed in the thermal design.

The space vacuum environment essentially allows only one means of energy transport to and from the spacecraft, that of radiative heat transfer. The overall energy balance is therefore completely defined by the solar and planetary heat input, internally generated heat, and the radiative energy transfer properties that are determined by the spacecraft configuration and materials. The source and sink temperatures (from the sun with a characteristic blackbody temperature of 5780 K and dark space at 3 K, respectively) for radiative transfer are extreme.

Under these conditions, extremes of both temperature and temperature gradient are common. Thermally isolated portions of an Earth-orbiting spacecraft can experience temperature variations from roughly 200 K during darkness to about 350 K in direct sunlight. One has only to consider such everyday experiences as the difficulty of starting a car in very cold weather, with battery and lubrication problems, or very hot weather (which may cause carburetor vapor lock) to appreciate that most machinery functions best at approximately the same temperatures as humans.

If appropriate internal conduction paths are not provided, temperature differences between the sunlit and dark sides of a spacecraft can be almost as severe as the extremes cited earlier. This results in the possibility of damage or misalignment due to differential expansion in the material. Space vehicles are sometimes rolled slowly about an axis normal to the sun line to minimize this effect. When this is impractical, and other means to minimize thermal gradients are not suitable, special materials having a very low coefficient of thermal expansion (such as Invar or graphite-epoxy) may need to be employed.

The fatiguing effect on materials of repeated thermal cycling between such extremes is also a problem and has resulted in many spacecraft component and subsystem failures (e.g., Landsat-D solar cell harness connections).

Thermal system design in vacuum is further complicated by the need for special care in ensuring good contact between bolted or riveted joints. Good thermal conductivity under such conditions is difficult to obtain, hard to quantify, and inconsistent in its properties. Usually a special thermal contact grease or pad is required to obtain consistently good conductive heat transfer.

The lack of free convection has been mentioned in connection with the 0-g environment; it is of course equally impossible in vacuum. Heat transfer internal to a spacecraft is therefore by means of conduction and radiation, in contrast to ground applications in which major energy transport is typically due to both free and forced convection. This results in the need for careful equipment design to ensure appropriate conduction paths away from all internal hot spots and detailed analytical verification of the intended design. This may sometimes be avoided by hermetically sealing an individual package, or, as is common for Soviet spacecraft, by sealing the whole vehicle. The disadvantage here is obviously that a single leak can result in loss of the mission.

The atmospheric entry thermal environment is the most severe normally encountered by a spacecraft, and vehicles designed for this purpose employ a host of special features to achieve the required protection. This is discussed in more detail in Chapters 6 and 10.

Planetary Environments

Interplanetary spacecraft designers face environmental problems that may be unique even for what is, after all, a rather specialized field. Flyby spacecraft, such as Pioneers 10 and 11 and Voyagers 1 and 2, may encounter radiation environments greatly exceeding those in near-Earth space. Flights to Mercury require the capability to cope with a factor of 10 increase in solar heating compared to Earth orbit, whereas Voyager 2 at Neptune in 1989, received only about 1% of the illumination at Earth. In addition to these considerations, planetary landers face possible hazards such as sulfuric acid in the Venusian atmosphere and finely ground windblown dust on Mars. Spacecraft intended for operation on the lunar surface must be designed to withstand alternating hot and cold soaks of 2-wk duration and a range of 200 K.

It is beyond the scope of this text to discuss in detail the environments of each extraterrestrial body, even where appropriate data exist. Spacecraft system designers involved in missions where such data are required must familiarize themselves with what is known and must usually include safety margins in all design calculations.

References

[1]Engels, R. C., Craig, R. R., and Harcrow, H. W., "A Survey of Payload Integration Methods," *Journal of Spacecraft and Rockets*, Vol. 21, Sept.–Oct. 1984, pp. 417–424.

[2]U.S. Standard Atmosphere, National Oceanic and Atmospheric Administration, NOAA S/T 76-1562, U.S. Government Printing Office, Washington, DC, 1976.

[3]May, T. C., and Woods, M. H., "Alpha-Particle-Induced Soft Errors in Dynamic Memories," *IEEE Transactions on Electron Devices*, Vol. ED-26, No. 1, Jan. 1979, pp. 2–9.

[4]Cunningham, S. S., "Cosmic Rays, Single Event Upsets and Things that Go Bump in the Night," *Proceedings of the AAS Rocky Mountain Guidance and Control Conference*, Paper AAS-84-05, 1984.

[5]Cunningham, S. S., Banasiak, J. A., and Von Flowtow, C. S., "Living with Things that Go Bump in the Night," *Proceedings of the AAS Rocky Mountain Guidance and Control Conference*, Paper AAS-85-056, 1985.

[6]Bouquet, F. L., and Koprowski, K. F., "Radiation Effects on Spacecraft Materials for Jupiter and Near-Earth Orbiters," *IEEE Transactions on Nuclear Science*, Vol. NS-29, No. 6, Dec. 1982, pp. 1629–1632.

[7]Frisch, B., "Composites and the Hard Knocks of Space," *Astronautics and Aeronautics*, Vol. , pp. 33–38.

[8]"Meteroid Environment Model-1969," NASA SP-8013.

[9]Cour-Palais, B., "Hypervelocity Impact in Metals, Glass, and Composites," *International Journal of Impact Engineering*, Vol. 5, 1987, pp. 221–237.

[10]Kessler, D. J., and Cour-Palais, B. G., "Collision Frequency of Artificial Satellites: The Creation of a Debris Belt," *Journal of Geophysical Research*, Vol. 83, No. A6, June 1978, pp. – .

[11]Kessler, D. J., and Cour-Palais, B. G., "Orbital Debris Environment and Spacecraft Shielding," NASA Johnson Space Center, 1988.

4
ASTRODYNAMICS

Astrodynamics is the study of the motion of man-made objects in space subject to both natural and artificially induced forces. It is the latter factor that lends a design element to astrodynamics that is lacking in its parent science, celestial mechanics. The function of the astrodynamicist is to synthesize trajectories that, within the limits imposed by physics and launch vehicle performance, accomplish desired mission goals. Experience gained since the dawn of the space age in 1957, together with the tremendous growth in the speed and sophistication of computer analyses, have allowed the implementation of mission designs not foreseen by early pioneers in astronautics. This trend was discussed briefly in Chapter 2 and shows every sign of continuing. The extrapolation of halo orbits[1] for the International Sun-Earth Explorer (ISEE) missions, the development of space colony concepts using the Earth-moon Lagrangian points,[2] together with the analysis by Heppenheimer[3] of "achromatic" trajectories to reach these points from the moon, and the extensive modern use of gravity-assist maneuvers[4] for interplanetary missions are but a few examples.

Astrodynamics, through its links to classical astronomy, has its roots in the very origins of the scientific revolution. The mathematical elegance of the field exceeds that found in any other area of astronautics. Problems posed in celestial mechanics have been a spur to the development of both pure and applied mathematics since Newton's development of the calculus (which he used, among other things, to derive Kepler's laws of planetary motion and to show that a spherically symmetric body acts gravitationally as if its mass were concentrated at a point at its center). Hoyle[5] comments on this point and notes that it has not been entirely beneficial; the precomputer emphasis on analytical solutions led to the development of many involved methods and "tricks" useful in the solution of celestial mechanics problems. Many of these methods persist to the present as an established part of mathematics education, despite having little relevance in an era of computational sophistication.

Because of the basic simplicity of the phenomena involved, it is possible in astrodynamics to make measurements and predictions to a level of accuracy exceeded in few fields. For example, it is not unusual to measure the position of an interplanetary spacecraft (relative to its tracking stations) to an accuracy of less than a kilometer. Planets such as Mars and Venus which have been orbited by spacecraft can now be located to within several tens of meters out of hundreds of millions of kilometers. Such precision is attained only at the price of extensive data processing and considerable care in modeling the solar system environment.

We shall not engage in detailed consideration of the methods by which the highest possible degree of precision is attained. Although it is true that the most accurate methods of orbit prediction and determination are desirable in the actual execution of a mission, such accuracy is rarely needed at the levels of mission definition appropriate to spacecraft design. What is required is familiarity with basic orbit dynamics and an understanding of when and why more complex calculations are in order. We take the view that spacecraft design requires a level of competence in astrodynamics approximately defined by the range of methods suitable for hand calculation. Any analysis absolutely requiring a computer for its completion is in general the province of specialists.

4.1 FUNDAMENTALS OF ORBITAL MECHANICS

Two-Body Motion

The basis of astrodynamics is Newton's law of universal gravitation:

$$F = -GMm/r^2 \qquad (4.1)$$

which yields the force between the two point masses M and m separated by a distance r and directed along the vector r (see Fig. 4.1) between them. It

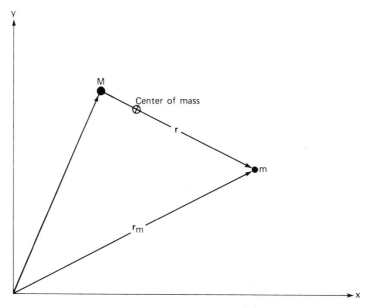

Fig. 4.1   Two-body motion in inertial space.

is a sophomore-level exercise in physics[6] to show that M and m may be extended spherically symmetric bodies without affecting the validity of Eq. (4.1).

A necessary and sufficient condition that F be a conservative (nondissipative, path-independent) force is that it be derivable as the negative (by convention) gradient of a scalar potential. This is the case for the gravitational force law, as seen by differentiating the potential function (per unit mass of m) given by

$$U = -GM/r \qquad (4.2)$$

U has dimensions of energy per unit mass and is thus the potential energy of mass m due to its position relative to mass M. Note that the singularity at the origin is excluded from consideration, since M and m cannot be coincident, and that the potential energy is taken as zero at infinity, an arbitrary choice, since Eq. (4.2) could include an additive constant with no change in the force law. With the negative-gradient sign convention indicated earlier, the potential energy is always negative.

Using Newton's second law,

$$F = ma \qquad (4.3)$$

in an inertial frame and equating the mutual force of each body on the other leads to the familiar inverse square law equation of motion

$$d^2r/dt^2 + G(M + m)/r^3 r = 0 \qquad (4.4)$$

where r is defined as shown in Fig. 4.1. Several key results may be obtained[7] for a universe consisting of only the two masses M and m:

1) The center of mass of the two-body system is unaccelerated and thus may serve as the origin of an inertial reference frame.

2) The angular momentum of the system is constant; as a result, the motion is in a plane normal to the angular momentum vector.

3) The masses M and m follow paths that are conic sections with their center of mass as one focus; thus, the possible orbits are a circle, an ellipse, a parabola, or a hyperbola.

We note that the two-body motion described by Eq. (4.4) is mathematically identical to the motion of a particle of reduced mass:

$$m_r = Mm/(M + m) \qquad (4.5)$$

subject to a radially directed force field of magnitude GMm/r^2. The two-body and central-force formulations are thus equivalent, which leads to the practice of writing Eq. (4.4) as

$$d^2r/dt^2 + \mu/r^3 r = 0 \qquad (4.6)$$

where $\mu = G(M + m)$. In nearly all cases of interest in astrodynamics, $m \ll M$, which leads to $m_r \simeq m$, $\mu \simeq GM$, and a blurring of the physical distinction between two-body and central-force motion. The system center of mass is then in fact the center of mass of the primary body M. For example, a satellite in Earth orbit has no measurable effect on the motion of the Earth, which appears as the generator of the central force. This approximation is still quite valid in the description of planetary motion, although careful measurements can detect the motion of the sun about the mutual center, or barycenter, of the solar system. Interestingly enough, the Earth-moon pair provides one of the few examples in the solar system where the barycenter of the two masses is sufficiently displaced from the center of the "primary" to be readily observable.

The formulation of Eq. (4.6) is especially convenient in that μ is determinable to high accuracy through observation of planetary or space-craft trajectories, whereas G is itself extremely difficult to measure accurately. As an aside, recent theoretical and experimental work[8] suggests that G may not be a constant, but decreases gradually over cosmologically significant time scales.

We now proceed to quantify the results cited earlier. Figure 4.2 depicts the possible orbits, together with the parameters that define their geometric properties. Note that the different conic sections are distinguished by a single parameter, the eccentricity e, which is related to the parameters a and b or a and p as shown in Fig. 4.2. It is also clear from Fig. 4.2 that a polar coordinate representation provides the most natural description of conic orbits, as a single equation,

$$r = p/(1 + e \cos\theta) \qquad (4.7)$$

accounts for all possible orbits. The angle θ is the true anomaly (often v in the classical celestial mechanics literature) measured from periapsis, the point of closest approach of M and m, as shown in Fig. 4.2. The parameter, or semilatus rectum p, is given by

$$p = a(1 - e^2) \qquad (4.8)$$

It may be useful to combine Eqs. (4.7) and (4.8) to yield

$$r = r_p(1 + e)/(1 + e \cos\theta) \qquad (4.9)$$

where

$$r_p = a(1 - e) \qquad (4.10)$$

is the periapsis radius, obtained at $\theta = 0$ in Eq. (4.7). Elliptic orbits also have a well-defined maximum or apoapsis radius at $\theta = \pi$ given by

$$r_a = a(1 + e) \qquad (4.11)$$

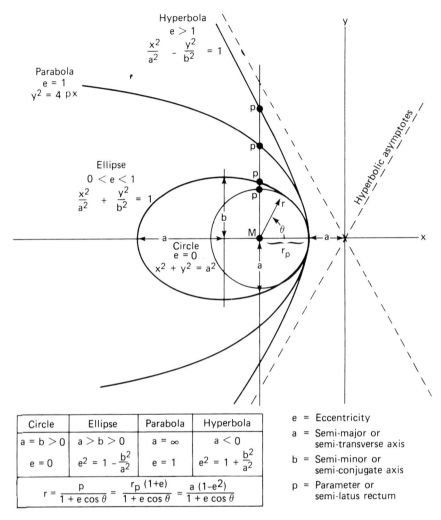

Fig. 4.2   Conic section parameters.

Circle	Ellipse	Parabola	Hyperbola
$a = b > 0$	$a > b > 0$	$a = \infty$	$a < 0$
$e = 0$	$e^2 = 1 - \dfrac{b^2}{a^2}$	$e = 1$	$e^2 = 1 + \dfrac{b^2}{a^2}$

$$r = \frac{p}{1 + e \cos \theta} = \frac{r_p(1+e)}{1 + e \cos \theta} = \frac{a(1-e^2)}{1 + e \cos \theta}$$

e = Eccentricity
a = Semi-major or semi-transverse axis
b = Semi-minor or semi-conjugate axis
p = Parameter or semi-latus rectum

Combining Eqs. (4.10) and (4.11) yields the following useful relationships for elliptic orbits only:

$$a = (r_a + r_p)/2 \tag{4.12}$$

and

$$e = (r_a - r_p)/(r_a + r_p) \tag{4.13}$$

It remains only to specify the relationships between the geometric parameters a, e, and p, and the physical variables energy and angular momentum. The solution of Eq. (4.6) establishes the required connection; this solution is given in a variety of texts[9] and will not be repeated here. We summarize the results below.

The total energy is simply the sum of kinetic and potential energy for each mass. Since the two-body center of mass is unaccelerated and we are assuming that $m \ll M$, the energy of the larger body is negligible; thus, the orbital energy is due to body m alone and is

$$E_t = T + U = V^2/2 - \mu/r \qquad (4.14)$$

where T is the kinetic energy per unit mass. In polar coordinates, with velocity components V_r and V_θ given by r and $r\, d\theta/dt$, respectively,

$$E_t = [r^2 + (r\, d\theta/dt)^2]/2 - \mu/r \qquad (4.15)$$

which is constant due to the previously discussed conservative property of the force law.

The polar coordinate frame in which V_r and V_θ are defined is referred to as the perifocal system. The Z axis of this system is perpendicular to the orbit plane with the positive direction defined such that the body m orbits counterclockwise about Z when viewed from the $+Z$ direction. The origin of coordinates lies at the barycenter of the system, and the X axis is positive in the direction of periapsis. The Y axis is chosen to form a conventional right-handed set. As discussed earlier, this axis frame is inertially fixed; however, it should not be confused with other inertial frames to be discussed in Sec. 4.1 (Coordinate Frames).

One of the more elegant features of the solution for central-force motion is the result that

$$E_t = -\mu/2a \qquad (4.16)$$

i.e., the energy of the orbit depends only on its semimajor (or semitransverse) axis. From Eqs. (4.14) and (4.16) we obtain

$$V^2 = \mu(2/r - 1/a) \qquad (4.17)$$

which is known as the vis-viva or energy equation.

The orbital angular momentum per unit mass of body m is

$$\boldsymbol{h} = \boldsymbol{r} \times d\boldsymbol{r}/dt = \boldsymbol{r} \times \boldsymbol{V} \qquad (4.18)$$

with magnitude given in terms of polar velocity components by

$$h = rV_\theta = r^2\, d\theta/dt \qquad (4.19)$$

and is constant for the orbit, as previously discussed. This is a consequence of the radially directed force law; a force normal to the radius vector is required for a torque to exist, and in the absence of such a torque, angular momentum must be conserved. From the solution of Eq. (4.6), it is found that

$$h^2 = \mu p \qquad (4.20)$$

Thus, the orbital angular momentum depends only on the parameter, or semilatus rectum p. It is also readily shown that the angular rate of the radius vector from the focus to the body m is

$$d\theta/dt = h/r^2 = h(1 + e \cos\theta)^2/p^2 \qquad (4.21)$$

Equations (4.16) and (4.20) may be combined with the geometric result (4.8) to yield the eccentricity in terms of the orbital energy and angular momentum,

$$e^2 = 1 + 2E_t(h/\mu)^2 \qquad (4.22)$$

This completes the summary of results from two-body theory that are applicable to all possible orbits. In subsequent sections, we consider specialized aspects of motion in particular orbits.

Circular and Escape Velocity

From Eq. (4.17), and noting that for a circular orbit $r = a$, we find that the required velocity at radius r,

$$V_{cir} = \sqrt{\mu/r} \qquad (4.23)$$

where V_{cir} is circular velocity. If $E_t = 0$, we have the condition for a parabolic orbit, which is the minimum-energy escape orbit. From Eq. (4.14),

$$V_{esc} = \sqrt{2\mu/r} = \sqrt{2}V_{cir} \qquad (4.24)$$

where V_{esc} is escape velocity. Of course, circular velocity can have no radial component, whereas escape velocity may be in any direction not intersecting the central body.

Circular and parabolic orbits are interesting limiting cases corresponding to particular values of eccentricity. Such exact values cannot be expected in practice; thus, in reality all orbits are either elliptic or hyperbolic, with $E_t < 0$ or $E_t > 0$. Nonetheless, near-circular or near-parabolic orbits may be used as reference trajectories for the actual motion, which is seen as a perturbation of the reference orbit. We will consider this topic in more detail later; for the present, we examine the features of motion in elliptic and hyperbolic orbits.

Motion in Elliptic Orbits

Figure 4.3 defines the parameters of interest in elliptic orbit motion. The conic section results given earlier are sufficient to describe the size and shape of the orbit, but do not provide the position of body m as a function of time. Because it is awkward to attempt a direct solution of Eq. (4.21) to

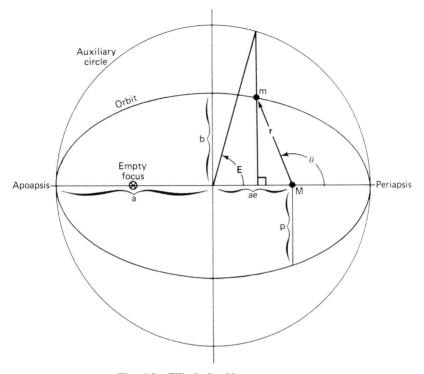

Fig. 4.3 Elliptical orbit parameters.

yield θ (and hence r) as a function of time, the auxiliary variable E, the eccentric anomaly, is introduced. The transformation between true and eccentric anomaly is

$$\tan(\theta/2) = [(1 + e)/(1 - e)]^{1/2} \tan(E/2) \qquad (4.25)$$

It is found[10] that E obeys the transcendental equation

$$f(E) = E - e \sin E - n(t - t_p) = 0 \qquad (4.26)$$

where

n = mean motion = $\sqrt{(\mu/a^3)}$
t_p = time of periapsis passage

which is known as Kepler's equation. The mean motion n is the average orbital rate, or the orbital rate for a circular orbit having the same semimajor axis as the given elliptic orbit. The mean anomaly

$$M \equiv n(t - t_p) \qquad (4.27)$$

is thus an average orbital angular position and has no physical significance unless the orbit is circular, in which case $n = d\theta/dt$ exactly, and $E = \theta = M$ at all times.

When E is obtained, it is often desired to know r directly, without the inconvenience of computing θ and solving the orbit equation. In such a case, the result

$$r = a(1 - e \cos E) \tag{4.28}$$

is useful. The radial velocity in the orbit plane is

$$rV_r = na^2 e \sin E = e(\mu a)^{1/2} \sin E \tag{4.29}$$

The tangential velocity V_θ is found from

$$V_\theta = r \, d\theta/dt = h/r \tag{4.30}$$

If $E = \theta = 2\pi$, then $(t - t_p) = \tau$, the orbital period. Equation (4.26) then gives

$$\tau = 2\pi\sqrt{a^3/\mu} \tag{4.31}$$

which is Kepler's third law.

The question of extracting E as a function of time in an efficient manner is of some interest. A numerical approach is required because no closed-form solution for $E(M)$ exists. At the same time, Eq. (4.26) is not a particularly difficult specimen; existence and uniqueness of a solution are easy to show.[9] Any common root finding method such as Newton's method or the modified false position method[11] will serve. All such methods are based on solving the equation in the "easy" direction (i.e., guessing E, computing M, and comparing with the known value) and employ a more or less sophisticated procedure to choose updated estimates for E. The question of choosing starting values for E to speed convergence to the solution has received considerable attention.[12] However, the current availability of programmable calculators, including some with built-in root finders, renders this question somewhat less important than in the past, at least for the types of applications stressed in this book.

When the orbit is nearly circular, $e \simeq 0$, and approximate solutions of adequate accuracy are available that yield θ directly in terms of mean anomaly. By expanding in powers of e, Eqs. (4.25) and (4.26) can be reduced to the result[10]

$$\theta \simeq M + 2e \sin M + (5e^2/4) \sin 2M + \cdots \tag{4.32}$$

In many cases of interest for orbital operations, the orbits will be nearly circular, and Eq. (4.32) can be used to advantage. For example, an Earth orbit of 200 km × 1000 km, quite lopsided by parking orbit standards, has an eccentricity of 0.0573, which implies that, in using Eq. (4.32), terms of order 2×10^{-4} are being neglected. For many purposes, such an error is unimportant.

When the orbit is nearly parabolic, numerical difficulties are encountered in the use of Kepler's equation and its associated auxiliary relations. This can be seen from consideration of Eq. (4.25) for $e \simeq 1$, where there is considerable loss of numerical accuracy in relating eccentric to true anomaly. The difficulty is also seen in the use of Kepler's equation near periapsis, where E and $e \sin E$ will be almost equal for near-parabolic orbits. Battin[13] and others have developed so-called universal formulas that avoid the difficulties in time-of-flight computations for Keplerian orbits. However, in spacecraft design the problems of numerical inaccuracy for nearly parabolic orbits are more theoretical than practical and will not concern us here.

Motion in Hyperbolic Orbits

The study of hyperbolic orbits is accorded substantially more attention in astrodynamics than it traditionally receives in celestial mechanics. In celestial mechanics, only comets pursue escape orbits, and these are generally almost parabolic; hence, orbit prediction and determination methods tend to center around perturbations to parabolic trajectories. In contrast, all interplanetary missions follow hyperbolic orbits, both for Earth departure and upon arrival at possible target planets. Also, study of the gravity-assist maneuvers mentioned earlier requires detailed analysis of hyperbolic trajectories.

Figure 4.4 shows the parameters of interest for hyperbolic orbits. Because a hyperbolic orbit of mass m and body M is a one-time event (possibly terminated by a direct atmospheric entry or a propulsive or atmospheric braking maneuver to effect orbital capture), the encounter is often referred to as hyperbolic passage. Although described by the same basic conic equation as for an elliptic orbit, hyperbolic passage presents some significant features not found with closed orbits.

In this section, we consider only the encounter between m and M, i.e., the two-body problem; hence, motion of M is ignored. Thus, if m is a spacecraft and M a planetary flyby target, then the separate motion of both m and M in solar orbit is neglected. This is equivalent to regarding the influence of M as dominating the encounter and ignoring that of the sun. This is of course the same approximation we have used in the preceding discussions, and, when m and M are relatively close, it is not a major source of error in the analysis of interest here. For example, the gravitational influence of the sun on a spacecraft in low Earth orbit will not usually be of significance in preliminary mission design and analysis.

Hyperbolic passage is fundamentally different. Although the actual encounter can indeed be modeled as a two-body phenomenon to the same fidelity as before, the complete passage must usually be examined in the context of the larger reference frame in which it takes place. This external frame provides the "infinity conditions" and orientation for the hyperbolic passage, and it is in this frame that gravity-assisted velocity changes must be analyzed. For the present, however, we consider only the actual two-

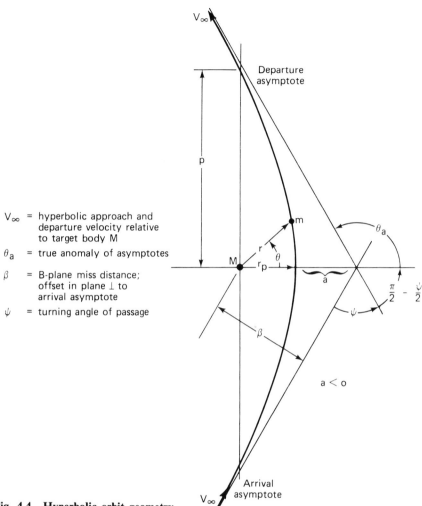

Fig. 4.4 Hyperbolic orbit geometry.

body encounter. The results will be useful within a so-called sphere of influence (actually not a sphere and not sharply defined) about the target body M. Determination of spheres of influence will be addressed in a subsequent section.

When m is "infinitely" distant from M, the orbit equation may be solved with $r = \infty$ to yield the true anomaly of the asymptotes. From Eq. (4.7),

$$\theta_a = \cos^{-1}(-1/e) \qquad (4.33)$$

and due to the even symmetry of the cosine function, asymptotes at $\pm\theta_a$ are obtained, as required for a full hyperbola. Since $r > 0$, values of true anomaly for finite r are restricted to the range $[-\theta_a, \theta_a]$, and so the orbit is concave toward the focus occupied by M.

Knowledge of θ_a serves to orient the orbit in the external frame discussed earlier. The hyperbolic arrival or departure velocity V_∞ is the vector difference, in the external frame, between the velocity of m and that of M. The condition that it must lie along an asymptote determines the orientation of the hyperbolic passage with respect to the external frame. This topic will be considered in additional detail in a subsequent section.

The magnitude of the hyperbolic velocity V_∞ is found from the vis-viva equation with $r = \infty$ to be

$$V_\infty^2 = -\mu/a = 2E_t \qquad (4.34)$$

where it is noted that the semimajor axis a is negative for hyperbolas. Since V_∞ is usually known for the passage from the infinity conditions, in practice one generally uses Eq. (4.34) to solve for a. Conservation of energy in the two-body frame requires V_∞ to be the same on both the arrival and departure asymptotes. However, the vector velocity V_∞ is altered by the encounter due to the change in its direction. This alteration of V_∞ is fundamental to hyperbolic passage and is the basis of gravity-assist maneuvers.

The change in direction of V_∞ is denoted by Ψ, the turning angle of the passage. If the motion of m were unperturbed by M, the departure asymptote would have a true anomaly of $-\theta_a + \pi$, whereas due to the influence of M the departure is in fact at a true anomaly:

$$\theta_a = -\theta_a + \pi + \Psi \qquad (4.35)$$

Hence,

$$\Psi/2 = \theta_a - \pi/2 = \sin^{-1}(1/e) \qquad (4.36)$$

where the second equality follows from Eq. (4.33). The velocity change seen in the external frame due to the turning angle of passage is, as seen from Fig. 4.5,

$$\Delta V = 2V_\infty \sin\Psi/2 = 2V_\infty/e \qquad (4.37)$$

The eccentricity may be found from Eq. (4.10) with the semimajor axis a given by Eq. (4.34), yielding

$$e = 1 + V_\infty^2 r_p/\mu \qquad (4.38)$$

This result is useful in the calculation of a hyperbolic departure from an initial parking orbit, or when, as is often the case for interplanetary exploration missions, the periapsis radius at a target plane is specified. However, Eq. (4.38) is inappropriate for use in pre-encounter trajectory correction maneuvers, which are conventionally referred to as the so-called B plane, the plane normal to the arrival asymptote. Pre-encounter trajec-

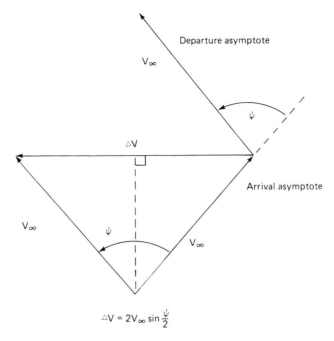

Fig. 4.5 Velocity vector change during hyperbolic passage.

tory corrections will generally be applied at an effectively "infinite" distance from the target body and will thus, almost by definition, alter the placement and magnitude of V_∞ relative to M. The B plane is therefore a convenient reference frame of such maneuvers.

The orbital angular momentum is easily evaluated in terms of the B-plane miss distance β, yielding

$$h = \beta V_\infty \tag{4.39}$$

which follows readily from the basic vector definition of Eq. (4.18) applied at infinity in rectangular coordinates. Using this result plus Eq. (4.34) in Eq. (4.22), we find

$$e^2 = 1 + (\beta V_\infty^2/\mu)^2 \tag{4.40}$$

From Eqs. (4.38) and (4.40), the periapsis radius is given directly in terms of the approach parameters β and V_∞ as

$$r_p/\beta = -(\mu/\beta V_\infty^2) + \sqrt{1 + (\mu/\beta V_\infty^2)^2} \tag{4.41}$$

while the B-plane offset required to obtain a desired periapsis is

$$\beta = r_p[1 + 2\mu/r_p V_\infty^2]^{1/2} \tag{4.42}$$

Equations (4.33–4.42), together with the basic conic section results given in Sec. 4.1 (Two-Body Motion), suffice to describe the spatial properties of hyperbolic passage. It remains to discuss the evolution of the orbit in time. As with elliptic orbits, the motion is most easily described via a Kepler equation,

$$f(F) = e \, \sinh F - F - n(t - t_p) = 0 \qquad (4.43)$$

where

n = mean motion = $[\mu/(-a)^3]^{1/2}$
t_p = time of periapsis passage

and again it is recalled that $a < 0$ for hyperbolic orbits. The hyperbolic anomaly F, as with the eccentric anomaly E, is an auxiliary variable defined in relation to a reference geometric figure, in this case an equilateral hyperbola tangent to the actual orbit at periapsis.[14] The details are not of particular interest here, since the analysis has even less physical significance than was the case for the eccentric anomaly. The transformation between true and hyperbolic anomaly is given by

$$\tan(\theta/2) = [(e + 1)/(e - 1)]^{1/2} \tanh(F/2) \qquad (4.44)$$

Analogously to Eqs. (4.28) and (4.29), it is found that

$$r = a(1 - e \, \cosh F) \qquad (4.45)$$

and

$$rV_r = na^2 e \, \sinh F = e(-a\mu)^{\frac{1}{2}} \sinh F \qquad (4.46)$$

with the tangential velocity again given by Eq. (4.30).

Motion in Parabolic Orbits

Parabolic orbits may be viewed as a limiting case of either elliptic or hyperbolic orbits as eccentricity approaches unity. This results in some mathematical awkwardness, as seen from Eq. (4.8),

$$a = \lim_{e \to 1}[p/(1 - e^2)] = \infty \qquad (4.47)$$

The semimajor axis is thus undefined for parabolic orbits. The result is of somewhat limited concern, however, and serves mainly to indicate the desirability of using Eq. (4.9), from which the semimajor axis has been eliminated, for parabolic orbits. Thus,

$$r = 2r_p/(1 + \cos\theta) \qquad (4.48)$$

and by comparison with Eq. (4.8), it is seen that

$$p = 2r_p \qquad (4.49)$$

It may be shown that the motion in time is given by

$$D^3/6 + D = M = n(t - t_p) \qquad (4.50)$$

The parabolic anomaly D is an auxiliary variable defined as

$$D = \sqrt{2} \tan \theta/2 \qquad (4.51)$$

and the mean motion is

$$n = \sqrt{\mu/r_p^3} \qquad (4.52)$$

As with hyperbolic orbits, n has no particular physical signficance. In terms of parabolic anomaly,

$$r = r_p(1 + D^2/2) \qquad (4.53)$$

while

$$rV_r = (\mu r_p)^{\frac{1}{2}}D = nr_p^2 D \qquad (4.54)$$

As always, the tangential velocity V_θ is given by Eq. (4.30). It should be noted that the exact definition of D varies considerably in the literature, as does the form of Kepler's equation [Eq. (4.50)]. Care should be taken in using analytical results from different sources for parabolic orbits.

Keplerian Orbital Elements

Orbital motion subject to Newtonian laws of motion and gravitational force results in a description of the trajectory in terms of second-order ordinary differential equations, as exemplified by Eq. (4.6). Six independent constants are thus required to determine a unique solution for an orbit; in conventional analysis, these could be the initial conditions consisting of the position and velocity vectors r and V at some specified initial time t_0, often taken as zero for convenience. In fact, however, any six independent constants will serve, with the physical nature of the problem usually dictating the choice.

In classical celestial mechanics, position and velocity information are never directly attainable. Only the angular coordinates (right ascension and declination) of objects on the celestial sphere are directly observable. Classical orbit determination is essentially the process of specifying orbital position and velocity given a time history of angular coordinate measurements. Direct measurement of r and V (or their relatively simple calculation from given data) is possible in astrodynamics, where ground tracking stations and/or onboard guidance systems may, for example, supply position and velocity vector estimates on a nearly continuous basis.

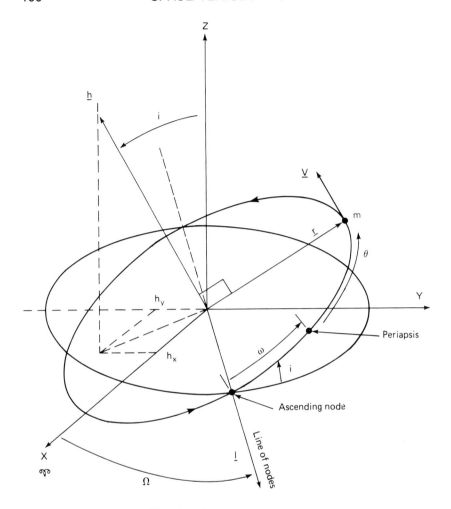

Fig. 4.6   Orbital elements.

However, even when **r** and **V** are obtained, information in this form is of mathematical utility only, since it conveys no physical "feel" for or geometric "picture" of the orbit. It is thus customary, when maximum accuracy is not essential, to describe the orbit in terms of six other quantities plus an epoch, a time t_0 at which they apply. These quantities, chosen to provide a more direct representation of the motion, are the Keplerian orbital elements, defined graphically in Fig. 4.6 and listed below:

a or p = semimajor axis or semi-latus rectum
e = eccentricity
i = inclination of orbit plane relative to a defined reference plane

Ω = longitude or right ascension of ascending node, measured in the reference plane from a defined reference meridian

ω = argument of periapsis, measured counterclockwise from the ascending node in the orbit plane

θ_0, M_0, or t_p = true or mean anomaly at epoch, or time of relevant periapsis passage

As seen, there is no completely standardized set of elements in use, a circumstance in part due to physical necessity. For example, the semimajor axis a is undefined for parabolic orbits and observationally meaningless for hyperbolic orbits, requiring use of a more fundamental quantity, the semilatus rectum p, or occasionally the angular momentum h. Nonetheless, the geometric significance and convenience of the semimajor axis for circular and elliptical orbits will not be denied, and it is always employed when physically meaningful.

Other problems occur as well. For nearly circular orbits, ω is ill defined, as is Ω for orbits with near-zero inclination. In such cases, various convenient alternate procedures are used to establish a well-defined set of orbital elements. For example, when the orbit is nominally circular, $\omega \equiv 0$ is often adopted by convention. When $i \simeq 0$, accurate specification of Ω and hence ω is difficult, and the parameter Π, the so-called longitude of periapsis, is often used. Here, $\Pi = \Omega + \omega$, with Ω measured in the X-Y plane and ω measured in the orbital plane. In such cases Π may be accurately known even though Ω and ω are each poorly specified.

It may be seen that a (or p) and e together specify the size and shape or, equivalently, the energy and angular momentum, of the orbit, and i, Ω, and ω provide the three independent quantities necessary to describe the orientation of the orbit with respect to some external inertial reference frame. The final required parameter serves to specify the position of body m in its orbit at a particular time. As indicated, the true or mean anomaly at epoch, or the time of an appropriate periapsis passage, may also be used. For spacecraft orbits, conditions at injection into orbit or following a midcourse maneuver may also be employed, as may the true or mean anomaly at some particular time other than the epoch.

In a simple two-body universe, the orbital elements are constant, and their specification determines the motion for all time. In the real world, additional influences or perturbations are always present and result in a departure from purely Keplerian motion; hence, the orbital elements are not constant. Perturbing influences can be of both gravitational and nongravitational origin and may include aerodynamic drag, solar radiation, and solar wind; the presence of a third body; a nonspherically symmetric mass distribution in an attracting body; or, in certain very special cases, relativistic effects. One or more of these effects will be important in all detailed analyses, such as are necessary for the actual execution of a mission, and in many cases for preliminary analysis and mission design as well. Indeed, it is common practice in mission design to make use of certain special perturbations to achieve desired orbital coverage, as discussed briefly in Chapter 2 and in Sec. 4.2 (Aspherical Mass Distribution).

It commonly happens that a spacecraft or planetary trajectory is predominantly Keplerian, but that there exist perturbations that are significant at some level of mission design or analysis. When approximate analyses of such cases are carried out, it may be found that the perturbing influences alter various elements or combinations of elements in a periodic manner, or in a secular fashion with a time constant that is small compared to the orbit period. Such analyses may be used to provide relatively simple corrections to a given set of orbital elements describing the average motion, or to a set of elements accurately defined at some epoch.

The result of procedures such as those just described is a description of the orbit in terms of a set of osculating elements, which are time varying and describe a Keplerian orbit that is instantaneously tangent to the true trajectory. In this way increased accuracy can be obtained while still retaining a description of the motion in terms of orbital elements, i.e., without resorting to a numerical solution. These topics will be considered in more detail in a later section.

Coordinate Frames

It is seen in the preceding discussion that a particular orbit is defined through its elements in relation to some known inertial reference frame. However, it is appropriate here to remark briefly on the two major inertial reference frames of interest.

The X axis in all cases of interest in the solar system is defined as the direction of the vernal equinox, the position of the sun against the fixed stars on (presently) March 21, the first day of spring. More precisely, the X axis is the line from the center of the Earth to the center of the sun when the sun crosses the Earth's equatorial plane from the southern to northern hemisphere.

For orbits about or observations from Earth (or, when appropriate, any other planet), the natural reference plane is the planetary equator. The orbital inclination i is measured with respect to the equatorial plane, and the longitude of the ascending node Ω is measured in this plane. The positive Z direction is taken normal to the reference plane in the northerly direction (i.e., approximately toward the North Star, Polaris, for Earth). The Y axis is taken to form a right-handed set and thus lies in the direction of the winter solstice, the position of the sun as seen from Earth on the first day of winter.

The coordinate frame thus defined is referred to as the geocentric inertial (GCI) system. Though fixed in the Earth, it does not rotate with the planet. It is seen that, in labeling the frame as "inertial," the angular velocity of the Earth about the sun is ignored. Since the frame is defined with respect to the "infinitely" distant stars, the translational offsets of the frame throughout the year are also irrelevant, and any axis set parallel to the defined set is equally valid.

For heliocentric calculations, planetary equators are not suitable reference planes, and another choice is required. It is customary to define the Earth's orbital plane about the sun, the ecliptic plane, as the reference plane for the solar system. The Earth's orbit thus has zero inclination, by definition, whereas all other solar orbiting objects have some nonzero inclination. The Z axis of this heliocentric inertial (HCI) system is again normal to the reference plane in the (roughly) northern direction, and the Y axis again is taken to form a right-handed set. The Earth's polar axis is inclined at approximately 23.5 deg relative to the ecliptic, and so the transformation from GCI to HCI is accomplished via a coordinate rotation of 23.5 deg about the X axis. The relationship between these two frames is shown in Fig. 4.7. The position vector (X, Y, Z) may be obtained in the frame in which the orbital elements are defined by

$$X = r[\cos(\omega + \theta) \cos\Omega - \sin(\omega + \theta) \sin\Omega \cos i] \qquad (4.55a)$$

$$Y = r[\cos(\omega + \theta) \sin\Omega + \sin(\omega + \theta) \cos\Omega \cos i] \qquad (4.55b)$$

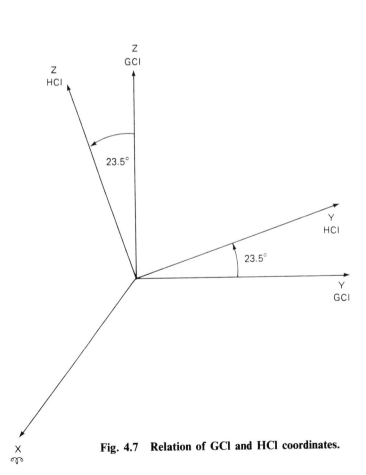

Fig. 4.7   Relation of GCI and HCI coordinates.

$$Z = r[\sin(\omega + \theta)\sin i] \tag{4.55c}$$

with r from Eq. (4.7) or (4.9) and $\theta(t)$ obtained from the appropriate Kepler equation and its auxiliary relations.

In either system, a variety of coordinate representations are possible in addition to the basic Cartesian (X,Y,Z) frame. The choice will depend in part on the type of equipment and observations employed. For example, a radar or other radiometric tracking system will produce information in the form of range to the spacecraft, azimuth angle measured from due North, and elevation angle above the horizon. Given knowledge of the tracking station location, such information is readily converted to standard spherical coordinates (r,θ,ϕ) and therefore to (X,Y,Z) or other coordinates.

When optical observations are made, as in classical orbit determination or when seeking to determine the position of an object against the background of fixed stars, range is not a suitable parameter. All objects appear to be located at the same distance and are said to be projected onto the celestial sphere. In this case, only angular information is available, and a celestial longitude-latitude system similar to that used for navigation on Earth is adopted. Longitude and latitude are replaced by right ascension and declination (α,δ). Right ascension is measured in the conventional trigonometric sense in the equatorial plane (about the Z axis), with 0 deg at the X axis. Declination is positive above and negative below the equatorial $(X\text{-}Y)$ plane, with a range of ± 90 deg. Figure 4.8 shows the

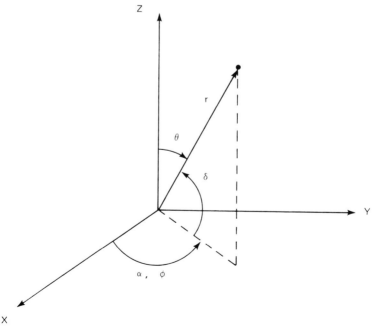

Fig. 4.8   Relationship between Cartesian, spherical, and celestial coordinates.

relationship between Cartesian, celestial, and spherical coordinates. Useful transformations are

$$X = r \sin\theta \cos\phi = r \cos\delta \cos\alpha \qquad (4.56a)$$

$$Y = r \sin\theta \sin\phi = r \cos\delta \sin\alpha \qquad (4.56b)$$

$$Z = r \cos\theta = r \sin\delta \qquad (4.56c)$$

$$r^2 = X^2 + Y^2 + Z^2 \qquad (4.56d)$$

$$\theta = \cos^{-1}[Z/(X^2 + Y^2 + Z^2)^{1/2}] \qquad (4.56e)$$

$$\phi = \alpha = \tan^{-1}(Y/X) \qquad (4.56f)$$

$$\delta = \sin^{-1}[Z/(X^2 + Y^2 + Z^2)^{\frac{1}{2}}] \qquad (4.56g)$$

Where line-of-sight vectors only are obtained, as with optical sightings, and r is assumed to be of unit length in Eqs. (4.56).

The Earth's spin axis is not fixed in space but precesses in a circle with a period of about 26,000 yr. This effect is due to the fact that the Earth is not spherically symmetric but has (to the first order) an equatorial bulge upon which solar and lunar gravitation act to produce a perturbing torque. The vernal equinox of course precesses at the same rate, with the result that very precise or very-long-term observations or calculations must account for the change in the "inertial" frames that are referred to the equinox. There is also a small deviation of about 9 arc-seconds over a 19-yr period due to lunar orbit precession that may in some cases need to be included. Specification of celestial coordinates for precise work thus includes a date or epoch (1950 and 2000 are in common current use) that allows the exact orientation of the reference frame with respect to the "fixed" stars (which themselves have measurable proper motion) to be computed.

In many spacecraft applications, corrections over the mission lifetime are small with respect to those relative to the years 1950 or 2000. For this reason, space missions are commonly defined with reference to true of date (TOD) coordinates, which have an epoch defined in a manner convenient to a particular mission.

Orbital Elements from Position and Velocity

As mentioned, knowledge of position and velocity at any single point and time in the orbit is sufficient to allow computation of all Keplerian elements. As an important example, it is possible given the position and velocity of the ascent vehicle at burnout to determine the various orbit injection parameters. A similar calculation would be required following a midcourse maneuver in an interplanetary mission.

As a matter of engineering practice, the single-measurement errors in vehicle position and velocity are of such magnitude as to result in a rather crude estimate of the orbit from one observation, and so rather elaborate filtering and estimation algorithms are employed in actual mission operations to obtain accurate results. However, for mission design and analysis such issues are unimportant, and it is of interest to know the orbital elements in terms of nominal position and velocity vectors. This information is also of use in sensitivity studies, in which the orbit dispersions that result from specified launch vehicle injection errors (see Chapter 5) are examined.

It is assumed that r and V are known in a coordinate system of interest, such as GCI for Earth orbital missions or HCI for planetary missions. In practice, one or more coordinate transformations must be performed to obtain data in the required form, since direct measurements will be made in coordinates appropriate to a ground-based tracking station or network. It is assumed here that i, j, and k are the unit vectors in the (X,Y,Z) directions for the appropriate coordinate system.

Given r and V in a desired coordinate system, the angular momentum is, from Eq. (4.18),

$$\boldsymbol{h} = \boldsymbol{r} \times \boldsymbol{V} \tag{4.57}$$

with magnitude

$$h = rv \sin(\pi/2 - \gamma) = rV \cos\gamma \tag{4.58}$$

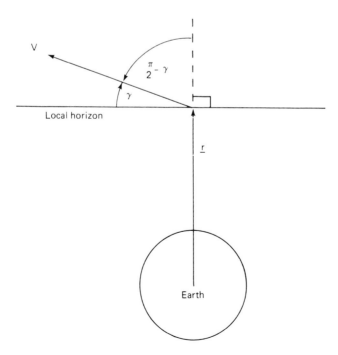

Fig. 4.9   Motion in orbit plane showing flight path angle.

where γ, the flight-path angle relative to the local horizon, is defined in Fig. 4.9. Thus,

$$\sin\gamma = \mathbf{r} \cdot \mathbf{V}/rV \tag{4.59}$$

Eccentricity may be found from Eq. (4.22), with E_t given by Eq. (4.14). Alternatively, it may be shown[9] that, in terms of flight-path angle γ,

$$e^2 = (rV^2/\mu - 1)^2 \cos^2\gamma + \sin^2\gamma \tag{4.60}$$

or

$$e^2 = [(rV^2/\mu - 1)^2 - 1] \cos^2\gamma + 1 \tag{4.61}$$

and

$$\tan\theta = \frac{(rV^2/\mu) \sin\gamma \cos\gamma}{(rV^2/\mu) \cos^2\gamma - 1} \tag{4.62}$$

Equation (4.62) avoids the ambiguity in true anomaly inherent in the use of the inverse cosine when Eq. (4.7) is used. Note that $\gamma = 0$ implies $\theta = 0$ if $rV^2/\mu > 1$, and $\theta = \pi$, if $rV^2/\mu < 1$. Thus, the spacecraft moves horizontally only at perigee or apogee if the orbit is elliptic. If $rV^2/\mu = 1$, the orbit is circular, and Eq. (4.59) will yield $\gamma = 0$, which implies that θ is undefined in Eq. (4.62). As discussed in Sec. 4.1.6, this difficulty is due to the fact that ω, and hence θ, are undefined for circular orbits. Defining $\omega = 0$ in this case will resolve the problem.

If defined, the semimajor axis is found from Eq. (4.8):

$$a = p/(1 - e^2) \tag{4.63}$$

or, if the orbit is nearly parabolic, we may use Eq. (4.20):

$$p = h^2/u \tag{4.64}$$

Since i is defined as the angle between $\mathbf{h}$ and $\mathbf{k}$,

$$\mathbf{h} \cdot \mathbf{k} = h \cos i = h_z \tag{4.65}$$

and

$$\cos i = h_z/h \tag{4.66}$$

where it is noted that 0 deg $\leqslant i \leqslant$ 180 deg.

The node vector $\mathbf{l}$ lies along the line of nodes between the equatorial and orbit planes and is positive in the direction of the ascending node. Thus,

$$\mathbf{l} = \mathbf{k} \times \mathbf{h} \tag{4.67}$$

Since Ω, the right ascension of the ascending node, is defined as the angle between i and l

$$l \cdot i = l \cos\Omega = l_x \qquad (4.68)$$

and

$$\cos\Omega = l_x / l \qquad (4.69)$$

where $l_y < 0$ implies $\Omega > 180$ deg. It may also be noted that

$$\tan\Omega = h_x / - h_y \qquad (4.70)$$

which avoids the cosine ambiguity in Eq. (4.69).

Finally, it is noted that ω is the angle from the node vector l to perigee, in the orbit plane, whereas θ is measured from perigee to r, also in the orbit plane. Thus,

$$l \cdot r = lr \cos(\omega + \theta) \qquad (4.71)$$

Hence, the argument of perigee is

$$\omega = \cos^{-1}(l \cdot r / lr) - \theta \qquad (4.72)$$

In Eq. (4.72), $k \cdot r > 0$ implies $0 \deg < \omega + \theta < 180 \deg$, and $k \cdot r < 0$ implies $180 \deg < \omega + \theta < 360 \deg$.

If desired, the periapsis time t_p may be found from the Kepler time-of-flight relations (4.26), (4.43), or (4.50), plus the auxiliary equations relating E, F, or D to true anomaly θ. As mentioned, Battin's universal formulas avoid the need to select from among several formulas during the calculation of orbital elements, but at the level of this text, this advantage is not important.

Effect of launch site on orbital elements. The location of the launch site (or, more accurately, the location and timing of actual booster thrust termination) is a determining factor in the specification of some orbital elements. Of these, the possible range of orbital inclinations is the most important. Qualitatively, it is clear that not all inclinations are accessible from a given launch site. For example, if injection into orbit does not occur precisely over the equator, an equatorial orbit is impossible, since the orbit plane must include the injection point. This is shown quantitatively by Bate et al.:[15]

$$\cos i = \sin\phi \cos\lambda \qquad (4.73)$$

where

ϕ = injection azimuth (North = 0 deg)
λ = injection latitude

If the boost phase is completed quickly so that injection is relatively close to the launch site, conditions for ϕ and λ as determined for the initial launch azimuth and latitude approximate those at vehicle burnout.

Equation (4.73) implies that direct orbits ($i < 90$ deg) require 0 deg $\leqslant \phi \leqslant 180$ deg, and furthermore that the orbital inclination is restricted to the range $|i| \geqslant |\lambda|$.

The longitude of the ascending node Ω also depends on the injection conditions, in this case the injection time. This is because launch is from a rotating planet, whereas Ω is defined relative to the fixed vernal equinox. When the choice of Ω_0 is important, as for a sun-synchronous orbit [see Sec. 4.2 (Aspherical Mass Distribution)], the allowable launch window can become quite small.

Orbit Determination

The preceding section gives a procedure for obtaining orbital elements when r and V are known in an inertial frame such as GCI or HCI. The problem of orbit determination then essentially consists of obtaining accurate values of r and V at a known time t. Doppler radar systems and certain other types of equipment allow this to be done directly (ignoring for the moment any coordinate transformations required to relate the observing site location to the inertial frame), whereas other types of radar or passive optical observations do not. Indeed, classical orbit determination may be thought of as the process of determining r and V given angular position (α, δ) observations in an inertial frame at known times.

Classical orbit determination remains important today. Radar observations are not possible for all targets and are seldom as accurate on any given measurement as are optical sightings. Given adequate observation time and sophisticated filtering algorithms (2 days of Doppler tracking followed by several hours of computer processing are required to provide the ephemeris data, accurate to within 10 m at epoch, used in the Navy Transit navigation system), radiometric measurement techniques are the method of choice in modern astrodynamics. But for accurate preliminary orbit determination, optical tracking remains unsurpassed. This is evident from the continued use and expansion of such systems, as for example in the Ground-Based Electro-Optical Deep Space Surveillance (GEODSS) program.[16]

As discussed, six independent pieces of time-tagged information are required to obtain the six orbital elements. Several types of observations have been used to supply these data:

1) Three position vectors $r(t)$ may be obtained at different known times. This case is applicable to radar systems that do not permit Doppler tracking to obtain velocity.

2) Three line-of-sight vectors (angular measurements) may be known at successive times. A solution due to Laplace gives position and velocity at the intermediate time.

3) Two position vectors $r(t)$ may be known and the flight time between them used to determine position and velocity. This is known as the

Gaussian problem, and remains useful today in part for its application to ballistic missile trajectory analysis and other intercept problems.

We note in passing that in few cases would an orbit be computed from only the minimum number of sightings. In practice, all data would be used, with the multiple solutions for the elements filtered or smoothed appropriately so as to allow a best solution to be obtained. However, practical details are beyond the scope of this text. Many references are available, e.g., Bate et al.,[15] Gelb,[17] Nahi,[18] and Wertz.[19] An excellent survey of the development of Kalman filtering for aerospace applications is given by Schmidt.[20]

We will not consider the details of orbit determination further in this text. Adequate references (e.g., Danby[10] and Bate et al.[15]) exist if required; however, such work is not generally a part of mission and spacecraft design.

Timekeeping Systems

Accurate measurement of time is crucial in astrodynamics. The analytic results derived from the laws of motion and presented here allow predictions to be made for the position of celestial objects at specific past or future times. Discrepancies between predictions and observations may, in the absence of a priori information, legitimately be attributed either to insufficient fidelity in the dynamic model or to inaccuracy in the measurement of time. Obviously, it is desirable to reduce uncertainties imposed by timekeeping errors, so that differences between measured and predicted motion can be taken as evidence for, and a guide to, needed improvements in the theoretical model.

The concept of time as used here is that of absolute time in the Newtonian or nonrelativistic sense. In this model, time flows forward at a uniform rate for all observers and provides, along with absolute or inertial space, the framework in which physical events occur. In astrodynamics and celestial mechanics, this absolute time is referred to as "ephemeris time," to be discussed later in more detail.

Special relativity theory shows this concept of time (and space) to be fundamentally in error. Perception of time is found to be dependent on the motion of the observer, and nonlocal observers are shown to be incapable of agreeing on the exact timing of events. General relativity shows further that the measurement of time is altered by the gravity field in which such measurements occur. These results notwithstanding, relativistic effects are important in astrodynamics only in very special cases, and certainly not for the topics of interest in this book. For our purposes, Newtonian concepts of absolute space and time are preferable to Einsteinian theories of spacetime.

Measurement of time. Measurement of absolute time is essentially a counting process in which the fundamental counting unit is derived from some observable, periodic phenomenon. Many such phenomena have been used throughout history as timekeeping standards, always with a progression toward the phenomena that demonstrate greater precision in their periodicity. In this sense, precision is obtained when the fundamental period is relatively insensitive to changes in the physical environment. Thus, pendulum clocks allowed a vast improvement in timekeeping standards compared to sundials, water clocks, etc. However, the period of a pendulum is a function of the local gravitational acceleration, which has a significant variation over the Earth's surface. Thus, a timekeeping standard based on the pendulum clock is truly valid only at one point on Earth. Moreover, the clock must be oriented vertically, and hence is useless, even on an approximate basis, for shipboard applications where accurate timekeeping is essential to navigation.

Until Jan. 1, 1958, the Earth's motion as measured relative to the fixed stars formed the basis for timekeeping in physics. By the end of the nineteenth century, otherwise unexplainable differences between the observed and predicted position of solar system bodies led to the suspicion that the Earth's day and year were not of constant length. Although this was not conclusively demonstrated until the mid-twentieth century, the standard or ephemeris second was nonetheless taken as 1/31,556,925.9747 of the tropical year (time between successive vernal equinoxes) 1900.

Since Jan. 1, 1958, the basis for timekeeping has been international atomic time (TAI), defined in the international system (SI) of units[21] as 9,292,631,770 periods of the hyperfine transition time for the outer electron of the Ce^{133} atom in the ground state. Atomic clocks measure the frequency of the microwave energy absorbed or emitted during such transitions and yield an accuracy of better than one part in 10^{14}. (For comparison, good quartz crystal oscillators may be stable to a few parts in 10^{13}, and good pendulum clocks and commonly available wristwatches may be good to one part in 10^{6}). The SI second was chosen to agree with the previously defined ephemeris second to the precision available in the latter as of the transition date.

Calendar time. Calendar time, measured in years, months, days, hours, minutes, and seconds, is computationally inconvenient but remains firmly in place as the basis for civil timekeeping in conventional human activities. Since spacecraft are launched and controlled according to calendar time, it is necessary to relate ephemeris time, based on atomic processes, to calendar time, which is forced by human conventions to be synchronous with the Earth's rotation.

The basic unit of convenience in human time measurement is the day, the period of time between successive appearances of the sun over a given meridian. The mean solar day (as opposed to the apparent solar day) is the average length of the day as measured over a year and is employed to remove variations in the day due to the eccentricity of the Earth's orbit and its inclination relative to the sun's equator.

Since "noon" is a local definition, a reference meridian is necessary in the specification of a planet-wide timekeeping (and navigation) system. The reference meridian, 0 deg longitude, runs through a particular mark at the former site of the Royal Observatory in Greenwich, England. Differences in apparent solar time between Greenwich and other locations on Earth basically reflect longitude differences between the two points, the principle that is the basis for navigation on Earth's surface. Twenty-four local time zones, each nominally 15 deg longitude in width, are defined relative to the prime meridian through Greenwich and are employed so that local noon corresponds roughly to the time when the sun is at zenith.

The mean solar time at the prime meridian is defined as Universal Time (UT), also called Greenwich Mean Time (GMT) or Zulu Time (Z). A variety of poorly understood and essentially unpredictable effects (e.g., variations in the accumulation of polar ice from year to year) alter the Earth's rotation period; thus, UT does not exactly match any given rotation period, or day. Coordinated Universal Time (UTC) includes these corrections and is the time customarily broadcast over the ratio. "Leap seconds" are inserted or deleted from UTC as needed to keep it in synchrony with Earth's rotation as measured relative to the fixed stars. Past corrections, as well as extrapolations of such corrections into the future, are available in standard astronomical tables and almanacs.

Ephemeris time. From earlier discussions, it is clear that ephemeris time is the smoothly flowing, monotonically increasing time measure relevant in the analysis of dynamic systems. In contrast, universal time is based on average solar position as seen against the stars and thus includes variations due to a number of dynamic effects between the sun, Earth, and moon. The resulting variation in the length of the day must be accounted for in computing ephemeris time from universal time. The required correction is

$$ET = UT + \Delta T \simeq UT + \Delta T(A) \qquad (4.74)$$

where ΔT is a measured (for the past) or extrapolated (for the future) correction, and $\Delta T(A)$ is an approximation to ΔT given by

$$\Delta T(A) = TAI - UT + 32.184 \text{ s} \qquad (4.75)$$

Hence,

$$ET \simeq TAI + 32.184 \text{ s} \qquad (4.76)$$

where $\Delta T = 32.184$ s was the correction to UT on Jan. 1, 1958, the epoch for international atomic time. For comparison, $\Delta T(A) = 50.54$ s was the correction for Jan. 1, 1980. The 18.36-s difference that has accumulated between TAI (which is merely a running total of SI seconds) and UT from 1958 to 1980 is indicative of the corrections required.

Julian dates. Because addition and subtraction of calendar time units are inconvenient, the use of ephemeris time is universal in astronomy and astrodynamics. The origin for ephemeris time is noon on Jan. 1, 4713 B.C., with time measured in days since that epoch. This is the so-called Julian

Day. For example, the Julian Day for Dec. 31, 1984 (also, by definition, Jan. 0, 1985) at 0 h is 2,446,065.5, and at noon on that day it is 2,446,066. Noon on Jan. 1, 1985, is then JD 2,446,067, etc. Tables of Julian date are given in standard astronomical tables and almanacs, as well as in navigation handbooks. Fliegel and Van Flandern[22] have published a very clever and widely implemented equation for the determination of any Julian date that is suited for use in any computer language (e.g., FORTRAN or PL/1) with integer-truncation arithmetic:

$$JD = K - 32075 + 1461 * [I + 4800 + (J - 14)/12]/4$$

$$+ 367 * \{J - 2 - [(J - 14)/12] * 12\}/12$$

$$- 3 * \{[I + 4900 + (J - 14)/12]/100\}/4 \qquad (4.77)$$

where

I = year
J = month
K = day of month

The Julian Day system has the advantage that practically all times of astronomical interest are given in positive units. However, in astrodynamics and spacecraft work in general, times prior to 1957 are of little interest, and the large positive numbers associated with the basic JD system are somewhat cumbersome. Accordingly, the Julian Day for Space (JDS) system is defined with an epoch of 0 h UT on Sept. 17, 1957. Thus,

$$JDS = JD - 2,436,099.5 \qquad (4.78)$$

Similarly, the Modified Julian Day (JDM) system has an epoch defined at 0 h UT, Jan. 1, 1950. These systems have the additional advantage for practical spacecraft work of starting at 0 h rather than at noon, which is convenient in astronomy for avoiding a change in dates during nighttime observations.

Sidereal time.   Aside from ephemeris time, the systems discussed thus far are all based on the solar day, the interval between successive noons or solar zenith appearances. It is this period that is the basic 24-h day. However, because of the motion of the Earth in its orbit, this "day" is not the true rotational period of the Earth as measured against the stars, a period known as the sidereal day, 23 h 56 m 4 s. There is exactly one extra sidereal day per year.

Sidereal time is of no interest for civil timekeeping but is important in astronautics for both attitude determination and control and astrodynamics, where the orientation of a spacecraft against the stars is considered. For example, it is the sidereal day that must be used to compute the period for a geosynchronous satellite orbit. Sidereal time is also needed to establish the instantaneous relationship of a ground-based observation station or launch site to the GCI or HCI frame. The local sidereal time is given by

$$\theta = \theta_G + \Omega_E \qquad (4.79)$$

where

θ = local sidereal time (angular measure)
θ_G = Greenwich sidereal time
Ω_E = east longitude of observing site

The Greenwich sidereal time is given in terms of its value at a defined epoch, t_0, by

$$\theta_G = \theta_{G_0} + \omega_e(t - t_0) \qquad (4.80)$$

where ω_e is the inertial rotation rate of Earth and was 7.292116×10^{-5} rad/s for 1980. Again, tables of θ_{G_0} for convenient choices of epoch are provided in standard astronomical tables and almanacs.

4.2 NON-KEPLERIAN MOTION

We have to this point reviewed the essential aspects of two-body orbital mechanics, alluding only briefly to the existence of perturbing influences that can invalidate Keplerian results. Such influences are always present and can often be ignored in preliminary design. However, this is not always the case, nor is it indeed always desirable; mission design often involves deliberate use of non-Keplerian effects. Examples include the use of atmospheric braking to effect orbital maneuvers and the use of the Earth's oblateness to specify desired (often sun-synchronous) rates of orbital precession. These and other aspects of non-Keplerian orbital dynamics are discussed in the following sections.

Sphere of Influence

Of the various possible perturbations to basic two-body motion, the most obvious are those due to the presence of additional bodies. Such bodies are always present and cannot be easily included in an analysis, particularly at elementary levels. It is then necessary to determine criteria for the validity of Keplerian approximations to real orbits when more than two bodies are present.

If we consider a spacecraft in transit between two planets, it is clear that when close to the departure planet, its orbit is primarily subject to the influence of that planet. Far away from any planet, the trajectory is essentially a heliocentric orbit, whereas near the arrival planet, the new body will dominate the motion. There will clearly be transition regions where two bodies will both have significant influence on the spacecraft motion. The location of these transition regions is determined by the so-called sphere of influence of each body relative to the other, a concept originated by Laplace.

For any two bodies, such as the sun and a planet, the sphere of influence is defined by the locus of points at which the sun's and the planet's gravitational fields have equal influence on the spacecraft. The term "sphere of influence" is somewhat misleading; every body's gravitational field extends to infinity, and in any case, the appropriate equal-influence boundary is not exactly spherical. Nonetheless, the concept is a useful one in preliminary design.

Although the relative regions of primary influence can be readily calculated for any two bodies, the concept is most useful when, as mentioned earlier, one of the masses is much greater than the other. In such a case, the so-called classical sphere of influence about the small body has the approximate radius

$$r \simeq R(m/M)^{\frac{2}{5}} \qquad (4.81)$$

where

r = sphere of influence radius
R = distance bewteen primary bodies m and M
m = mass of small body
M = mass of large body

A spacecraft at a distance less than r from body m will be dominated by that body, but at greater distances, it will be dominated by body M. Table 4.1 gives sphere-of-influence radii from Eq. (4.81) for the planets relative to the sun and the moon relative to the Earth.

Other sphere-of-influence definitions are possible. A popular alternate to Eq. (4.81) is the Bayliss sphere of influence, which replaces the 2/5 power

Table 4.1 Planetary spheres of influence

Planet	Mass ratio (sun planet)	Sphere of influence, km
Mercury	6.0236×10^6	1.12×10^5
Venus	4.0852×10^5	6.16×10^5
Earth	3.3295×10^5	9.25×10^5
Mars	3.0987×10^6	5.77×10^5
Jupiter	1.0474×10^3	4.88×10^7
Saturn	3.4985×10^3	5.46×10^7
Uranus	2.2869×10^4	5.18×10^7
Neptune	1.9314×10^4	8.68×10^7
Pluto	3×10^6	1.51×10^7
Moon[a]	81.30	66,200

[a]Relative to Earth.

with 1/3. This defines the boundary where the direct acceleration due to the small body equals the perturbing acceleration due to the gradient of the larger body's gravitational field.

Away from the sphere-of-influence boundary, Keplerian orbits are reasonably valid. Near the boundary, a two-body analysis is invalid, and alternate methods are required. In some cases, it may be necessary to know only the general characteristics of the motion in this region. The results of restricted three-body analysis, discussed later, may then be useful. Generally, however, more detailed information is required. If so, the mission analyst has two choices. Accurate trajectory results may be obtained by a variety of methods, all requiring a computer for their practical implementation. Less accurate, preliminary results may be obtained by the method of patched conics, discussed in Sec. 4.4. The use of this method is consistent with the basic sphere-of-influence calculation of (4.81) and thus does not properly account for motion in the transition region between two primary bodies.

Restricted Three-Body Problem

The general case of the motion of three massive bodies under their mutual gravitational attraction has never been solved in closed form. Sundman[23] developed a general power series solution; however, useful results are obtained only for special cases or by direct numerical integration of the basic equations. A special case of particular interest involves the motion of a body of negligible mass (i.e., a spacecraft) in the presence of two more massive bodies. This is the restricted three-body problem, first analyzed by Lagrange, and is applicable to many situations of interest in astrodynamics.

Earlier it was found[10] that, in contrast to the simple Keplerian potential function of Eq. (4.2), the restricted three-body problem yields a complex potential function with multiple peaks and valleys. This is shown in Fig. 4.10, where contours of equal potential energy are plotted. Along these contours, particles may move with zero relative velocity.

The five classical Lagrangian points are shown in Fig. 4.10. L_1, L_2, and L_3 are saddle points, i.e., positions of unstable equilibrium. Objects occupy-

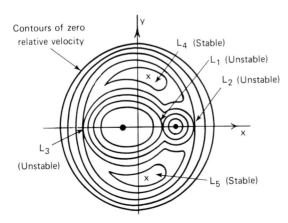

Fig. 4.10 Lagrange points for restricted three-body problem.

ing these positions will remain stationary only if they are completely unperturbed. Any disturbances will result in greater displacement from the initial position. L_1 and L_3 are positions where the centrifugal force of the spacecraft in the rotating Earth-moon system is balanced by their combined colinear gravitational force. L_2 is a similar point where the lunar gravitational force plus the centrifugal force are balanced by the opposing gravitational force from the Earth. Spacecraft can occupy these positions for extended periods, but only if a supply of stationkeeping fuel is provided to overcome perturbing forces.

L_4 and L_5 are positions of stable equilibrium; objects displaced slightly from these positions will experience a restoring force toward them. Small bodies can occupy stable orbits about L_4 or L_5, a fact that is observationally confirmed by the presence of the Trojan asteroids that occupy the stable Lagrange points 60 deg ahead of and behind Jupiter in its orbit about the sun. The properties of L_4 and L_5 have led to considerable analysis of their use as sites, in the Earth-moon system, for permanent space colonies and manufacturing sites that would be supplied with raw materials form lunar mining sites.[3]

Aspherical Mass Distribution

As indicated in Sec. 4.1 (Two-Body Motion), an extended body such as the Earth acts gravitationally as a point mass provided its mass distribution is spherically symmetric. That is, the density function $\rho(r,\theta,\phi)$ in spherical coordinates must reduce to a function $\rho = \rho(r)$; there can be no dependence on latitude θ or longitude ϕ. Actually, the Earth is not spherically symmetric, but more closely resembles an oblate spheroid with a polar radius of 6357 km and an equatorial radius of 6378 km. This deformation is due to the angular acceleration produced by its spin rate (and is quite severe for the large, mostly gaseous outer planets such as Jupiter and Saturn), but numerous higher-order variations exist and are significant for most Earth orbital missions.

The Earth is approximately spherically symmetric; thus, it is customary[24] to describe its gravitational potential in spherical coordinates as a perturbation to the basic form of Eq. (4.2), e.g.,

$$U(r,\theta,\phi) = -\mu/r + B(r,\theta,\phi) \tag{4.82}$$

where $B(r,\theta,\phi)$ is a spherical harmonic expansion in Legendre polynomials $P_{nm}(\cos\theta)$. The complete solution[25] includes expansion coefficients dependent on both latitude and longitude. Both are indeed necessary to represent the observable variations in the Earth's field. However, low-orbiting spacecraft having short orbital periods are not sensitive to the longitudinal, or tesseral, variations because, being periodic, they tend to average to zero. Spacecraft in high orbits with periods too slow to justify such an averaging assumption are too high to be influenced significantly by what are, after all, very small effects. Thus, for the analysis of the most significant orbital perturbations, only the latitude-dependent, or zonal, coefficients are important. (An exception is a satellite in a geostationary orbit; since it hovers over a particular region on Earth, its orbit is subjected to a continual perturbing force in the same direction, which will be important.) If the longitude-dependent variations are ignored, the gravitational potential is[25]

$$U = -(\mu/r)\left[1 - \sum_{n=2}^{\infty} (R_e/r)^n J_n P_n(\cos\theta)\right] \qquad (4.83)$$

where

R_e = radius of Earth
r = radius vector to spacecraft
J_n = nth zonal harmonic coefficient
$P_n(x) = P_{n0}(x) = n$th Legendre polynomial

It is seen in Table 4.2 that, for Earth, J_2 dominates the higher-order J_n, which themselves are of comparable size, by several orders of magnitude. This is to be expected, because J_2 accounts for the basic oblateness effect, which is the single most significant aspherical deformation in the Earth's figure. Furthermore, since $(R_e/r) < 1$, it is to be expected that the second-order term in Eq. (4.83) will produce by far the most significant effects on the spacecraft orbit.

This is in fact the case. The major effect of Earth's aspherical mass distribution is a secular variation in the argument of perigee ω and the longitude of the ascending node Ω. Both of these depend to the first order only on the Earth's oblateness, quantitatively specified by J_2. The results are

$$d\omega/dt \simeq (3/4)nJ_2(R_e/a)^2(4 - 5\sin^2 i)/(1 - e^2)^2 \qquad (4.84)$$

$$d\Omega/dt \simeq -(3/2)nJ_2(R_e/a)^2 \cos i/(1 - e^2)^2 \qquad (4.85)$$

where

n = mean motion, $\sqrt{(\mu/a^3)}$
a = semimajor axis
i = inclination
e = eccentricity

Variations in a, e, i, and n also occur but are periodic with zero mean and small amplitude. In Eqs. (4.84) and (4.85) we have approximated these

Table 4.2 Zonal harmonics for Earth

J_2	1.082×10^{-3}
J_3	-2.54×10^{-6}
J_4	-1.61×10^{-6}

parameters by their values for a Keplerian orbit, whereas in fact they depend on J_2. Errors implicit in this approximation may be as much as 0.1%.[19]

Of interest here is the fact that, for direct ($i < 90$ deg) orbits, $d\Omega/dt < 0$, whereas for retrograde orbits it is positive. When the rotation rate equals 360 deg/yr, the orbit is said to be sun-synchronous, since its plane in inertial space will precess to remain fixed with respect to the sun. Such orbits are frequently used to allow photography or other Earth observations to take place under relatively fixed viewing and lighting conditions. Practical sun-synchronous orbits are usually nearly circular, slightly retrograde (96 deg $< i < 100$ deg) and have a mean altitude in the range of 200–1000 km.

From Eq. (4.84) we note that $d\omega/dt = 0$ for $\sin^2 i = 4/5$, or $i = 63.435$ deg; thus, in this case there is no perturbation to the argument of perigee. For $i < 63.435$ deg, the rotation of the line of apsides is in the direction of the orbit, whereas for larger inclinations, it rotates in the direction opposite to the motion.

The computed perturbations due to the Earth's oblateness and the higher-order geopotential variations are most important for relatively low orbits and may not be the dominant gravitational disturbances for higher orbits. For example, considering only the J_2 term in Eq. (4.83), it is easily found that, at $r \simeq 15,000$ km, the maximum possible perturbing acceleration is about 10^{-3} m/s^2. This is comparable to or exceeded by the perturbing acceleration produced by the sun on a spacecraft in such an orbit. At higher altitudes, oblateness effects are even smaller and may well rank behind lunar perturbations (typically two orders of magnitude smaller than solar effects) in significance. The exception, again, can be satellites in geostationary orbits, which are subject to essentially constant perturbing forces due to geopotential variations.

The results of this section are of course not restricted to Earth-orbiting spacecraft and can be used to establish orbits with particular characteristics about other planets. During the Viking missions to Mars, for example, the Viking Orbiters were inserted into highly elliptic orbits with a periapsis of a few hundred kilometers and a period of 24.68 h (1 Martian day). This was initially done to optimize data relay from the landers. However, this orbit minimized fuel usage during the injection maneuver, allowed both high- and low-altitude photography, and, because of the rotation of the line of apsides, allowed the periapsis to occur over a substantial range of latitudes. This allowed most of the planet to be photographed at both small and large scale.

It is interesting to note that the results presented in this section may be used in reverse order; that is, orbital tracking may be used to establish the values of the J_n coefficients (and, in the full-blown expansion, the values of the tesseral harmonic coefficients as well).[26] Tracking of the early Vanguard satellites was used to establish the pear-shaped nature of the Earth's figure, shown by the fact that J_3 is nonzero. Today satellite geodesy is the method that produces the most accurate determination of the harmonic coefficients. Proposals have been advanced for placing two spacecraft in the same relatively low orbit and using their differential acceleration as an even more sensitive indicator of variations in the Earth's figure[21] (see Table 4.2).

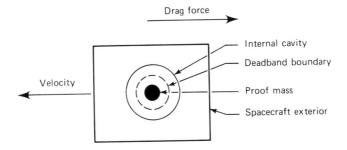

Fig. 4.11   Drag compensation concept.

The use of the potential function expansion approach relies for its computational practicality on the assumption that the Earth is nearly spherical and that its potential may be approximated by small perturbations to the simple point-mass form. For the Earth this is true, but this is not always so elsewhere in the solar system. The moon is both substantially ellipsoidal and possessed of regions of significant mass concentration, or mascons, discovered during the Apollo era by tracking lunar orbiting spacecraft. In such cases, the oblateness term may not be the most significant source of orbital perturbations.

Satellite geodesy imposes unique spacecraft design requirements for its accomplishment. Low-orbiting spacecraft yield the most useful results, yet at low altitudes atmospheric drag dwarfs the minor perturbations that it can be desired to measure. Great care is required to develop analytical models for these effects so that the biases they introduce can be removed from the data. The problem can also be circumvented by the use of an onboard drag-compensation system. The system operates by enclosing a small free-floating proof mass inside the spacecraft (see Fig. 4.11). The proof mass is maintained in a fixed position relative to the spacecraft body by means of very small thrusters operating in a closed-loop control system.[21] This drag-compensation system removes all nongravitational forces on the proof mass, and hence the spacecraft, and thus forces it to follow a purely gravitational trajectory. The measured departures from simple Keplerian motion are then due to perturbations in the Earth's potential field.

This approach has advantages for navigation satellites as well as for geodetic research. First implemented in the U.S. Navy TRIAD program in 1972, the drag-free (also called DISCOS, or Disturbance Compensation System) concept[28] has allowed the measurement of along-track gravitational perturbations down to a level of $5 \times 10^{-12} g$. Its use has been proposed for other applications where removal of all nongravitational effects is important.

Atmospheric Drag

The influence of atmospheric drag is important for all spacecraft in low Earth orbit, both for attitude control (see Chapter 7) and in astrodynamics.

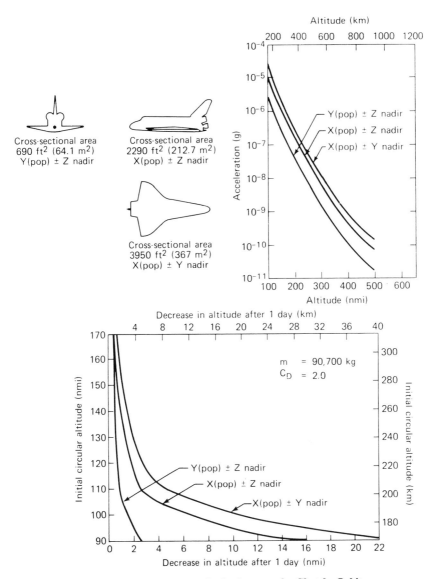

Fig. 4.12   Effects of atmospheric drag on the Shuttle Orbiter.

Such spacecraft will eventually re-enter the atmosphere due to the cumulative effect of atmospheric drag unless provided with an onboard propulsion system for periodic reboost. The period of time required for this to occur is called the orbital lifetime of the spacecraft and depends on the mass and aerodynamic properties of the vehicle, its orbital altitude and eccentricity, and the density of the atmosphere.

An example of the effect of atmospheric drag on a spacecraft in low Earth orbit is given in Fig. 4.12, which shows the predicted decay for the Space Shuttle as a function of orbital height and vehicle attitude.[29]

Because of the difficulty of modeling both the environmental and spacecraft aerodynamic properties in this flight regime, quantitative results for orbital perturbations due to atmospheric drag are difficult to obtain. In the following sections we discuss the approximate results that can be obtained and supply guidelines for more detailed modeling effects where appropriate.

Atmosphere models. As will be seen, satellite orbital lifetime depends strongly on the variation of the upper atmosphere density with altitude. Although the gross behavior of the atmospheric density is well established, exact properties are difficult to determine and highly variable. It is this factor, more than any other, that makes the determination of satellite lifetimes so uncertain and renders even the most sophisticated analysis of more academic than practical interest. "Permanent" satellites will have lifetimes measured in years, with an uncertainty measured at least in months and quite possibly also in years.

As a spacecraft approaches re-entry, this uncertainty decreases; however, predicted lifetimes may be in error by one or more orbital periods even on the day of re-entry. Errors of about 10% of the remaining lifetime represent the best obtainable precision for orbit decay analyses. Difficulties in orbital decay analysis were graphically illustrated during the final months and days prior to the re-entry of Skylab in 1979. Kaplan et al.[30] present an excellent summary of the analytical and operational effort expended in a largely unsuccessful attempt to effect a controlled re-entry of that vehicle.

Standard models exist[31] for the variations in average atmospheric properties with altitude (see Chapter 3). For properties at altitudes above 100 km, which are of principal interest in spacecraft dynamics, these models are based primarily on the work of Jacchia.[32] As indicated in Chapter 5, the basic form of the Earth's atmospheric density profile is obtained by requiring hydrostatic equilibrium in conjunction with a specified temperature profile, determined via a combination of analytical and empirical means. The temperature profile is modeled as a sequence of layers having either a constant temperature or a constant temperature gradient.

These assumptions result in a density profile having the form of Eqs. (5.18a) or (5.18b) in each layer, i.e., an exponential or power law dependence. In practice, there may be little difference in the density as predicted by each of these forms, and it is common and analytically convenient to assume an exponential form:

$$\rho = \rho_0 \exp - \beta(r - r_0) \qquad (4.86)$$

where

$1/\beta$ = scale height
ρ_0 = density at reference altitude
r_0 = reference altitude
r = altitude with respect to the Earth's center

The reference altitude may be the top or bottom of a specified layer. If a single layer is assumed, the atmosphere is referred to as strictly exponential, and the reference level may be sea level. It may also be desirable to use a reference level at some altitude, such as 100 km, appropriate to the analysis. This allows accuracy to be maintained at orbital altitudes, at the possible sacrifice of sea level results, which are irrelevant in this situation. The exact form of β is[33]

$$\beta = (g/R_{gas} + dT/dr)/T \qquad (4.87)$$

where

R_{gas} = specific gas constant
T = temperature
g = gravitational acceleration

The ability to obtain the integrated result of Eq. (4.86), and consequently its accuracy, depends on the assumption of constant scale height, and thus on the assumption that the g and T are both constant. Clearly this is not strictly true. Gravitational acceleration varies by about 4% over the altitude range from sea level to 120 km.

Rather large local temperature gradients, on the order of 10 K/km at 120 km, may exist. Scale heights for Earth below about 120 km range from 5 to 15 km, with a mean value of about 7.1 km. In the worst case, when $1/\beta = 5$ km, examination of temperature profile data for Earth indicates that the gravitational term is approximately 0.14 km^{-1} and the temperature gradient term is approximately 0.04 km^{-1}. However, for the relatively small range of altitudes of importance in satellite aerodynamic drag analyses, it is often acceptable to assume β to be a constant given by its mean value in the range of interest. Shanklin et al.[34] studied the sensitivity of near-term orbit ephemeris predictions to differences among various atmosphere models and concluded that the simple exponential model discussed earlier yielded, for the four cases studied, results indistinguishable from those using more complex models.

Standard atmosphere data were presented in Figs. 3.21 and 3.22 and Table 3.3. Extensive variations from average atmosphere properties exist. Fluctuations are observed on a daily, 27-day, seasonal, yearly, and 11-yr basis, as well as with latitude. Of these, the 11-year cycle is the most pronounced, due to the variations in solar flux with the sunspot cycle. Density fluctuations of a factor of 10 at 350 km and a factor of 5 at 800 km are observed.

Higher solar fluxes produce greater atmospheric density at a given altitude. Solar flux is commonly reported in units of 10^4 Janskys (1 Jansky = 10^{-26} W · m^{-2} · Hz^{-1}), with typical 11-yr minima of 80 and maxima of 150, with peaks of 250 not unrealistic. Jacchia[35] shows the effects of such peaks. For example, at 500 km with a solar flux of 125, a nominal density of 1.25×10^{-12} kg/m^3 is observed, whereas a solar flux of 160 will produce the same density at 600 km.

Effects of drag on orbital parameters. Drag is defined as the component of force antiparallel to the spacecraft velocity vector relative to the atmosphere. We may often ignore the velocity component due to the rotation of the atmosphere, since it is small compared to the spacecraft orbital velocity. There is then no component of force normal to the orbit plane, and, to a first approximation, atmospheric drag thus has no effect on the elements ω, Ω, and i, which determine the orientation of the orbit in space.

As discussed earlier, atmospheric density varies exponentially with height above the Earth, with a scale height on the order of tens of kilometers, and is always small at orbital altitudes. Thus, if the orbit is even slightly elliptic, the principal drag occurs at perigee and may be modeled approximately as an impulsive reduction in velocity. It will be seen in Sec. 4.3 that the result of such an impulsive ΔV maneuver is to reduce the orbital apogee while leaving the perigee altitude essentially unchanged. Thus, atmospheric drag reduces the semimajor axis and the eccentricity of the orbit, tending to circularize it. A low orbit that is nearly circular experiences a significant continuing drag force and may have a lifetime of only days or hours.

The above model is of course only approximate, and aerodynamic drag does include variations in all orbital elements.[36] However, more detailed analysis does confirm the principal features of this model, namely, that drag tends to circularize the orbit at an altitude very near that of the original perigee. Analysis of the motion of a satellite in the upper regions of a planetary atmosphere is extremely complex. The classical treatment in this area is that of King-Hele,[37] whereas that of Vinh et al.[33] provides an excellent recent text. We include here the results of importance in preliminary mission design.

It is found that the process of orbit decay is most conveniently represented by specifying eccentricity as the independent variable. In this way, the geometric properties of the decaying orbit can be specified without reference to the spacecraft dynamic characteristics. In nondimensional form, it is found that the semimajor axis decays as

$$a/a_0 = 1 + \varepsilon h_1(\alpha) + \varepsilon^2 h_2(\alpha) + \varepsilon^3 h_3(\alpha) + \varepsilon^4 h_4(\alpha) + \cdots \qquad (4.88)$$

where the subscript 0 implies initial orbit conditions and

$$\alpha = \beta a_0 e$$

$$\varepsilon = 1/\beta a_0$$

$$h_1 = B - B_0$$

$$h_2 = 2(A - A_0) + (A - 3)(B - B_0)$$

$$h_3 = 7[\alpha^2 - (\beta a_0 e_0)^2]/2 - (4A_0 - 13)(A - A_0)/2$$

$$+ (2\alpha^2 - 4A - A^2 + 13)(B - B_0)$$

$$+ (\alpha^2 + A - A^2 + 3)(B - B_0)^2/2$$

$$h_4 = [\alpha^2 - (\beta a_0 e_0)^2](7A - 4A_0 - 35)/2$$

$$+ (A - A_0)(12\alpha^2 + 16A_0^2 + 42A_0 + 4A_0 A - 9A - 8A^2 + 213)/6$$

$$+ (B - B_0)(2A^3 + 7A^2 + 46A - 25\alpha^2 - 138)/2$$

$$+ (B - B_0)^2(A^3 + 6A^2 - \alpha^2 A - 7\alpha^2 - 35)/2$$

$$+ 2(A - A_0)(B - B_0)(\alpha^2 + A - A^2 + 3)$$

$$+ (B - B_0)^3(2A^3 - 3A^2 - 2\alpha^2 A + \alpha^2 - 6)/6$$

$$A = \alpha I_0(\alpha)/I_1(\alpha)$$

$$B = \ell n[\alpha I_1(\alpha)]$$

where I_n is the imaginary Bessel function, nth order, first kind.[11]
The decay in semimajor axis a/a_0 found through Eq. (4.88) is then used to determine the changes in other parameters. It is found that the perigee radius decays as

$$r_p/r_{p_0} = (1 - \varepsilon\alpha)(a/a_0)/(1 - e_0) \qquad (4.89)$$

and the apogee radius is found from

$$r_a/r_{a_0} = (1 + \varepsilon\alpha)(a/a_0)/(1 + e_0) \qquad (4.90)$$

and the decrease in orbital period is

$$\tau/\tau_0 = (a/a_0)^{3/2} \qquad (4.91)$$

The preceding results are used as follows:
1) Given the initial orbit and the atmosphere model, find ε.
2) Choose e in the range $0 < e < e_0$ and compute α.
3) Compute the h_i and solve Eqs. (4.88–4.91).
4) Repeat steps 2 and 3 as needed for a range of eccentricities.
This procedure allows the evolving shape of the orbit to be predicted during the decay process. In spacecraft design applications, however, it will more commonly be desired to study the evolution of the orbit in time. To do this, the orbital eccentricity must be known as a function of time. King-Hele[37] finds, and the higher-order analysis of Vinh et al.[33] confirms, that eccentricity varies with time as

$$(e/e_0)^2 = 1 - T/T_L \qquad (4.92)$$

where T is the nondimensional time and T_L the nondimensional orbital lifetime given in terms of the initial orbit parameters by

$$T_L = (e_0/\varepsilon)^2[1 - 5e_0/6 + 23e_0^2/48 + 7\varepsilon/8 + \varepsilon e_0/6 + 9\varepsilon^2/16e_0]/2 \qquad (4.93)$$

If the orbit is initially nearly circular ($e_0 < 0.02$), T_L is given more accurately by

$$T_L = [1 - (9\beta^2 a_0^2 e_0^2/20 - 1)\varepsilon/2](e_0^2/2\varepsilon^2) \qquad (4.94)$$

T contains all of the spacecraft parameters and is defined as

$$T = 2\pi(SC_D/m)\rho_{p_0}f\beta^2 a_0^3 e_0 I_1(\beta a_0 e_0) \exp(-\beta a_0 e_0)t/\tau_0 \qquad (4.95)$$

The parameter f provides a latitude correction, usually less than 10%, to the orbital speed as it appears in the drag model

$$D = \tfrac{1}{2}\rho V^2 f S C_D \qquad (4.96)$$

where

D = drag force
C_D = drag coefficient
S = projected area normal to flight path
ρ_{p_0} = initial periapsis density

and is given by

$$f = [1 - (\omega_e r_{p_0}/V_{p_0}) \cos i]^2 \qquad (4.97)$$

where

ω_e = angular velocity of the Earth (7.292×10^{-5} rad/s)
V_{p_0} = initial periapsis velocity

The term (m/SC_D) in Eq. (4.95) is the ballistic coefficient and is a measure of the ability of an object to penetrate atmospheric resistance. Typical values for spacecraft will be on the order of $10-100$ kg/m². The projected area can be complicated to compute for other than simple spacecraft shapes, but this presents no fundamental difficulties. Determination of the drag coefficient is more complex. This issue is addressed in the next section.

If only the orbital lifetime is required, then Eq. (4.93) or (4.94) alone is sufficient, and the more complex procedure for analyzing the evolution of the orbit due to drag is unnecessary. This is often the case in preliminary design work.

Table 4.3   Newtonian flow drag coefficients

Body	C_D
Sphere	1.0
Circular cylinder	1.3
Flat plate at angle α	$1.8 \sin^3\alpha$
Cone of half-angle δ	$2 \sin^2\delta$

Drag coefficient. As seen from Eq. (4.95), the orbit lifetime depends directly on C_D. For analysis of orbit decay, we are concerned with the so-called free molecular flow regime. In this regime, the flow loses its continuum nature and appears to a first approximation as a stream of independent particles (i.e., Knudsen number $\gg 1$). Furthermore, the flow is at very high speed, about Mach 25 or more than 7.5 km/s in most cases, and thus is hypersonic. This implies that any pressure forces produced by random thermal motion are small in comparison to those due to the directed motion of the spacecraft through the upper atmosphere.

The high-speed, rarefied flow regime permits the use of Newtonian flow theory, in which the component of momentum flux normal to a body is assumed to be transferred to the body by means of elastic collisions with the gas molecules. The tangential component is assumed unchanged. The net momentum flux normal to the body surface produces a pressure force that, when integrated over the body, yields the drag on the body. Geometric shadowing of the flow is assumed, which implies that only the projected area of the spacecraft can contribute to the drag force. Subject to these assumptions, and assuming Mach $25 \simeq \text{Mach}\infty$, the pressure coefficient

$$C_p \equiv (p - p_\infty)/\tfrac{1}{2}\rho V^2 \qquad (4.98)$$

is given by

$$C_p = 1.84 \sin^2\theta \qquad (4.99)$$

where θ is the local body angle relative to the flow velocity vector.

The drag coefficient is obtained by integrating the streamwise component of the pressure coefficient over the known body contour. For simple shapes the required integrations can be performed analytically, yielding useful results for preliminary design and analysis. Table 4.3 gives values of C_D for spheres, cylinders, flat plates, and cones subject to the preceding assumptions.

Newtonian flow theory allows only approximate results at best. Even at the high speeds involved in orbital decay analysis, random thermal motion is important, as in the exact, nonelastic nature of the interaction of the gas molecules with the body surface.

A more accurate treatment results from the assumption that atmospheric molecules striking the spacecraft are in Maxwellian equilibrium, having both random and directed velocity components. Some of the molecules that strike the surface are assumed to be re-emitted inelastically, with a Maxwellian distribution characteristic of the wall temperature T_w. This model, due to Shaaf and Chambre,[38] results in a pressure force on the wall of

$$p = 2p_i - \sigma_n(p_i - p_w) \qquad (4.100)$$

where

p_i = pressure due to incident molecular flux
p_w = pressure due to wall re-emissions
σ_n = normal momentum accommodation coefficient

With Maxwellian distributions assumed, p_i and p_w may be computed and the net pressure force obtained. A similar analysis by Fredo and Kaplan[39] yields the shear force and introduces a dependence on the tangential momentum accommodation coefficient σ_t.

The accommodation coefficients σ_n and σ_t characterize the type of interaction the gas particles make with the surface. The equation $\sigma_n = \sigma_t = 0$ implies specular reflection, as in Newtonian flow, whereas $\sigma_n = \sigma_t = 1$ implies diffuse reflection, i.e., total accommodation ("sticking") of the particles to the surface followed by subsequent Maxwellian re-emission at the wall temperature. It is traditional in orbit decay studies to assume total accommodation as an improvement on the known deficiencies of the Newtonian model. In practice this is never true, with $\sigma_n = 0.9$ about the maximum value observed, and σ_t somewhat less. Furthermore, both co-efficients are strongly dependent on incidence angle.[40] The work of Fredo and Kaplan[39] shows these effects to be important, particularly for complex, asymmetric shapes. However, this approach requires a large computer for its implementation and is of limited usefulness in preliminary design calculations.

Solar Radiation Pressure

Observed solar radiation intensity at the Earth's orbit about the sun is closely approximated (to within 0.3%) by

$$I_s = 1358 \ \text{W/m}^2/(1.0004 + 0.0334 \cos D) \qquad (4.101)$$

where

I_s = integrated intensity (in W/m^2) on the area normal to the sun
D = phase of year [$D = 0$ on July 4 (aphelion)]

given by Smith and Gottlieb.[41] Note that 1358 W/m^2 is the mean intensity observed at a distance of 1 A.U.; solar radiation intensity for planets at other distances from the sun is computed from the inverse square law. For practical purposes in spacecraft design, the observed intensity of Eq. (4.101) is essentially that due to an ideal blackbody at 5780 K.[42] This assumption is especially convenient when it is necessary to consider analytically the spectral distribution of radiated solar power. It is of course erroneous in that it does not account for the many absorption lines in the solar spectrum due to the presence of various elements in the sun's atmosphere.

We note that intensity has dimensions of power per unit area, and that power is the product of force and the velocity at which the force is applied. Solar radiation thus produces an effective force per unit area, or pressure, given by

$$p_s = I_s/c = 4.5 \times 10^{-6} \text{ N/m}^2 \qquad (4.102)$$

where c is the speed of light in vacuum.

As we have seen, the passage of a satellite through the upper atmosphere produces a drag force resulting from the dynamic pressure, $\frac{1}{2}\rho V^2$, due to the upper atmosphere density ρ and the orbital velocity V. With $V \simeq 7.6$ km/s, it is found that a density of $\rho \simeq 1.55 \times 10^{-13}$ kg/m^3 will produce a dynamic pressure equal to the solar radiation pressure from Eq. (4.102). In the standard atmosphere model, this density is found at an altitude of roughly 500 km. At 1000 km, aerodynamic drag produces only about 10% of the force due to solar radiation pressure. Thus, at many altitudes of interest for Earth orbital missions, solar radiation pressure exerts a perturbing force comparable to or greater than atmospheric drag.

Solar radiation pressure is fundamentally different from aerodynamic drag in that (for a symmetric satellite) the force produced is in the antisolar direction, rather than always opposite the spacecraft velocity vector. The resulting effects may for some orbits average nearly to zero over the course of an orbit but are generally not confined to only variations in the eccentricity and semimajor axis, as with aerodynamic drag. Depending on the orbit and the orientation and symmetry properties of the spacecraft, changes in all orbital elements are possible. Of course, solar radiation pressure does not act on a spacecraft during periods of solar occultation by the Earth or other bodies.

The perturbing effects of solar radiation pressure can be deliberately enhanced by building a spacecraft with a large area-to-mass ratio. This is the so-called solar sail concept. If the center of mass of the vehicle is maintained near to or in the plane of the sail in order to minimize "weathervane" or "parachute" tendencies, then the sail can provide significant propulsive force normal to the sun vector, allowing relatively sophisticated orbital maneuvers. Solar sails have been proposed by many authors for use in planetary exploration.[43] The solar sail concept enables some missions, such as a Mercury Orbiter, to be performed in a much better fashion than with currently available or projected chemical boosters. Optimal use of solar radiation pressure to maximize orbital energy and angular momentum has been studied by Van der Ha and Modi.[44]

Satellites in high orbits, such as geosynchronous communications or weather satellites, must be provided with stationkeeping fuel to overcome the long-term perturbations induced by solar radiation pressure. Planetary missions must similarly take these effects into account for detailed trajectory calculations. And, as will be seen in Chapter 7, the force produced by solar radiation can be important in attitude control system design.

Solar radiation produces a total force upon a spacecraft given by

$$F_s = KSp_s \qquad (4.103)$$

where

K = accommodation coefficient
S = projected area normal to sun

The K is an accommodation coefficient characterizing the interaction of the incident photons with the spacecraft surface. It is in the range $1 \leqslant K \leqslant 2$, where $K = 1$ implies total absorption of the radiation (ideal blackbody), and $K = 2$ implies total specular reflection back along the sun line (ideal mirror).

Solar radiation is to be distinguished from the solar wind, which is a continuous stream of particles emanating from the sun. The momentum flux in the solar wind is small compared with that due to solar radiation.

4.3 BASIC ORBITAL MANEUVERS

Many spacecraft, especially those intended for unmanned low Earth orbital missions, pose only loose requirements on orbit insertion accuracy and need no orbit adjustments during the mission. In other cases, nominal launch vehicle insertion accuracies are inadequate, or the desired final orbit cannot be achieved with only a single boost phase, and postinjection orbital maneuvers will be required. Still other missions will involve frequent orbit adjustments in order to fulfill basic objectives. In this section, we consider simple orbital maneuvers including plane changes, one- and two-impulse transfers, and combined maneuvers.

In the following we assume impulsive transfers; i.e., the maneuver occurs in a time interval that is short with respect to the orbital period. Since orbit adjust maneuvers typically consume a few minutes at most and orbital periods are 100 min or longer, this is generally a valid approximation. In such cases, the total impulse (change in momentum) per unit mass is simply equal to the change in velocity ΔV.

This quantity is the appropriate measure of maneuvering capability for spacecraft that typically are fuel-limited. To see this, note that, if a thruster produces a constant force F on a spacecraft with mass m for time interval Δt, we obtain upon integrating Newton's second law

$$\Delta V = (F/m)\Delta t \qquad (4.104)$$

Since F/m will be essentially constant during small maneuvers, and since the total thruster on-time is limited by available fuel, the total ΔV available

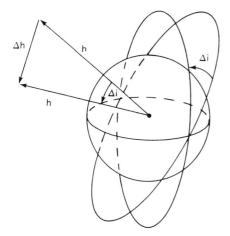

Fig. 4.13 Plane rotation.

for spacecraft maneuvers is fixed and is a measure of vehicle maneuver performance capability.

Plane Changes

Most missions, in cases where any orbit adjustment at all is required, will require some adjustment of the orbital plane. This plane (see Fig. 4.13) is perpendicular to the angular momentum vector h, which for a Keplerian orbit is permanently fixed in space. A pure plane rotation alters h in direction but not in magnitude and thus requires an applied torque normal to h. This in turn requires the application of a force on the spacecraft, e.g., a thruster firing, parallel to h. For pure plane rotation we see from the geometry of Fig. 4.13 and the law of cosines that the change in angular momentum is related to the rotation angle v by

$$\Delta h = h[2(1 - \cos v)]^{\frac{1}{2}} = 2h \sin(v/2) \qquad (4.105)$$

and since $h = rV_\theta$, the required impulse is

$$\Delta V = 2V_\theta \sin(v/2) \simeq V_\theta v \qquad (4.106)$$

where the last equality is valid for small v.

This impulse is applied perpendicular to the initial orbit plane. There results a node line between the initial and final orbits running through the point where the thrust is applied. Since, in fact, the location of this node line is usually determined by mission requirements, the timing of the maneuver is often fixed. Maximum spatial separation between the initial and final orbits occurs ± 90 deg away from the point of thrust application.

Note that, if the impulse is applied at the line of nodes of the original orbit (i.e., in the equatorial plane of the coordinate system in use), then the plane rotation will result in a change of orbital inclination only, with $\Delta i = v$. If the impulse is applied at $\omega + \theta = 90$ deg, i.e., at a point in the

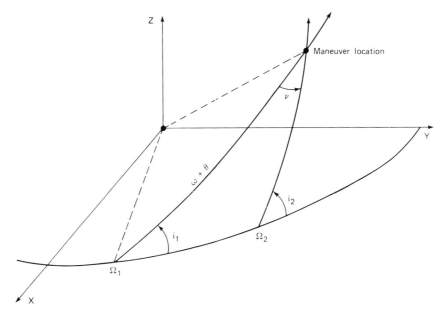

Fig. 4.14   General plane change maneuver.

orbit where the radius vector is perpendicular to the line of nodes, the orbit will precess by the amount $\Delta\Omega = v$ without altering its inclination. Maneuvers initiated at other points will alter both i and Ω.

In practice, adjustments for both i and Ω may be required. Adjustments to Ω are required to compensate for timing errors in orbit insertion (which may not be errors at all and may be, as for interplanetary missions, unavoidable due to the existing planetary configuration). Alterations to inclination angle are required to compensate for azimuthal heading [see Sec. 4.1 (Effect of Launch Site on Orbital Elements)] errors at injection. Adjustments to Ω and i can be done separately, but the typical maneuver is a single plane rotation executed at the node line between the initial and desired orbits. Figure 4.14 shows the spherical triangle that is applicable to this case. From the spherical triangle law of cosines, the required plane rotation is

$$\cos v = \cos i_1 \cos i_2 + \sin i_1 \sin i_2 \cos(\Omega_2 - \Omega_1) \qquad (4.107)$$

and from the spherical triangle law of sines, the maneuver is performed at a true anomaly in the initial orbit found from

$$\sin(\omega + \theta) = \sin i_2 \sin(\Omega_2 - \Omega_1)/\sin v \qquad (4.108)$$

Two locations in true anomaly θ are possible, corresponding to the choice of $\pm v$ at the two nodes. To minimize ΔV requirements, the maneuver should be executed at the node with the largest radius vector.

Interestingly, a plane rotation executed with a single impulsive burn at the line of nodes is not always a minimum-energy maneuver. Lang[45] shows that if the desired node line and the orbital eccentricity are such that

$$e > |\cos\omega^*| \tag{4.109}$$

where ω^* is the central angle from the desired node location to periapsis in the initial orbit, then a two-impulse transfer is best, with the maneuvers occurring at the minor axis points. In this case, the total impulse is given by

$$\Delta V = v|\sin\omega^*|\sqrt{\mu/a} \tag{4.110}$$

applied in the ratio

$$\Delta V_1/\Delta V_2 = -\sin(\omega^* + \theta_2)/\sin(\omega^* + \theta_1) \tag{4.111}$$

Differences between the one- and two-impulse transfers can be significant if eccentricity is large. Lang shows that, for $e = 0.7$ and $\omega^* = 90$ or 270 deg, a 29% savings is realized using the optimal maneuver. For $e < 0.1$, savings are always less than 10%.

Equation (4.106) shows that plane changes are expensive; a 0.1-rad rotation for a 200-km circular parking orbit requires a ΔV of approximately 0.78 km/s. The relative expense of plane changes compared with, for example, perigee adjustments, has produced considerable interest in the use of aerodynamic maneuvers in the upper atmosphere to effect plane rotations.[46] Such possibilities are clearly enhanced for vehicles such as the Space Shuttle, which have a significant lift vector that can be rotated out of the plane of the atmospheric entry trajectory.

Mission requirements specifying node location may well be in conflict with the desire to minimize fuel expenditure. It may be noted that ΔV requirements are minimized by executing the maneuver when V_θ is smallest, i.e., at the apoapsis of an elliptic orbit. Unless the required node line location coincides with the line of apsides of the initial orbit, this minimum-impulse maneuver cannot be achieved.

A simple example is found in the deployment of a communications satellite into a geostationary, hence equatorial, orbit. As seen in Sec. 4.1 (Effect of Launch Site on Orbital Elements), the minimum inclination orbit for launch from Cape Canaveral is 28.5 deg. The satellite will either be injected directly into a highly elliptic transfer orbit with apoapsis at geostationary altitude, or into a nearly circular parking orbit and then later into the transfer orbit. Plane rotation must be done over the equator, and it is highly desirable that it be done at apogee. Thus, the initial launch (or the maneuver into the transfer orbit) must be timed to cause the apogee to be so placed. This is most easily accomplished from a circular parking orbit, which is one reason why initial injection into such an orbit is typically a part of more complex mission sequences.

However, such freedom is not always available. Interplanetary missions provide a ready example. Although most of the planets lie close to the plane of Earth's orbit (ecliptic plane), none is exactly in it; and hence, heliocentric orbit plane changes are always required for interplanetary transfer, unless the mission can be timed to allow the target planet to be intercepted when it is at its heliocentric line of nodes (i.e., in the ecliptic plane). This is not often the case.

As will be seen in Sec. 4.4, a spacecraft on an outer planetary mission will at the time of intercept be at, or at least closer to, the transfer orbit apoapsis than it was at departure from the Earth. As discussed, this is the more desirable position for a plane change. However, since the plane rotation maneuver must occur 90 deg prior to intercept (if maximum effect is to be gained from the maneuver), the node line is thus specified and will not be at the optimal aphelion point. The magnitude of the required plane change is equal (if the maneuver occurs 90 deg prior to intercept) to the

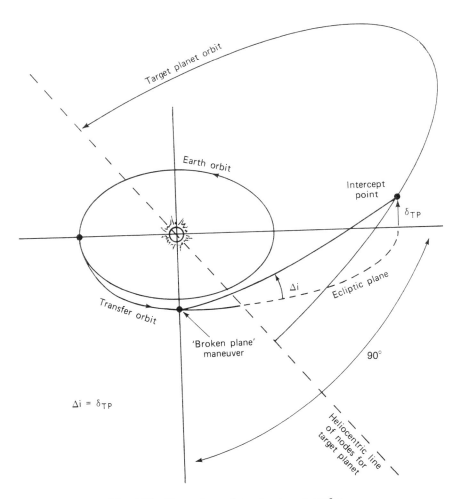

Fig. 4.15   Noncoplanar interplanetary transfer.

ecliptic latitude of the target planet at the time of the encounter. Figure 4.15 shows the geometry. A similar situation exists for inner planetary missions, except that it is desirable to change planes as near to Earth as possible, when heliocentric velocity is least.

The plane change may be done closer to the target planet than the optimal point 90 deg prior to encounter. In such a case, a larger rotation angle is required since there will not be time prior to encounter for the maneuver to take maximum effect. It may be necessary to balance this loss against a gain due to performing the maneuver farther from the sun, where the tangential velocity V_θ is smaller. Appropriate use of these "broken plane" maneuvers can yield acceptable planetary transfers when no direct transfer is possible.[47]

Coplanar Transfers

We now consider maneuvers that leave the orientation of the orbit unchanged but that may alter the elements a, e, and ω and the period τ. Since the direction of h is to remain fixed, all maneuvers must produce torques parallel to h and are thus confined to the orbit plane.

Single-impulse transfer. The geometry of a general single-impulse orbit transfer is shown in Fig. 4.16. An impulsive maneuver is executed at

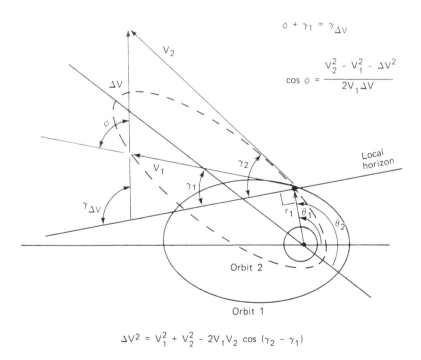

$$\phi + \gamma_1 = \gamma_{\Delta V}$$

$$\cos \phi = \frac{V_2^2 - V_1^2 - \Delta V^2}{2V_1 \Delta V}$$

$$\Delta V^2 = V_1^2 + V_2^2 - 2V_1 V_2 \cos(\gamma_2 - \gamma_1)$$

Fig. 4.16 Single-impulse transfer between two intersecting orbits.

some position (r_1, θ_1) in the orbit plane, with velocity and flight-path angle (V_1, γ_1) yielding a new orbit (which must in the single-impulse case always intersect the old) with velocity and flight-path angle (V_2, γ_2) and true anomaly θ_2. Since the maneuver is impulsive, the radius vector r_1 cannot change during its execution.

Orbit 1 is assumed to be known. In typical situations of interest it may be required to 1) determine the ΔV and heading for the maneuver, given the desired new orbit and the specified transfer point; or 2) determine V_2 and all other characteristics of Orbit 2, given the impulse ΔV and the heading for the maneuver. The heading may be the pitch angle ϕ relative to the orbital velocity vector V_1, or it may be specified by the flight-path angle $\gamma_{\Delta V}$ relative to the local horizontal. In either case, the maneuver satisfies the vector equation

$$V_2 = V_1 + \Delta V \qquad (4.112)$$

However, for preliminary calculations it may be more convenient to use the law of cosines with the velocity vector diagram of Fig. 4.16 to determine the required information.

In case 1, where the new orbit is known and the transfer point (r_1, θ_1) is specified, V_2 is immediately found from Eq. (4.17), and γ_2 is found from Eq. (4.58). Then the impulse and heading are

$$\Delta V^2 = V_1^2 + V_2^2 - 2V_1 V_2 \cos(\gamma_2 - \gamma_1) \qquad (4.113)$$

$$\cos\phi = (V_2^2 - V_1^2 - \Delta V^2)/2V_1 \Delta V \qquad (4.114)$$

or, from the law of sines,

$$\phi = \pi - \sin^{-1}[(V_2/\Delta V)\sin(\gamma_2 - \gamma_1)] \qquad (4.115)$$

with

$$\phi = \gamma_1 - \gamma_{\Delta V} \qquad (4.116)$$

In case 2, the new orbit is to be found given ϕ or $\gamma_{\Delta V}$. With ϕ computed from Eq. (4.116) if need be, Eq. (4.114) is used to solve for V_2, and then Eq. (4.113) is used to find γ_2. The results of Sec. 4.1 (Orbital Elements from Position and Velocity) are then applied to find h, p, e, a, θ, and ω.

The most important special case for a single-impulse transfer occurs when the maneuver ΔV is applied tangent to the existing velocity vector V_1. Then $\phi = 0$ or π, and Eq. (4.114) yields $V_2 = V_1 \pm \Delta V$. (A negative root for V_2 exists but is not operationally sensible.) The tangential ΔV application thus allows the maximum possible change in orbital energy for a given fuel expenditure, adding to or subtracting from the existing velocity in a scalar fashion. Moreover, from Eq. (4.113), $\gamma_2 = \gamma_1$; hence, the flight-path angle is not altered at the point of maneuver execution.

At apogee or perigee, $V_r = 0$, $V_\theta = V_1$, and $\gamma_1 = 0$. A tangentially applied impulse alters only V_θ, leaving $V_r = 0$, and does not change the perigee or

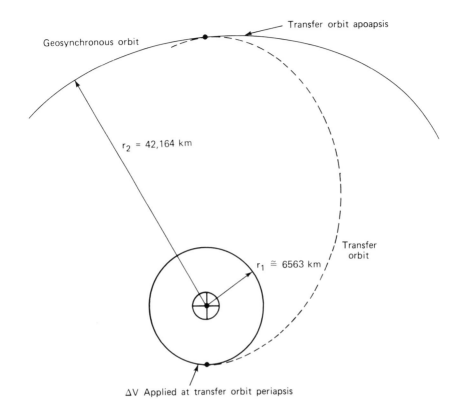

Fig. 4.17 Apogee raising maneuver from circular parking orbit.

apogee radius. Since angular momentum is conserved,

$$h = rV_\theta = r_a V_a = r_p V_p \qquad (4.117)$$

an increase or decrease in V_p with r_p constant must result in an increase or decrease in r_a. Thus, a tangential maneuver at one of the apsides takes effect at the opposite apsis. Since γ is not altered at the maneuver point, the line of apsides does not rotate. An example is shown in Fig. 4.17, which illustrates an apogee-raising maneuver for a satellite initially in a low circular parking orbit and intended for a circular geostationary orbit. The high orbit is attained through injection into an intermediate, highly elliptic geosynchronous transfer orbit.

Two-impulse transfer. Two maneuvering impulses are required for transfer between two nonintersecting orbits. Figure 4.18 shows the required geometry. Analysis of this case requires two successive applications of the results in the previous section, first for a maneuver from Orbit 1 to the transfer orbit, and then for a maneuver into Orbit 2 from the transfer orbit.

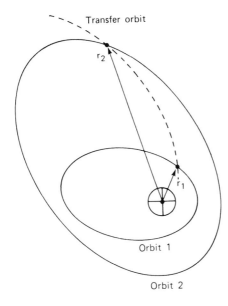

Fig. 4.18 Two-impulse transfer between nonintersecting orbits.

Again, it may be required to assess the results of particular maneuvers, or to determine the maneuvers required for a particular transfer.

Maneuvers defined by two (or more) impulses involve no particular analytical difficulty. From the trajectory design viewpoint, however, an entirely new order of complexity, and thus flexibility, is introduced. In the second case above, the transfer orbit must be known to conduct the maneuver analysis. What, indeed, should the transfer orbit be in order to perform a particular mission? The problem of determining a suitable transfer orbit between specified end conditions, subject to appropriate constraints, is the essential problem of astrodynamics. As mentioned, the design element is what distinguishes astrodynamics from its parent field, classical celestial mechanics. It is this activity that is the main concern of the professional astrodynamicist.

It is worth noting that, conceptually, trajectory design and orbit determination may be viewed as essentially the same problem. If the specified end conditions for the transfer are taken to be observations of an orbiting body, then determination of the "transfer orbit," if it is unique, between these positions is precisely the task of orbit determination. If the given sightings do not uniquely determine the orbit (i.e., determine r and V at a known epoch), then additional information must be obtained. To the trajectory designer, the possible lack of uniqueness between endpoints offers the freedom to apply other constraints, such as fuel usage or transfer time.

Lambert problem. The classical two-impulse trajectory design problem is the so-called Lambert, Gaussian, or "time-of-flight" problem. The Gaussian problem was discussed briefly in Sec. 4.1 (Orbit Determination).

It was pointed out that the specification of two position vectors r_1 and r_2 together with the flight time between them is sufficient for orbit determination. To the trajectory designer, this is equivalent to the statement that, for fixed endpoints, the possible trajectories are parameterized according to time of flight. Typically a range of solutions having different energy requirements is available, with the shorter flight times generally (but not always) associated with higher energy requirements.

This property of conic trajectories is referred to as Lambert's theorem, which states that the transfer time is a function of the form

$$t = t(r_1 + r_2, c, a) \tag{4.118}$$

where c is the chord length between the position vectors r_1 and r_2 and a is the transfer orbit semimajor axis. Since a and E_t are related by Eq. (4.16), the rationale for the statements in the previous paragraph is clear.

Kaplan[9] and Bate et al.[15] give excellent introductory discussions of the time-of-flight problem. Modern practical work in the field is oriented toward trajectory design and is primarily the work of Battin[48] and Battin and co-workers.[49,50]

Hohmann transfer.   Figure 4.17 shows an important special case for transfer between two coplanar nonintersecting orbits. The transfer orbit is shown with an apogee just tangent to the desired geosynchronous orbit, whereas the perigee is tangent to the initial circular parking orbit. From the results of the previous section, any higher transfer orbit apogee would also allow the geosynchronous orbit to be reached. However, the transfer orbit that is tangent to both the arrival and departure orbits, called the Hohmann transfer, has the property that it is the minimum-energy, two-impulse transfer between two coplanar circular orbits.

In practice Hohmann transfers are seldom used, in part because given departure and arrival orbits are rarely both circular and coplanar. Also, Hohmann orbits are slow, a factor that may be significant for interplanetary missions.

Because of the physical constraints defining the Hohmann transfer, its ΔV requirements are easily computed. ΔV_1 for departure from a circular orbit at r_1 is

$$\Delta V_1 = V_{p_{TO}} - \sqrt{\mu/r_1} \tag{4.119}$$

(where subscript TO denotes transfer orbit), whereas upon arrival at the circular orbit at r_2,

$$\Delta V_2 = \sqrt{\mu/r_2} - V_{a_{TO}} \tag{4.120}$$

and

$$\Delta V = \Delta V_1 + \Delta V_2 \tag{4.121}$$

The transfer orbit properties are easily found; since

$$a_{TO} = (r_1 + r_2)/2 \qquad (4.122)$$

the vis-viva equation yields

$$V^2_{P_{TO}} = \mu(2/r_1 - 1/a_{TO}) \qquad (4.123)$$

and

$$V^2_{a_{TO}} = \mu(2/r_2 - 1/a_{TO}) \qquad (4.124)$$

or by noting from Eq. (4.117) that

$$V_{a_{TO}} = V_{P_{TO}}(r_1/r_2) \qquad (4.125)$$

Inbound and outbound transfers are symmetric; thus, no loss of generality is incurred by considering only one case. A simple example serves to illustrate the method.

Example

Compute the mission ΔV for a Hohmann transfer from a 185-km circular Space Shuttle parking orbit to a geosynchronous orbit at an altitude of 35,786 km above the Earth.

Solution

For Earth

$$\mu = 3.986 \times 10^5 \text{ km}^3/\text{s}^2$$

$$R_e = 6378 \text{ km}$$

Thus,

$$r_1 = 6563 \text{ km}$$

$$r_2 = 42,164 \text{ km}$$

$$a_{TO} = 24,364 \text{ km}$$

and from Eq. (4.119),

$$\Delta V_1 = 10.252 \text{ km/s} - 7.793 \text{ km/s} = 2.459 \text{ km/s}$$

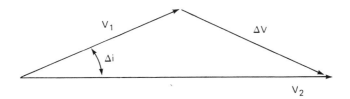

Fig. 4.19   Vector diagram for combined plane change and orbital energy adjustment.

Similarly, from Eq. (4.120),

$$\Delta V_2 = 3.075 \text{ km/s} - 1.596 \text{ km/s} = 1.479 \text{ km/s}$$

Hence, for the mission,

$$\Delta V = 3.938 \text{ km/s}$$

Combined Maneuvers

It is often possible to combine a required in-plane and out-of-plane maneuver and effect a fuel savings. A practical example is that of the previous section, in which the ΔV requirement for a Hohmann transfer from a parking orbit to geosynchronous orbit was computed. To attain a geostationary orbit, a plane change of 28.5 deg (assuming a due-east launch from Cape Canaveral) is required.

The situation is as shown in Fig. 4.19. As in previous sections, V_1 is the existing velocity vector, V_2 is the desired vector, and

$$\Delta V^2 = V_1^2 + V_2^2 - 2 V_1 V_2 \cos\Delta i \qquad (4.126)$$

gives the velocity increment required for the transfer to effect the combined plane change and alteration of the in-plane elements a, e, and ω.

In the preceding example, $V_1 = 1.596$ km/s, $V_2 = 3.075$ km/s, and $\Delta i = 28.5$ deg; thus we find $\Delta V = 1.838$ km/s. If the maneuvers are performed separately, with the plane change first, Eq. (4.106) yields $\Delta V_1 = 0.786$ km/s, and the circularization maneuver requires $\Delta V_2 = 1.479$ km/s, as before. This produces a total mission $\Delta V = 2.265$ km/s, clearly a less efficient approach.

More general combinations of maneuvers are possible that have concomitant fuel savings over sequential ΔV applications. Investigation of optimal orbit transfers is a perennial topic of research in astrodynamics. Recent works by Small[51] and Hulkower et al.[52] provide useful results in this area.

4.4   INTERPLANETARY TRANSFER

The results of the previous sections are sufficient for the analysis of all basic orbital transfers, including those for interplanetary missions. However, trajectory design for such missions is sufficiently complex to justify separate discussion.

Although trajectory design for advanced mission analysis or actual mission execution will invariably be accomplished using direct numerical integration of the equation of motion, such procedures are inappropriate for preliminary assessments. Not only are the methods time-consuming and highly specialized, but also, without a preliminary analytical solution, they offer no way to eliminate the many cases that are not at all close to the actual solution of interest.

For initial mission design and feasibility assessments, the so-called method of patched conics is universally employed. We consider the application of the method in some detail.

Method of Patched Conics

As the name implies, this approach uses a series of Keplerian orbits to define the trajectory. Each separate conic section is assumed to be solely due to the influence of the dominant body for that portion of the mission. The different segments are "patched" at the sphere of influence boundaries between different bodies given by Eq. (4.81). Thus, a spacecraft trajectory between Earth and Mars will be modeled for departures as a geocentric escape hyperbola, which at great distances from Earth becomes a heliocentric elliptic orbit, followed again by a hyperbolic approach to Mars under the influence of that planet's gravitational field.

Patched-conic trajectory designs are accomplished in three well-defined steps:

1) The heliocentric orbit from the departure planet to the target planet is computed, ignoring the planet at either end of the arc.

2) A hyperbolic orbit at the departure planet is computed to provide the "infinity" (sphere of influence boundary) conditions required for the departure end of the heliocentric orbit in step 1.

3) A hyperbolic approach to the target is computed from the infinity conditions specified by the heliocentric orbit arrival.

The statement of the patched-conic procedure shows that it cannot yield truly accurate results. A spacecraft in an interplanetary trajectory is always under the influence of more than one body, and especially so near the sphere of influence boundaries. Transitions from one region of dominance to another are gradual and do not occur at sharply defined boundaries. Keplerian orbit assumptions in these regions are incorrect, yet conveniently applicable analytical results do not exist, even for the restricted three-body problem. And other perturbations, such as solar radiation pressure, also occur and can have significant long-term effects.

The patched-conic method yields good estimates of mission ΔV requirements and thus allows quick feasibility assessments. Flight times are less well predicted, being in error by hours, days, or even weeks for lengthy interplanetary missions. Such errors are of no consequence for preliminary mission design but are unacceptable for mission execution. An encounter at the target planet must occur within seconds of the predicted time if a flyby or orbit injection maneuver is to be properly performed. For example, the heliocentric velocity of Mars in its orbit is roughly 24 km/s. If an orbit injection were planned to occur at a 500 km periapsis height, a spacecraft arriving even 10 s late at Mars would likely enter the atmosphere.

Patched-conic techniques are useful at the preliminary design level for hand calculation or for implementation in a relatively simple computer program. As stated, actual mission design and execution must employ the most accurate possible numerical integration techniques. The difference in accuracy obtainable from these two approaches can be a source of difficulty, even at the preliminary design level, for modern interplanetary missions involving application of multiple ΔV or planetary flybys and the imposition of targeting constraints and limitations on total maneuver ΔV. In such cases, the errors implicit in patched-conic approximations during early phases of the trajectory may invalidate subsequent results, and detailed numerical calculations are too cumbersome even on the fastest machines for use in preliminary analysis. Recent applications of constrained parameter optimization theory to the multiple encounter problem have resulted in relatively fast, efficient techniques for trajectory design that eliminate 90–99% of the error of simple conic methods.[47]

Heliocentric trajectory. This portion of the interplanetary transfer will usually be computed first, unless certain specific conditions required at an encounter with the target planet should require a particular value of V_∞ for the hyperbolic approach. As stated earlier, the calculation ignores the planet at each end of the transfer and thus gives the ΔV to go from the orbit of the departure planet to the orbit of the arrival planet. To be strictly correct, the departure and arrival should begin and end at the sphere of influence boundary for each planet; however, these regions are typically quite small with respect to the dimensions of the heliocentric transfer and are often ignored. Of course, calculations for Earth-moon missions cannot justifiably employ this assumption.

The heliocentric segment is not restricted to Hohmann transfers or even to coplanar transfers, though these are common assumptions in preliminary design. The assumption of coplanarity may cause serious errors in ΔV computations and should be avoided. However, given the overall accuracy of the method, the assumption of circular orbits at the departure and arrival planets is often reasonable and because of its convenience is used where possible. This assumption may not be justified for missions to planets with substantially elliptic orbits, e.g., Mercury or possibly Mars or, in the extreme case, Pluto.

The heliocentric trajectory design will usually be constrained by available launch energy, desired travel time, or both. When both are important, appropriate tradeoffs must be made, with the realization, however, that energy savings achieved through the use of near-Hohmann trajectories can be nullified by the increased mass and/or redundant systems required because of the longer flight times. Again, minimum-energy orbits (Hohmann-type doubly tangent transfer plus a heliocentric plane change

$$V_\infty = V_S - V_P$$

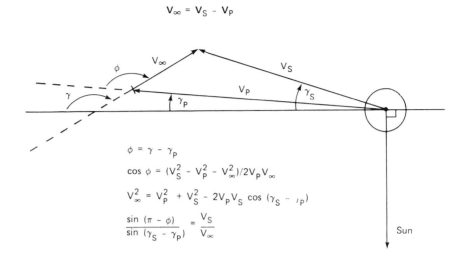

$$\phi = \gamma - \gamma_P$$

$$\cos \phi = (V_S^2 - V_P^2 - V_\infty^2)/2V_P V_\infty$$

$$V_\infty^2 = V_P^2 + V_S^2 - 2V_P V_S \cos (\gamma_S - \gamma_P)$$

$$\frac{\sin (\pi - \phi)}{\sin (\gamma_S - \gamma_P)} = \frac{V_S}{V_\infty}$$

Fig. 4.20   Approach velocity at target planet.

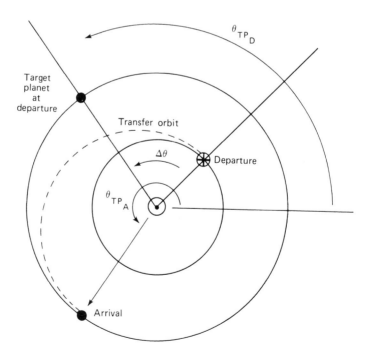

Fig. 4.21   Phasing for interplanetary transfer.

if required) are often assumed initially because of the computational convenience. If the flight times are unacceptable, a faster transfer must be used, with consequently higher ΔV requirements.

If a doubly tangent transfer orbit is assumed, then Eqs. (4.122–4.125) may be used to determine the transfer orbit characteristics. Equation (4.106) is used to compute any additional ΔV required to match the heliocentric declination of the target planet at encounter. This cannot be done until the actual timing of the mission is determined.

It will be required to know the arrival velocity of the spacecraft relative to the target planet. This is simply the vector difference

$$V_\infty = V_S - V_P \qquad (4.127)$$

between the heliocentric transfer orbit velocity and the velocity of the target planet in its heliocentric orbit, as shown in Fig. 4.20. If the minimum-energy transfer is used, this reduces to the difference in their scalar speeds. If a faster transfer is required, the heliocentric velocity vector will not be tangent to the target planet orbit at encounter. The arrival velocity relative to the target planet will then be given by the methods of Sec. 4.3 (Coplanar Transfers). Equations (4.113–4.116) become

$$V_\infty^2 = V_P^2 + V_S^2 - 2V_P V_S \cos(\gamma_S - \gamma_P) \qquad (4.128)$$

$$\cos\phi = (V_S^2 - V_P^2 - V_\infty^2)/2V_P V_\infty \qquad (4.129)$$

or, from the law of sines,

$$\phi = \pi - \sin^{-1}[(V_S/V_\infty) \sin(\gamma_S - \gamma_P)] \qquad (4.130)$$

with

$$\gamma = \gamma_P - \phi \qquad (4.131)$$

It will always be advantageous for the transfer orbit to be tangent to that of the departure planet. This may not be possible, however, when gravity-assist maneuvers are used at intermediate planets between the departure planet and the ultimate target. The departure conditions from the intermediate planet are determined by the hyperbolic encounter with that planet, as will be seen in Sec. 4.4 (Gravity-Assist Trajectories).

Once a trial orbit has been assumed and the heliocentric transfer time computed, it is necessary and possible to consider the phasing or relative angular position of the departure and arrival planets for the mission. The geometric situation is shown in Fig. 4.21. Clearly, departure must occur when the relative planetary positions are located such that, as the spacecraft approaches the target planet orbit in the transfer trajectory, the planet is there also. Assuming the transfer time has been found, the difference in true anomaly between departure and arrival planets at launch is

$$\Delta\theta = (\theta_{TO_A} - \theta_{TO_D}) - (\theta_{TP_A} - \theta_{TP_D}) \qquad (4.132)$$

where, from Eq. (4.7), the true anomaly is

$$\theta = \cos^{-1}[(p/r - 1)/e] \qquad (4.133)$$

for any conic orbit. If the departure or arrival planet orbit is circular, $\theta(t) = n(t - t_0)$, whereas if near-circularity can be assumed, $\theta(t)$ may be obtained from Eq. (4.32).

If coplanar circular planetary orbits are assumed, then no difference in ΔV requirements is found between missions executed at different calendar times. In fact, substantial advantages exist for missions that can be timed to encounter the target planet near its heliocentric line of nodes (implying minimum plane change requirements) or when the combination of Earth and target planet perihelion and aphelion phasing is such as to minimize the semimajor axis of the required transfer orbit. For example, an Earth-Mars minimum-energy transfer orbit can have a semimajor axis from 1.12 A.U. to 1.32 A.U, resulting in a ΔV difference at Earth departure of about 500 m/s.

Departure hyperbola. When the heliocentric transfer has been computed, the required spacecraft velocity in the neighborhood of the departure planet is found from the vis-viva equation. This velocity is in heliocentric inertial coordinates; the departure planet will itself possess a considerable velocity in the same frame. In the patched-conic method, the heliocentric transfer is assumed to begin at the sphere of influence boundary between the departure planet and the sun. This boundary (see Table 4.1) may be assumed to be at infinity with respect to the planet. As indicated by Eq. (4.127), the planetary departure hyperbola must therefore be designed to supply V_∞, the vector difference between the spacecraft transfer orbit velocity V_S and the planetary velocity V_P. Again, when V_S is parallel to V_P, V_∞ is their simple scalar difference.

We assume for convenience in this discussion that departure is from Earth. If the departure maneuver is executed with zero geocentric flight-path angle γ, the maneuver execution point will define the periapsis location for the outbound half of a hyperbolic passage, discussed in Sec. 4.1 (Motion in Hyperbolic Orbits).

The geocentric hyperbola must be tangent at infinity to the heliocentric transfer orbit; hence, the orientation of the departure asymptote is known in the heliocentric frame. The excess hyperbolic velocity V_∞ is also known from Eq. (4.127), and r_p is usually fixed by parking orbit requirements (r_p is typically about 6600 km for Earth). Therefore, θ_a, e, ΔV, and the departure location in the heliocentric frame are specified by Eqs. (4.36–4.38). Equation (4.39) allows the offset β for the passage to be computed if desired. Figure 4.22 shows the geometry for hyperbolic departure.

Encounter hyperbola. The determination of V_∞ relative to the target planet from the heliocentric transfer orbit has been discussed. The encounter orbit at the target planet is shown in Fig. 4.23 in a frame centered in the

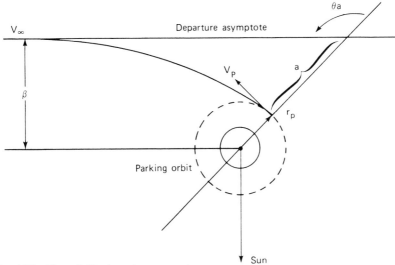

Fig. 4.22   Hyperbolic departure geometry.

planet. The results of Sec. 4.1 (Motion in Hyperbolic Orbits) again allow the required parameters to be found.

Operational requirements for the encounter will usually differ somewhat from those for departure. The periapsis radius is found from the approach parameters β and V_∞ from Eq. (4.41). This will be of interest for flyby and orbital-injection missions, where a particular periapsis altitude may be appropriate for photography or to attain a desirable orbit about the planet. For passages where a gravity assist is required to allow a continuation of the mission to another planet or moon, the turning angle Ψ of the passage will be critical, and β and r_p will be adjusted accordingly.

Note that impact is achieved for $r_p \leq R$, the planetary radius. From Eq. (4.42), the β-plane offset for impact is then given by

$$\beta_{\text{impact}} \leqslant R[1 + 2\mu/RV_\infty^2]^{\frac{1}{2}} \qquad (4.134)$$

The term in brackets will be somewhat greater than unity; thus, a large targeting area at "infinity" funnels down to a considerably smaller planetary area. This is of course due to the attractive potential of the planet and is referred to as its collision cross section.

Atmospheric entry and braking without direct planetary impact will require targeting for a small annulus in the atmosphere above the planet, such that

$$R + h_{\text{min}} < r_p < R + h_{\text{max}} \qquad (4.135)$$

The minimum acceptable entry height h_{min} is usually determined by the maximum acceptable dynamic loading due to atmospheric deceleration.

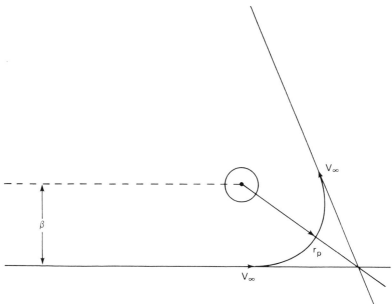

Fig. 4.23 Hyperbolic encounter in target planet frame of reference.

The maximum limit $h_{\max}$ will often be fixed by entry heating constraints. These topics are discussed more fully in Chapter 6.

From Eq. (4.42), it is seen that the β-plane annular width, which maps into an annular region near periapsis, is given by

$$\Delta\beta = (\beta/r_p - \mu/\beta V_\infty^2)\Delta r_p \qquad (4.136)$$

Equation (4.136) may be used to determine the β-plane targeting requirement for the encounter to ensure hitting the entry corridor of Eq. (4.135).

Gravity-Assist Trajectories

Upon completion of the hyperbolic passage at a target planet, the approach velocity V_∞ relative to the planet will have been turned through an angle Ψ. In the heliocentric inertial frame, the encounter thus produces the result of Fig. 4.24. As seen by applying Eq. (4.127) on arrival and departure, V_p and V_∞ do not change during the passage, but V_∞ does, with the result that the spacecraft velocity V_{S_D} in the inertial frame is altered with respect to V_{S_A}.

It will be required to know the values of V_{S_D} and γ_{S_D} upon exit from the hyperbolic passage. Examination of Fig. 4.24, with ΔV and Ψ known from Eq. (4.37), yields

$$V_{S_D}^2 = V_{S_A}^2 + \Delta V^2 - 2V_{S_A}\,\Delta V\cos v \qquad (4.137)$$

$$v = 3\pi/2 + \Psi/2 - \phi_A + \gamma_{S_A} - \gamma_P \qquad (4.138)$$

$$\phi_A = \pi - \sin^{-1}[(V_{S_A}/V_\infty)\,\sin(\gamma_{S_A} - \gamma_P)] \qquad (4.139)$$

$$\gamma_{S_D} = \gamma_{S_A} + \sin^{-1}[(\Delta V/V_{S_D})\,\sin\nu] \qquad (4.140)$$

Figure 4.24 depicts a situation in which the heliocentric energy of the spacecraft is increased (at the infinitesimal expense of that of the planet) as a result of the hyperbolic passage. This would be applicable to missions such as Voyager 1 and 2, in which encounters at Jupiter were used to direct the two spacecraft toward Saturn (and, for Voyager 2, subsequently to Uranus and Neptune) much more efficiently than by direct transfer from Earth.

Energy-loss cases are equally possible. One application is to inner planetary missions such as Mariner 10, which reached Mercury via a pioneering gravity-assist maneuver performed at Venus. Similarly, the Ulysses spacecraft will be directed toward the sun by means of a energy-loss maneuver at Jupiter. Figure 4.25 shows a typical situation involving heliocentric energy loss following the encounter.

It will be noted that in Fig. 4.24 the planetary-relative approach vector V_{∞_A} is rotated through angle Ψ in a counterclockwise or positive sense to obtain V_{∞_D}, whereas the opposite is true in Fig. 4.25. These are typical, though not required, situations producing energy gain at outer planets and energy loss at inner planets. Consideration of the encounter geometry will show that clockwise rotation of Ψ occurs for spacecraft passage between the target body and its primary (e.g., between a planet and the sun), whereas counterclockwise or positive rotation results from passage behind the target body as seen from its primary. This is shown conceptually in Fig. 4.26. Spacecraft heliocentric energy gain or loss as a result of the encounter depends on the orientation of V_{S_A} and the rotation angle Ψ of the relative approach velocity V_{∞_A} during encounter.

Equation (4.37) shows that the maximum-energy gain or loss occurs if $\Psi = 180$ deg; in this case scalar addition of V_∞ to V_P results. This is an idealized situation requiring $r_p = 0$ for its implementation. Actually, the maximum heliocentric ΔV obtainable from an encounter occurs for a grazing passage with $r_p = R$. Note that a Hohmann trajectory yields a

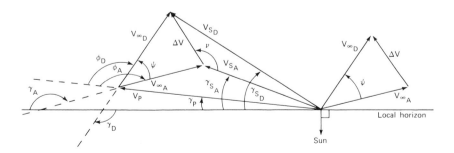

Fig. 4.24   Spacecraft energy gain in inertial frame during hyperbolic passage.

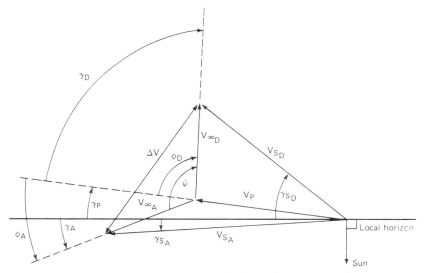

Fig. 4.25   Spacecraft energy loss in inertial frame during hyperbolic passage.

transfer orbit that is tangent to the target body orbit; V_P and V_{S_A} are thus colinear. Examination of Fig. 4.24 or 4.25 shows that in this case energy can only be gained (for outbound transfers) or lost (for inbound transfers), regardless of the sign of the rotation angle Ψ. Non-Hohmann transfers allow a wider range of encounter maneuvers.

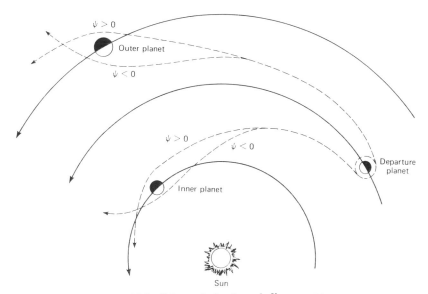

Fig. 4.26   Schematic for hyperbolic passage.

The gravity-assist maneuver for planetary exploration has arrived as a mature technique since its initial use on the Mariner 10 mission. The Galileo Jupiter orbital mission is designed around extensive use of gravity assists in the Jovian system to allow the spacecraft to be directed from one moon to another. The first spacecraft to encounter the tail of a comet, ISEE-3 (renamed the International Cometary Explorer), was directed to the comet Giacobini-Zinner in late 1983 via a gravity assist from the moon.

Lunar Transfer

The problem of calculating lunar transfer trajectories is conceptually similar to that of interplanetary transfer analysis, and, indeed, the method of patched conics can be used for preliminary assessment of mission requirements. However, the results obtained are considerably less satisfactory than for interplanetary transfers due to a number of complicating factors.

The masses of the Earth and moon are more nearly equal than for any other primary and satellite (possibly excluding Pluto and Charon) in the solar system. The moon's sphere of influence is therefore large with respect to the Earth-moon separation, and a spacecraft in transit between the two spends much of its time close to the sphere-of-influence boundary. Also, the sun's influence on the trajectory is significant. For these reasons, the patched-conic method is less accurate than for heliocentric transfers and yields truly useful results only for outbound ΔV calculations.

Accurate results are also obtained with considerably more trouble than for interplanetary trajectories, because of the size of the lunar sphere of influence in relation to the transfer orbit dimensions. This implies that the location of the point where the spacecraft crosses the boundary is important in determining the characteristics of the transfer orbit, a fact that adds considerable complexity to the numerical procedures.

We do not include an extended treatment of lunar transfer calculations here. Cursory mission requirements can be assessed by the methods of Sec. 4.4 (Method of Patched Conics); more detailed analysis must be done via numerical integration of the equations of motion, possibly using the patched-conic solution as an initial guess. Bate et al.[15] give an excellent discussion of the use of patched-conic techniques in lunar transfer calculations. Their treatment includes non-coplanar transfer analyses, important in this case because of the relatively large lunar orbit inclination (which is in fact not constant, but varies between 18.2 and 28.5 deg over an 18.6-yr period) in the GCI frame.

4.5   PERTURBATION METHODS

We have on several occasions mentioned that truly Keplerian orbits are essentially nonexistent and have given methods for analyzing some of the perturbations to Keplerian orbits that are important in spacecraft and mission design. Perturbation theory forms an elaborate structure in astrodynamics and classical celestial mechanics and, indeed, comprises much of the current literature in the latter subject. Such topics are completely beyond the scope of this text. However, many of the results cited earlier are due to perturbation theory, and a brief outline of this topic is in order.

Perturbation methods are broadly divided into special and general theories. Special perturbation theory is ultimately characterized by the direct numerical integration of the equations of motion due to a dominant acceleration and one or more small perturbing accelerations. As with all numerical analysis, results are unique to the given case, and it is often unclear how to extrapolate the results of one situation to another case of interest.

General perturbation analysis, historically the first approach to be developed, proceeds as given earlier, except that the perturbing accelerations are integrated analytically, at least to some given order of accuracy. Since closed-form integration of given perturbing accelerations will rarely be possible, series expansion to a desired order of accuracy is used to represent the perturbation, and the series integrated term by term. Analytical results are thus available, and broader applications and more general conclusions are possible. Nearly all important results have been obtained through general perturbation methods; on the other hand, special perturbation techniques are more applicable to practical mission design and execution.

Common special perturbation techniques are the methods of Cowell and Encke and the method of variation of parameters. Cowell's method is conceptually the simplest, at least in an era of digital computers, and consists of directly integrating the equations of motion, with all desired perturbing accelerations included, in some inertial frame. The method is uncomplicated and readily amenable to the inclusion of additional perturbations if a given analysis proves incomplete. The primary pitfalls are those associated with the use of numerical integration schemes by the unwary. Reference to appropriate numerical analysis texts and other sources[53] is recommended even if standard library procedures are to be used. Cowell's method is relatively slow, a factor that is in the modern era often irrelevant; the computer costs of a slow method may be more than compensated by the reduction of engineering manpower costs attendant to a simple approach. The speed of the method is increased substantially, with only slight complexity, by employing spherical coordinates (r,θ,ϕ) instead of Cartesian coordinates.

Encke's method antedates Cowell's, which is not surprising since the latter poses formidable implementation requirements in a precomputer era. Encke's method also employs numerical integration techniques, but proceeds by integrating the difference between a given reference orbit (often called the osculating or tangent orbit) and the true orbit due to the perturbing acceleration. Since the perturbation is assumed small (a possible pitfall in the application of Encke's method), the difference between the true and reference orbits is presumably small, and larger integration step sizes can be used for much of the orbit. Encke's method, depending on the situation, executes from 3 to 10 times faster than Cowell's.

The method of variation of parameters is conceptually identical to that of general perturbation analysis, with the exception that the final step of series expansion and term by term integration is skipped in favor of direct numerical integration. In this sense, it is something of a compromise method between special and general perturbations. For example, the effects due to nonspherical primary mass distributions discussed in Sec. 4.2 are analyzed by the variation of parameters method. The results yield analytical rather than numerical forms for the variation of the parameters or elements (Ω and ω in this case) by obtaining $d\Omega/dt$, $d\omega/dt$, etc. This allows more interesting general conclusions to be drawn than with a purely numerical approach; however, complete analysis of the final effects must still be done numerically.

4.6 ORBITAL RENDEZVOUS

Orbital rendezvous and docking operations are essential to the execution of many current and planned space missions. First proven during the manned Gemini flights of 1965 and 1966, rendezvous and docking was a required technique for the Apollo lunar landing missions and the Skylab program. It is essential for Space Shuttle missions involving satellite retrieval, inspection, or repair. Unmanned, ground-controlled rendezvous and docking procedures have been demonstrated on many Soviet flights and have been proposed as an efficient technique for an unmanned Mars sample return mission.[54] In this section, we discuss rendezvous orbit dynamics and procedures. More detailed consideration of guidance algorithms, sensors, and docking procedures is included in Chapter 8.

Equations of Relative Motion

Preliminary rendezvous maneuvers, often called phasing maneuvers, may well be analyzed in an inertial frame such as GCI and carried out using the methods of Sec. 4.3. However, the terminal phase of rendezvous involves the closure of two vehicles separated by distances that are small (e.g., tens or hundreds of kilometers) relative to the dimensions of the orbit. It is then expected that the difference in acceleration experienced by the two vehicles is relatively small and thus that their differential motion might easily be obscured by their gross orbital motion. Also, guidance algorithms are generally described in terms of the position and velocity of one vehicle relative to another. For these reasons, a description of the orbital motion and maneuvers in a planetary centered reference frame is often inappropriate for rendezvous analysis. Instead, it is customary to define a target vehicle (TV) and a chase vehicle (CV) and to describe the motion of the chase vehicle in a noninertial coordinate frame fixed in the target vehicle. In this way, one obtains the equations of relative motion between the vehicles.

The coordinate frame for the analysis is shown in Fig. 4.27. It is assumed that the two orbits are in some sense "close," having similar values of the elements a, e, i, and Ω. $\boldsymbol{R}_T$ and $\boldsymbol{R}_C$ are the inertial vectors to the target and chase vehicles, respectively, and $\boldsymbol{r}$ is their separation vector,

$$\boldsymbol{r} = \boldsymbol{R}_C - \boldsymbol{R}_T \tag{4.141}$$

initially assumed to be small. The frame in which $\boldsymbol{r}$ is expressed is centered in the TV, and it is convenient to use the rotating local vertical system

(r,s,z) shown in Fig. 4.27, where r is parallel to the TV radius vector $\mathbf{R}_T$, s is normal to r in the orbit plane, and z is perpendicular to the TV orbit plane. In this system, the vector equations of motion for a CV maneuvering with acceleration $\boldsymbol{a}$ and a nonmaneuvering TV are, from Eq. (4.6),

$$\mathrm{d}^2\mathbf{R}_C/\mathrm{d}t^2 + (\mu/R_C^3)\mathbf{R}_C = \boldsymbol{a} \qquad (4.142)$$

$$\mathrm{d}^2\mathbf{R}_T/\mathrm{d}t^2 + (\mu/R_T^3)\mathbf{R}_T = 0 \qquad (4.143)$$

Equations (4.142) and (4.143) are then differenced and combined with Eq. (4.141), using several simplifying vector identities that assume small r, to yield the equation of motion of the CV in the TV frame. A variety of linearized equations may then be obtained, depending on the simplifying assumptions used.

If the TV and CV orbits are both nearly circular with similar semimajor axes and orbital inclinations, the basic relative motion equations first given by Hill in 1878 and subsequently rediscovered by Clohessy and Wiltshire[55] apply:

$$\mathrm{d}^2 r/\mathrm{d}t^2 - 2n\,\mathrm{d}s/\mathrm{d}t - 3n^2 r = a_r$$

$$\mathrm{d}^2 s/\mathrm{d}t^2 + 2n\,\mathrm{d}r/\mathrm{d}t = a_s$$

$$\mathrm{d}^2 z/\mathrm{d}t^2 + n^2 z = a_z \qquad (4.144)$$

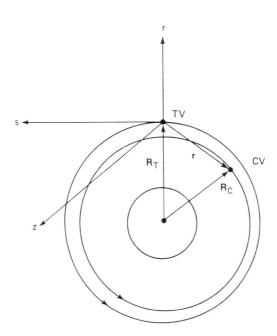

Fig. 4.27   Rotating local vertical coordinates.

where n is the mean TV orbital rate and $n \simeq d\theta/dt$ by assumption. Note that, although small separation was initially assumed, the downtrack range s does not explicitly appear above and is thus not restricted. In the circular orbit case, the important criteria for orbital separation are the radial and out-of-plane components. If orbits of nonzero eccentricity are allowed, as below, restrictions on downtrack separation will again appear. Note also that the out-of-plane component decouples from the other two; for small inclinations, the motion normal to the orbit plane is a simple sinusoid.

The circular orbit approximation is common and often realistic, since many rendezvous operations can be arranged to occur, at least in the final stages, between CV and TV in nominally circular orbits. Also, as discussed in Sec. 4.1 (Motion in Elliptic Orbits), most practical parking orbits are of nearly zero eccentricity. However, Jones[56] has shown that both zero eccentricity and small eccentricity approximations can yield significant errors (see Fig. 4.28) in some cases compared with results obtained using Stern's equations.[57] These equations are linear and thus retain the assumption of small displacements between TV and CV but are valid for arbitrary eccentricity. We have

$$d^2r/dt^2 - 2(d\theta/dt)(ds/dt) - [(d\theta/dt)^2 + 2\mu/R_T^3]r - (d^2\theta/dt^2)s = a_r$$
$$(4.145a)$$

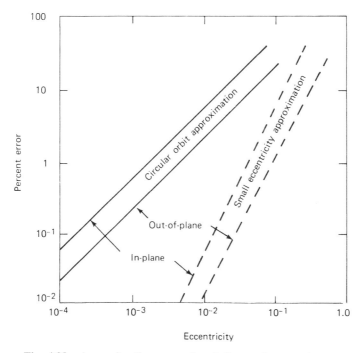

Fig. 4.28   Approximation errors in relative motion equations.

$$d^2s/dt^2 + 2(d\theta/dt)(dr/dt) - [(d\theta/dt)^2 - \mu/R_T^3]s + (d^2\theta/dt^2)r = a_s$$
(4.145b)

$$d^2z/dt^2 + (\mu/R_T^3)z = a_z$$
(4.145c)

where θ and R_T are the true anomaly and radius of the TV in the inertial frame.

It is seen that Eq. (4.145) reduces to Eq. (4.144) when the TV orbit is circular. Dunning[58] gives equations of intermediate complexity between the above two sets, in which the second-derivative terms in true anomaly θ are omitted. These and equivalent results obtained by Jones[55] give first-order corrections for eccentricity compared with the Clohessy-Wiltshire equations.[55]

Care should be exercised in the choice of formulation used. Jones[56] finds approximately 5% error using the Hill equations compared to results obtained using Stern's equations for $e = 0.01$, and 10% error for $e = 0.05$. However, error estimates are only approximate and depend on the actual case of interest, and both sets of equations contain linearization errors that can be expected to dominate at longer CV to TV ranges. Classical guidance algorithms[59] for terminal rendezvous implicitly invoke the same assumptions as for the Hill equations; thus, when doubt exists, it is wise to study the sensitivity of the results obtained to the choice of orbital eccentricity assumed and the dynamics model employed.

The solution to the Hill equations in the case of unforced motion is easily obtained.[9] The rotating coordinate system used in the analysis produces what at first glance appear to be rather unusual trajectories. Figure 4.29 shows typical CV motion[65] for cases where it is above and below the orbit of the TV. Clearly, maneuvers to achieve rendezvous are facilitated if the CV is initially above and ahead or initially below and behind the TV. Rendezvous procedures are structured so as to attain this geometry prior to the initiation of the terminal phase closing maneuvers.

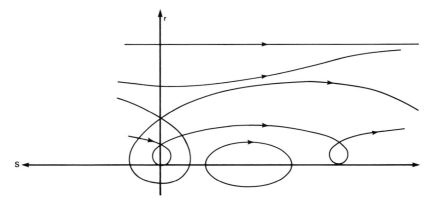

Fig. 4.29 Relative motion trajectories.

Rendezvous Procedures

A variety of rendezvous procedures have been implemented in the U.S. manned flight program, and others have been proposed for unmanned vehicles such as would be required for a planetary sample return program.[54] We consider here the basic operational scenario for U.S. manned rendezvous missions. All such missions have utilized essentially circular target vehicle orbits. The baseline procedure that was developed during the Gemini program[60] and implemented operationally on Apollo[61] is the so-called concentric flight plan (CFP) approach. The CFP procedure involves five basic steps:

1) Any out-of-plane component in the CV orbit is removed by waiting until $z = 0$, in the notation of Eq. (4.144), and thrusting with acceleration a_z to yield $dz/dt = 0$. In the formulation of the Hill equations, this is equivalent to a small plane change at the line of nodes, as given by Eq. (4.106). This maneuver was not required to be, and typically was not, the first in the sequence and in operational cases often was not needed at all.

2) A waiting period is needed to allow proper phasing to develop between the two vehicles, as discussed for interplanetary transfer in Sec. 4.4 (The Heliocentric Trajectory). The "above and ahead" or "below and behind" geometry must be attained, together with the requirement imposed by Eq. (4.132). An adjustment to the CV orbit (typically a perigee raising maneuver for the usual case of the CV below the TV) is made to achieve the desired phasing prior to the next step.

3) Upon attainment of proper vehicle phasing, the so-called constant differential height (CDH) or coelliptic maneuver was performed. Assuming the TV to occupy the higher orbit, this maneuver is done at the CV orbit apogee and places the chase craft in an orbit that is concentric (or coelliptic) with that of the target but several tens of kilometers lower. A coelliptic separation of 15 n.mi. was used for Apollo lunar rendezvous. A strong advantage of the CFP approach is that the CDH phase allows the next, or terminal, phase to be relatively insensitive to the timing and execution of earlier operations.

4) When proper vehicle-to-vehicle phasing is obtained, the terminal phase initiation maneuver is executed. In Gemini and Apollo, the nominal final transfer maneuver was a two-impulse, non-Hohmann trajectory requiring a transfer angle of 130 deg. This value was selected on the basis of simulations showing a relative lack of sensitivity of the arrival conditions to errors in TPI timing and impulse magnitude. Also, this transfer trajectory was shown to result in a minimal rotation rate in the CV-to-TV inertial line of sight during the closure process, a feature that is useful both in the design of guidance algorithms and as a piloting aid. Other final transfer trajectories are possible; Jezewski and Donaldson[62] have studied optimal maneuver strategies using the Clohessy-Wiltshire equations.[55]

5) Regardless of the terminal phase maneuver selected, as the range is reduced, a closed-loop terminal guidance scheme will be used to control the reduction of range and range rate to zero. Some form of proportional guidance[63] is generally employed; whatever the technique, the orbit dynamics of the closure trajectory are no longer central to the scheme. Small

corrections are applied, based on differences between actual and predicted velocity vs range during closure, to allow the nominal transfer trajectory to be maintained by the CV as the target is approached. As final closure occurs, braking maneuvers are performed to reduce any residual relative velocity to zero in the neighborhood of the target vehicle.

The rendezvous phase is complete when the chase and target vehicles are separated by a small distance, typically less than 100 m, and have essentially zero relative velocity. This defines the stationkeeping phase, in which the available CV acceleration a is often assumed to dominate the differential accelerations in Eq. (4.144) due to the orbital dynamics effects. In such a case, for example,

$$d^2r/dt^2 \simeq a_r \qquad (4.146)$$

and similarly for the other components. Thus, the motion in the near neighborhood of the TV is essentially rectilinear and dominated by the CV control maneuvers, provided transit times are kept small relative to the orbit period.

Note that stationkeeping is entirely possible even if the CV control authority is low; however, neglect of the orbital dynamics in local maneuvers is then not possible. Low-impulse stationkeeping control implies maneuvers having a duration significant with respect to the orbital period. Orbital dynamics effects will always be apparent in such cases. This is graphically demonstrated in the case of manned maneuvering unit activities in the vicinity of the Shuttle.[64]

It is worth noting that the only completely passive stationkeeping positions possible are directly ahead of or behind the target vehicle in its orbit. Radial or out-of-plane offsets will, in the absence of control maneuvers, result in oscillations of the CV about the target during the orbital period.

References

[1]Farquhar, R. W., Muhonen, D. P., Newman, C. R., and Heuberger, H. S., "Trajectories and Orbital Maneuvers for the First Libration Point Satellite," *Journal of Guidance and Control*, Vol. 3, No. 6, Nov.–Dec. 1980, pp. 549–554.

[2]O'Neill, G. K., *The High Frontier*, Morrow, New York, 1977.

[3]Heppenheimer, T. A., "Achromatic Trajectories and Lunar Material Transport for Space Colonization," *Journal of Spacecraft and Rockets*, Vol. 15, No. 3, May–June 1978, pp. 236–239.

[4]Flandro, G. A., "Solar Electric Low-Thrust Missions to Jupiter with Swingby Continuation to the Outer Planets," *Journal of Spacecraft and Rockets*, Vol. 5, Sept. 1968, pp. 1029–1032.

[5]Hoyle, F., *Astronomy and Cosmology — A Modern Course*, Freeman, San Francisco, CA, 1975.

[6]Halliday, D., and Resnick, R., *Physics*, 3rd ed., Wiley, New York, 1977.

[7]Goldstein, H., *Classical Mechanics*, 2nd ed., Addison-Wesley, Reading, MA, 1980.

[8]Dirac, P. A. M., *Directions in Physics*, Wiley-Interscience, New York, 1975.

[9]Kaplan, M. H., *Modern Spacecraft Dynamics and Control*, Wiley, New York, 1976.

[10]Danby, J. M. A., *Celestial Mechanics*, Macmillan, New York, 1962.

[11]Hamming, R. W., *Introduction to Applied Numerical Analysis*, McGraw-Hill, New York, 1971.

[12]Sheela, B. V., "An Empirical Initial Estimate for the Solution of Kepler's Equation," *Journal of the Astronautical Sciences*, Vol. 30, No. 4, Oct.–Dec. 1982, pp. 415–419.

[13]Battin, R. H., *An Introduction to the Mathematics and Methods of Astrodynamics*, AIAA, Washington, D.C., 1987.

[14]Herrick, S., *Astrodynamics*, Vol. 2, Van Nostrand Reinhold, London, 1971.

[15]Bate, R. D., Mueller, D. D., and White, J. E., *Fundamentals of Astrodynamics*, Dover, New York, 1971.

[16]"GEODSS Photographs Orbiting Satellite," *Aviation Week and Space Technology*, Vol. 119, No. 26, 28 Nov. 1983, pp. 146–147.

[17]Gelb, A. (ed.), *Applied Optimal Estimation*, MIT Press, Cambridge, MA, 1974.

[18]Nahi, N. T., *Estimation Theory and Applications*, Krieger, Huntingdon, NY, 1976.

[19]Wertz, J. R. (ed.), *Spacecraft Attitude Determination and Control*, Reidel, Boston, MA, 1978.

[20]Schmidt, S. F., "The Kalman Filter: Its Recognition and Development for Aerospace Applications," *Journal of Guidance and Control*, Vol. 4, Jan.–Feb. 1981, pp. 4–7.

[21]Mechtly, E. A., "The International System of Units," NASA SP-7012, U.S. Government Printing Office, Washington, DC, 1973.

[22]Fliegel, H. F., and Van Flandern, T. C., "A Machine Algorithm for Processing Calendar Dates," *Communications of the ACM*, Vol. 11, Oct. 1968, p. 657.

[23]Sundman, K. F., "Memoire sur le Probleme des Trois Corps," *Acta Mathematica*, Vol. 36, 1913, pp. 105–179.

[24]Kaula, W. M., *An Introduction to Planetary Physics: The Terrestrial Planets*, Wiley, New York, 1968.

[25]Arfken, G. A., *Mathematical Methods for Physicists*, 2nd ed., Academic, New York, 1970.

[26]Kaula, W. M., *Theory of Satellite Geodesy*, Blaisdell, Waltham, MA, 1966.

[27]Kershner, R. B., "Technical Innovations in the APL Space Department," *Johns Hopkins APL Technical Digest*, Vol. 1, No. 4, Oct.–Dec. 1980, pp. 264–278.

[28]"A Satellite Freed of All but Gravitational Forces," *Journal of Spacecraft and Rockets*, Vol. 11, Sept. 1974, pp. 637–644.

[29]*STS User's Handbook*, May 1982.

[30]Kaplan, M. H., Cwynar, D. J., and Alexander, S. G., "Simulation of Skylab Orbit Decay and Attitude Dynamics," *Journal of Guidance and Control*, Vol. 2, No. 6, Nov. 1979, pp. 511–516.

[31]U. S. Standard Atmosphere, National Oceanic and Atmospheric Administration, NOAA S/T 76-1562, U.S. Government Printing Office, Washington, DC, 1976.

[32]Jacchia, L. G., "Revised Static Models of the Thermosphere and Exosphere with Empirical Temperature Profiles," Smithsonian Astrophysical Observatory Special Rept. 332, 1971.

[33]Vinh, N. X., Busemann, A., and Culp, R. D., *Hypersonic and Planetary Entry Flight Mechanics*, Univ. of Michigan Press, Ann Arbor, MI, 1980.

[34]Shanklin, R. E., Lee, T., Samii, M., Mallick, M. K., and Capellari, J. O., "Comparative Studies of Atmospheric Density Models Used for Earth Orbit Estimation," *Journal of Guidance, Control, and Dynamics*, Vol. 7, March–April 1984, pp. 235–237.

[35]Jacchia, L. G., "Thermospheric Temperature, Density, and Composition: New Models," Smithsonian Astrophysical Observatory Special Rept. 375, March 1977.

[36]Liu, J. J. F., "Advances in Orbit Theory for an Artificial Satellite with Drag," *Journal of the Astronautical Sciences*, Vol. 31, No. 2, April–June 1983, pp. 165–188.

[37]King-Hele, D., *Theory of Satellite Orbits in an Atmosphere*, Butterworths, London, 1964.

[38]Schaaf, S. A., and Chambre, P. L., "Flow of Rarefied Gases," *Fundamentals of Gas Dynamics*, edited by H. W. Emmons, Princeton Univ. Press, Princeton, NJ, 1958.

[39]Fredo, R. M., and Kaplan, M. H., "Procedure for Obtaining Aerodynamic Properties of Spacecraft," *Journal of Spacecraft and Rockets*, Vol. 18, July–Aug. 1981, pp. 367–373.

[40]Knechtel, E. D., and Pitts, W. C., "Normal and Tangential Momentum Accommodation Coefficients for Earth Satellite Conditions," *Astronautica Acta*, Vol. 18, No. 3, 1973, pp. 171–184.

[41]Smith, E., and Gottlieb, D. M., "Possible Relationships Between Solar Activity and Meterological Phenomena," NASA Goddard Space Flight Center, Greenbelt, MD, NASA TR X-901-74-156, 1974.

[42]Siegel, R., and Howell, J. R., *Thermal Radiation Heat Transfer*, 2nd ed., Hemisphere, Washington, DC, 1981.

[43]Jacobson, R. A., and Thornton, C. L., "Elements of Solar Sail Navigation with Applications to a Halley's Comet Rendezvous," *Journal of Guidance and Control*, Vol. 1, Sept.–Oct. 1978, 365–371.

[44]Van der Ha, J. C., and Modi, V. J., "On the Maximization of Orbital Momentum and Energy Using Solar Radiation Pressure," *Journal of the Aeronautical Sciences*, Vol. 27, Jan. 1979, pp. 63–84.

[45]Lang, T. J., "Optimal Impulsive Maneuvers to Accomplish Small Plane Changes in an Elliptical Orbit," *Journal of Guidance and Control*, Vol. 2, No. 2, July–Aug. 1979, pp. 301–307.

[46]Ikawa, H., "Synergistic Plane Changes Maneuvers," *Journal of Spacecraft and Rockets*, Vol. 19, Nov. 1982, pp. 300–324.

[47]D'Amario, L. D., Byrnes, D. V., and Stanford, R. H., "Interplanetary Trajectory Optimization with Application to Galileo," *Journal of Guidance and Control*, Vol. 5, No. 5, Sept. 1982, pp. 465–468.

[48]Battin, R. H., "Lambert's Problem Revisited," *AIAA Journal*, Vol. 15, May 1977, pp. 703–713.

[49]Battin, R. H., Fill, T. J., and Shepperd, S. W., "A New Transformation Invariant in the Orbital Boundary-Value Problem," *Journal of Guidance and Control*, Vol. 1, Jan.–Feb. 1978, pp. 50–55.

[50]Battin, R. H., and Vaughan, R. M., "An Elegant Lambert Algorithm," *Journal of Guidance, Control, and Dynamics*, Vol. 7, Nov.–Dec. 1984, pp. 662–666.

[51]Small, H. W., "Globally Optimal Parking Orbit Transfer," *Journal of the Astronautical Sciences*, Vol. 31, No. 2, April–June 1983, pp. 251–264.

[52]Hulkower, N. D., Lau, C. O., and Bender, D. F., "Optimum Two-Impulse Transfers for Preliminary Interplanetary Trajectory Design," *Journal of Guidance, Control, and Dynamics*, Vol. 7, July–Aug. 1984, pp. 458–462.

[53]Hamming, R. W., *Numerical Methods for Scientists and Engineers*, 2nd ed., McGraw-Hill, New York, 1973.

[54]Tang, C. C. H., "Co-Apsidal Autonomous Terminal Rendezvous in Mars Orbit," *Journal of Guidance and Control*, Vol. 3, Sept.–Oct. 1980, pp. 472–473.

[55]Clohessy, W. H., and Wiltshire, R. S., "Terminal Guidance System for Satellite Rendezvous," *Journal of Aerospace Sciences*, Vol. 27, 1960, pp. 653–658.

[56]Jones, J. B., "A Solution of the Variational Equations for Elliptic Orbits in Rotating Coordinates," AIAA Paper 80-1690, Aug. 1980.

[57]Stern, R. G., "Interplanetary Midcourse Guidance Analysis," MIT Experimental Astronomy Laboratory Rept. TE-5, Cambridge, MA, 1963.

[58]Dunning, R. S., "The Orbital Mechanics of Flight Mechanics," NASA SP-325, U.S. Government Printing Office, Washington, DC, 1975.

[59]Chiarappa, D. J., "Analysis and Design of Space Vehicle Flight Control Systems: Volume VIII—Rendezvous and Docking," NASA Contractor Rept. CR-827, U.S. Government Printing Office, Washington, DC, 1967.

[60]Parten, R. P., and Mayer, J. P., "Development of the Gemini Operational Rendezvous Plan," *Journal of Spacecraft and Rockets*, Vol. 5, Sept. 1968, pp. 1023–1026.

[61]Young, K. A., and Alexander, J. D., "Apollo Lunar Rendezvous," *Journal of Spacecraft and Rockets*, Vol. 7, Sept. 1970, pp. 1083–1086.

[62]Jezewski, D. J., and Donaldson, J. D., "An Analytic Approach to Optimal Rendezvous using the Clohessy-Wiltshire Equations," *Journal of the Astronautical Sciences*, Vol. 27, No. 3, July–Sept. 1979, pp. 293–310.

[63]Nesline, F. W., and Zarchan, P., "A New Look at Classical vs. Modern Homing Missile Guidance," *Journal of Guidance and Control*, Vol. 4, Jan.–Feb. 1981, pp. 78–85.

[64]Covault, C., "MMU," *Aviation Week and Space Technology*, Vol. 120, No. 4, 23 Jan. 1984, pp. 42–56.

[65]Schneider, A. M., Prussing, J. E., and Timin, M. E., "A Manual Method for Space Rendezvous Navigation and Guidance," *Journal of Spacecraft and Rockets*, Vol. 6, Sept. 1969, pp. 998–1006.

5
PROPULSION

Probably no single factor constrains the design of a space vehicle and the execution of its mission more than does the state of the art in propulsion technology. Ascent propulsion capability, together with the physical limitations imposed by celestial mechanics, sets the limits on payload mass, volume, and configuration that bound the overall design. The economics of space flight and our progress in exploiting space are driven inexorably by the cost per kilogram of mass delivered to orbit. Mankind's reach in exploring interplanetary space is limited by the energy available from current upper stages. Though the advent of the Space Shuttle has expanded many of the boundaries of the spacecraft design environment, it is still true that the scope of most space missions is ultimately set by propulsion system limitations.

Yet, despite its importance, ascent propulsion is probably the factor over which the spacecraft designers have the least control. Except those involved directly in the areas of rocket engine, booster, or upper-stage design, most aerospace engineers will be in the position of customers with freight to be moved. A limited number of choices are available, and the final selection is seldom optimal for the given task, but is merely the least unsatisfactory. Rarely is a particular mission so important that a specific engine or launch vehicle will be designed to fit its needs. Indeed, there have been few boosters designed for space missions at all; most are converted Intermediate Range Ballistic Missiles (IRBMs) and Intercontinental Ballistic Missiles (ICBMs). The Saturn family, the Space Shuttle, and Ariane are conspicuous exceptions, but even as this book is written, a significant fraction of payloads reach space on various versions of the Atlas, the Delta, and the Titan.

We therefore take the view that ascent propulsion is essentially a "given" in the overall design. We do not explore in detail the multitude of considerations that go into the design of launch vehicles. The text by Sutton and Ross[1] is probably the best source for those seeking more detail in this area. Our purpose is to explore the factors that are involved in the selection of launch vehicles and upper stages for a given mission. We have tried to include a comprehensive discussion of the capabilities of the various vehicles, including those of both current and projected availability which could be of interest.

However, injection into a specified trajectory does not end the consideration of propulsion systems required by the spacecraft systems engineer. Low-orbit satellites may need vernier engines for drag compensation.

163

Satellites in geosynchronous Earth orbit (GEO) require a similar system for stationkeeping purposes. Many spacecraft will need substantial orbit adjustments or midcourse maneuvers. Attitude control systems will often employ small thrusters, either for direct control or for adjustment of spacecraft angular momentum. The future holds the prospect of the development of orbital transfer vehicles (OTVs) for operations in Earth orbit. Planetary landers, such as the Surveyor, Apollo, and Viking missions to the moon and Mars, involve the development of specialized descent propulsion systems. For these and other reasons, the spacecraft designer must be familiar with propulsion system fundamentals.

5.1 ROCKET PROPULSION FUNDAMENTALS

Thrust Equation

The fundamental equation for rocket engine performance is the thrust equation,[2]

$$T = \dot{m}V_e + (p_e - p_a)A_e \qquad (5.1)$$

where

T = thrust force
$\dot{m}$ = flow rate = $\rho_e V_e A_e$
ρ_e = fluid density at nozzle exit
V_e = exhaust velocity at nozzle exit
p_e = exhaust pressure at nozzle exit
p_a = ambient pressure
A_e = nozzle exit area

This equation is valid for reaction motors that generate thrust through the expulsion of a fluid stream without ingesting fuel or oxidizer from any source external to the vehicle. In aerospace applications, the working fluid is a gas, possibly nitrogen gas stored under pressure, water molecules resulting from the combustion of oxygen and hydrogen, or hydrogen gas superheated by passage through a nuclear reactor. If efficiency is at all important, the gas is made as hot as possible and expanded through a supersonic nozzle, as in Fig. 5.1, to increase V_e. (This of course also causes the term $p_e A_e$ to decrease, but the loss here is usually small compared to the gain in $\dot{m}V_e$.) In all large operational engines to date, heating of the gas has been accomplished chemically. The working fluid is thus composed of the products of the combustion cycle producing the heat.

Since the dominant term in the thrust equation is $\dot{m}V_e$, it is customary to rewrite Eq. (5.1) as

$$T = \dot{m}[V_e + (p_e - p_a)A_e/\dot{m}] = \dot{m}V_{eq} \qquad (5.2)$$

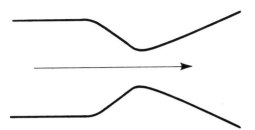

Fig. 5.1   Supersonic nozzle.

where V_{eq} is the equivalent exhaust velocity. The portion of V_{eq} due to the pressure term will nearly always be small relative to V_e and back-of-the-envelope performance calculations often make use of this fact by dropping V_{eq} for the more simply obtained V_e. However, the pressure term is by no means negligible when accurate results are desired, as later examples will show. Even rough calculations can sometimes require its inclusion. For example, the Space Shuttle main engine suffers about a 20% loss of thrust at sea level compared to vacuum conditions because of pressure effects.

Specific Impulse

The total change in momentum of the expelled propellant (and hence in the opposite direction of the rocket) is

$$I = \int T \, dt = \int \dot{m} V_{eq} \, dt \qquad (5.3)$$

If V_{eq} is constant over the burning time (never strictly true, even ignoring various transients for real motors, except for horizontal flight and flight in vacuum), Eq. (5.3) becomes

$$I = m_p V_{eq} \qquad (5.4)$$

where m_p is the mass of propellant consumed. Note that it is not necessary in Eq. (5.3) to assume $\dot{m}$ is constant to obtain this result, but only to assume that any throttling used is done in such a way as to leave V_{eq} unaffected (a desirable but seldom achievable condition).

To focus on the efficiency of the engine rather than its size or duration of operation (which determine m_p), it is convenient to define the specific impulse,

$$I_{sp} = I/m_p = V_{eq} = T/\dot{m} \qquad (5.5a)$$

This is a definition founded in basic physics that is, however, only recently gaining acceptance. It is far more customary in engineering circles to use instead the weight flow to normalize the total impulse, yielding

$$I_{sp} = I/m_p g = V_{eq}/g = T/\dot{m}g \qquad (5.5b)$$

With the definition of Eq. (5.5b), specific impulse is measured in seconds in any consistent system of units, a not inconsiderable advantage. Note, however, that Eq. (5.5a) reveals the fundamental physics of the situation; the specific impulse, or change in momentum per unit mass, is merely the equivalent exhaust velocity. Specific impulse is the most important single measure of rocket engine performance, because it relates in a fundamental way to the payload-carrying capability of the overall vehicle, as we shall see in later sections.

Nozzle Expansion

Returning to Eq. (5.1), it is of interest to explore some general considerations in the operation of rocket engines for maximum efficiency. It is clear from Eq. (5.1) that $p_e < p_a$ is undesirable, since the pressure term is then negative and reduces thrust. Also, this condition can be harmful when operating inside the atmosphere, since the exit flow will tend to separate from the walls of the nozzle, producing a region of recirculating flow that can under some conditions set up destructive vibrations due to unbalanced and shifting pressure distributions.

It is not immediately clear but is true that $p_e \gg p_a$ is also to be avoided. This situation basically indicates a failure to expand the exhaust nozzle as much as might be done, with a consequent loss of potentially available thrust. And again, within-the-atmosphere operation at very large exit-to-ambient pressure ratios can cause undesirable interactions of the exhaust plume with the external airflow.

It is thus ideal to have a close match between nozzle exit pressure and ambient pressure. There are practical limits to the degree to which this can be accomplished. An engine operating in vacuum would require an infinite exit area to obtain $p_e = 0$. Very large nozzles introduce a mass penalty that can obviate the additional thrust obtained. Large nozzles are more difficult to gimbal for thrust vector control and may be unacceptable if more than one engine is to be mounted on the same vehicle base. Furthermore, it is clear that engines intended for ascent propulsion cannot in any case provide an ideal pressure match at more than one altitude. An effective compromise is to operate the nozzle as much as possible in a slightly underexpanded condition, the effects of which are less harmful than overexpansion. Still, as a practical matter, overexpansion must often be tolerated. Shuttle main engines sized for sea level operation would be extremely inefficient for high-altitude operation. These engines operate with an exit pressure of about 0.08 atm and thus do not approach an underexpanded condition until an altitude of about 18 km is reached.

As noted earlier, choice of expansion ratio for conventional nozzle, fixed-area ratio engines is usually a compromise. For engines that operate from liftoff through the atmosphere and into space, as do most lower-stage engines, the compromise will be driven by performance and liftoff thrust requirements. Engines that provide most of the liftoff thrust and do not perform long in vacuum have lower optimal expansion ratios than those that operate solely in space. In this latter case, the highest practical expansion ratio is usually constrained by the available volume and increasing nozzle weight. Engines like the Space Shuttle main engine (SSME) and Atlas sustainer, which must operate at liftoff but perform most of their work in space, are the most difficult compromise, a trade between vacuum performance, sea level performance, and other factors. The maximum expansion ratio is often limited by the need to prevent flow separation in the nozzle, with its resulting asymmetric side loads.

Even some space engines are limited by this problem. In the case of the J-2, a 250,000-lb thrust LO_2/LH_2 engine used in the second and third stages of the Saturn 5, it was desired to test the engines in a sea level environment even though all flight operations would be in vacuum. This avoided the expense of the very large vacuum test facilities that would have been needed for each test of a higher-expansion-ratio engine. The J-2 was marginal in its ability to maintain full flow at the nozzle exit under these conditions and was frequently plagued by flow separation and sideloads during testing. The problem was most annoying during engine startup and at off-design operation.

A variety of unconventional nozzle concepts have been suggested that have as their goal the achievement of optimum expansion at all altitudes. The concept that these nozzles have in common is a free expansion and deflection of the jet. The most prominent examples are plug or spike nozzles and the expansion-deflection (ED) nozzles.

A plug or spike nozzle is depicted in Fig. 5.2. The combustion chamber is a torus or, more probably, an annular ring of a number of individual combustors. The nozzle through which the gases exit the combustion chamber will converge to a sonic throat and may be followed in some cases by a diverging supersonic section as in conventional nozzles. This expansion, if used, will be small relative to the engine design expansion ratio. The gas is directed inboard and slightly aft, the angle being defined by the overall characteristics of the engine. The gas impinges on the central plug, which is carefully contoured to turn the flow in the aft direction. The unconfined gas tends to expand, even as its momentum carries it inboard and along the plug. The boundary condition that must be satisfied by the outer sheath of gas is that it match the ambient pressure; the stream expands to achieve this condition. As the vehicle ascends and ambient pressure decreases, the stream expands accordingly, as shown in Fig. 5.3. Expansion ratio is thus always near-optimal, tending to infinity in vacuum.

In the initial concept, the central spike tapered to a point. It was quickly discerned that performance was equally good, and the mass much lower, if the point were truncated. The final variant of this concept is the Rocket-

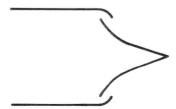

Fig. 5.2   Plug or spike nozzle.

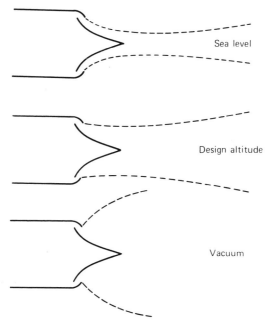

Fig. 5.3   Plug nozzle at various altitudes.

dyne Aerospike. In this case, the spike is still more truncated, but a substantial secondary flow (provided by the turbine exhaust in a complete engine) is fed through the bottom of the plug to help maintain the core flow shape and provide an adequate base pressure.

One disadvantage of the plug nozzle is that the heavy, high-pressure combustion section is the largest diameter of the engine rather than a small, compact cylinder, thus resulting in a heavy engine. The expansion-deflection concept was an attempt to obtain a pressure-compensating nozzle in a more nearly conventional overall shape. A central plug shaped like an inverted mushroom turns the flow from the combustion chamber outward and nearly horizontal, as in Fig. 5.4. The contoured outer skirt then turns the flow aft. The pressure compensation is supposed to come from the degree of expansion into the annular central space behind the plug. Without secondary flow this does not work, since the self-pumping action of the flow tends to close the flow behind the plug, creating a low-pressure area behind the plug base. This results in a conventional aerodynamic "pressure drag," which seriously inhibits engine performance. A large secondary flow from the turbopump system or from an ambient air bleed is required to obtain a workable engine.

A variation on these concepts is shown in Fig. 5.5. This is sometimes called the "linear plug." The combustors and the deflection surface form a linear array. This concept is especially well suited to streamlined, flying-wing-type vehicles. This application may see the first flight use of a pressure-compensated nozzle for space vehicles.

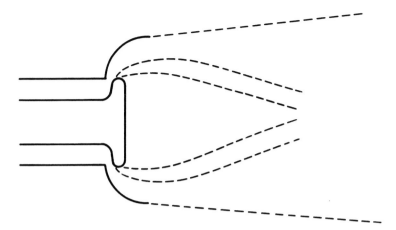

Fig. 5.4   Expansion—deflection nozzle.

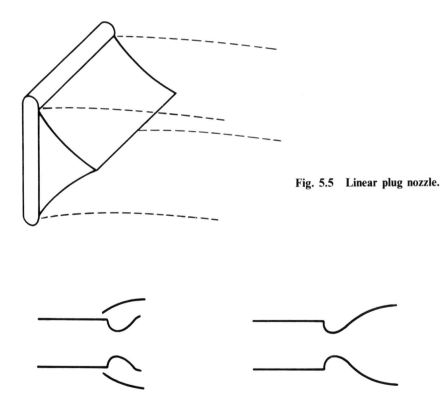

Fig. 5.5   Linear plug nozzle.

Fig. 5.6   Extendable nozzle.

Extendable exit cones (EECs) have become common as a solution to the problem of launching upper stages containing motors that must operate efficiently in vacuum without being so large as to pose packaging problems for launch. In the EEC concept, the exit nozzle is designed in two sections, as shown in Fig. 5.6, where the second section is translated from its stored to its operational position by springs or pneumatic plungers. This technique is primarily applied to designs where a radiatively cooled, dump cooled, or ablative exit cone is used; it is not suitable for engines with regenerative cooling.

Calculation of Specific Impulse

The exact calculation of rocket engine exhaust velocity and pressure, and hence specific impulse and thrust, is an exceedingly complex task requiring the numerical solution of a multidimensional coupled set of partial differential equations describing the fluid dynamic and chemical processes involved. However, surprisingly good results can be obtained by idealizing the rocket engine flowfield as a quasi-one-dimensional adiabatic, frictionless, shock-free flow of a calorically perfect gas having a fixed chemical composition determined by the combustion process. If this model is employed, the energy equation of gasdynamics,

$$c_p T_c = c_p T_e + V_e^2/2 \qquad (5.6)$$

may be combined with the isentropic pressure-temperature relation,

$$T_e/T_c = (p_e/p_c)^{(k-1)/k} \qquad (5.7)$$

to yield for the exhaust velocity

$$V_e^2 = k R_{gas} T_c [1 - (p_e/p_c)^{(k-1)/k}]/(k-1) \qquad (5.8)$$

where

k = ratio of specific heats, c_p/c_v
p_e = nozzle exit pressure
p_c = combustion chamber pressure
T_c = combustion chamber temperature
R_{gas} = exhaust flow specific gas constant $= \mathscr{R}/\mathscr{M}$
$\mathscr{R}$ = universal gas constant
$\mathscr{M}$ = exhaust gas molecular weight

We use the less conventional k for the ratio of specific heats, as opposed to the customary γ, in order to avoid confusion with the notation for flight-path angle used throughout this text.

The extent to which this result is useful to the designer is a matter for careful judgment. It establishes the parameters on which exhaust velocity, and hence specific impulse, depends. Thus, it is clear that high combustion temperature and low exhaust gas molecular weight are advantageous. High chamber pressure is also seen to be desirable, as is a low effective ratio of specific heats (k is always greater than unity).

However, additional complexities exist. Chamber temperature depends on the chemical reactions that take place during combustion; a highly energetic reaction such as

$$H_2 + F_2 \rightarrow 2HF$$

will produce higher temperatures than a reaction such as

$$CH_4 + 2O_2 \rightarrow CO_2 + 2H_2O$$

because of the inherent differences in bonding energy. However, the rate of energy release also affects the chamber temperature; the combustion process is essentially an equilibrium reaction whose rate depends on equilibrium "constants" that are functions of pressure. Reactions yielding products having a lower specific volume than the constituents, such as

$$2H_2 + O_2 \rightarrow 2H_2O$$

will proceed faster at higher pressure, releasing energy at a greater rate. The net result is a generally small but useful temperature increase due to chamber pressure, which is not seen in the above simplified model.

Also, most large, liquid-fueled rocket engines are regeneratively cooled, meaning that the thrust chamber and nozzle are cooled by the flow of propellant in a surrounding jacket prior to injection and combustion. This process, intended to protect the metal walls, removes little heat from the main flow (thus allowing the adiabatic flow assumption to be retained) but may raise the precombustion fuel temperature by several hundred degrees, thus raising the energy level of the propellant. This effect can add several seconds of I_{sp} compared to that predicted by Eq. (5.8).

The utility of k, as a parameter describing the gas is somewhat questionable in a chemically reacting flow. For simple diatomic gases at temperatures below roughly 500 K, it is both theoretically and observably true that $k = 7/5$. For example, this is true for air (neglecting CO_2 and other minor constituents). At higher temperatures c_p and c_v are not constant (the gas is not "calorically perfect") and do not allow the static enthalpy to be written in the form

$$h = c_p T \qquad (5.9)$$

as we did earlier in Eq. (5.6). The similarity parameter k thus has little meaning and does not appear in the basic equations that must be solved to obtain the exhaust velocity. However, it is commonly found that good results can be obtained using Eq. (5.8) and similar calorically perfect gas results, provided that an empirically determined "hot k" is used. This will be on the order of $k = 1.21 - 1.26$ for a wide range of fuels and oxidizers.

Table 5.1 Specific impulse for operational engines

Engine	Thrust	Fuel	Oxidizer	I_{sp}	Expansion ratio
Rocketdyne RS-27 (Delta)	207,000 lbf	RP-1	Liquid oxygen	262 (S.L.)	8:1
Atlantic Research Corp. 8096-39 (Agena)	17,000 lbf	UDMH	H.P. nitric acid	300 (Vac)	45:1
Aerojet AJ110	9,800 lbf	UDMH/N_2H_4	N_2O_4	320 (Vac)	65:1
TRW TR-201 (Delta)	9,900 lbf	UDMH/N_2H_4	N_2O_4	303 (Vac)	50:1
TRW MMPS (Spacecraft)	88 lbf	MMH	N_2O_4	305 (Vac)	180:1
TRW MRE-5	4 lbf	N_2H_4	—	226 (Vac)	?
Rocket Research					
MR 104C	129 lbf	N_2H_4	—	239 (Vac)	53:1
MR 50L	5 lbf	N_2H_4	—	225 (Vac)	40:1
MR 103A	0.18 lbf	N_2H_4	—	223 (Vac)	100:1
United Technologies					
Orbus 6	23,800 lbf	Solid		290 (Vac)	47:1
Orbus 21	58,560 lbf	Solid		296 (Vac)	64:1
Morton Thiokol					
STAR 48	17,210 lbf	Solid		293 (Vac)	55:1
STAR 37F	14,139 lbf	Solid		286 (Vac)	41:1
Pratt & Whitney					
RL-10	16,500 lbf	Liq. H_2	Liq. O_2	444 (Vac)	?

In computing specific impulse, as opposed to exhaust velocity, the pressure term must be accounted for. Again, it is typically not large, but is significant, since payload mass is highly sensitive to I_{sp}. Subject to the same approximations as previously, the exit area is given by

$$A_e/A^* = (1/M_e)\{2[1 + (k - 1)M_e^2/2]/(k + 1)\}^{\frac{1}{2}(k + 1)/(k - 1)} \qquad (5.10a)$$

where

$M_e = V_e/a_e = $ exit Mach number
$a_e^2 = kRT_e = $ exit speed of sound
$A^* = $ sonic throat area

It is also necessary to know $\dot{m}$ if the pressure effect on I_{sp} is to be assessed. Subject to the same calorically perfect approximations as previously noted, it is found that

$$\dot{m} = p_c A^* \{(k/RT_c)[2/(k + 1)]^{(k + 1)/(k - 1)}\}^{\frac{1}{2}} \qquad (5.10b)$$

As an indication of available engine performance, Table 5.1 gives actual specific impulse for a variety of engines with varying fuel/oxidizer combinations.

Nozzle Contour

A characteristic of all supersonic flow devices, including rocket engine thrust chambers, is the use of a convergent-divergent nozzle to achieve the transition from subsonic to supersonic flow. Various shapes are possible for the subsonic converging section, since the flow is not particularly sensitive to the shape in this region. A simple cone, faired smoothly into cylindrical combustion zone and the rounded throat, is usually satisfactory. For the divergent section, a cone is the most straightforward and obvious shape and indeed was used in all early rocket designs. The chosen cone angle was a compromise between excessive length, mass, and friction loss for a small angle vs the loss due to nonaxial flow velocity for larger angles.

The classic optimum conical nozzle tends to have about a 15-deg half-angle. Such an angle was generally satisfactory for low-pressure engines operating at modest expansion ratios. As chamber pressures and expansion ratios increased in the search for higher performance, conic nozzles became unsatisfactory. The increased length needed in such cases results in greater weight and high moment of inertia, which causes difficulty in gimballing the motor for thrust vector control. This gave rise to the so-called "bell" or contoured nozzle. This concept involves expanding the flow at a large initial angle and then turning it so that it exits in a nearly axial direction (most nozzles will still have a small divergence angle, e.g., 2 deg at the exit). Design of the nozzle to achieve this turning of the flow without producing undesired shock waves requires the application of the method of characteristics and is beyond the intended scope of this text. Bell or contour nozzles are often referred to by the percentage of length as compared to a 15-deg cone of the same expansion ratio. For example, an 80% bell has a length 80% of that of the equivalent conic nozzle.

The efficiency or thrust coefficient of bell and cone nozzles is essentially the same. Although the flow exiting the bell is more nearly axial, the losses involved in turning the flow tend to compensate for this advantage. Practical engine designs turn the flow quite rapidly after the sonic throat, a process that introduces various inefficiencies. Gradually contoured nozzles such as used in high-speed wind tunnels are possible but tend to be quite long and generally do not offer sufficient advantage to compensate for their weight, volume, and cost penalty.

Engine Cooling

A variety of cooling concepts have been proposed for use in rocket engines, many of which have seen operational use, often in combination. The most common approach for large engines with lengthy operating times is "regenerative cooling," mentioned earlier, where one of the propellants is passed through cooling passages in the thrust chamber and nozzle wall before being injected into the combustion chamber. This very effective and efficient approach is usually supplemented by film or boundary-layer cooling, where propellant is injected so as to form a cooler, fuel-rich zone near the walls. This is accomplished by the relatively simple means of altering the propellant distribution at the injector, which usually consists of a "shower head" arrangement of many small entrance ports for fuel and oxidizer. In some cases injector orifices may be oriented to spray directly on the engine wall. Probably the most extreme example occurred in the pioneering V-2, which had a series of holes drilled just above the throat to bleed in raw fuel to protect the combustion chamber, which was fabricated from mild steel. In most regenerative cooling designs, the fuel is used as the coolant, although oxidizer has been used and is increasingly suggested for high-mixture-ratio, high-pressure LO_2/LH_2 engines.

Ablative thrust chambers are commonly employed in engines designed for a single use, in cases where neither propellant is an efficient coolant, or for operational reasons such as when deep throttling or pulse mode operation is required. When a wide range of throttling is available, propellant (and hence coolant) flow at the lower thrust settings may become so sluggish that fluid stagnation and overheating may occur. When pulsed operation is desired, as, for example, in thrusters used for on-orbit attitude or translation control, the volume of the coolant passages is incompatible with the requirement for short, sharp pulses. Ablative thrust chamber endurance of several thousand seconds has been demonstrated. In cases where throat erosion is critical, refractory inserts have been used. Ablative chambers are especially sensitive to mixture ratio distribution in the flow, with hot streaks causing severe local erosion, especially if oxidizer-rich.

Radiation-cooled thrust chambers have been extensively used in smaller engine assemblies. Refractory metals or graphite have been commonly used in the fabrication of such motors, which tend to be simple and of reasonably low mass and have nearly unlimited life. However, these desirable features are sometimes offset by the nature of radiation cooling, which causes difficulty in some applications. The outer surface of the thrust chamber, which rejects heat at temperatures approaching 1500 K, must have an unimpeded view of deep space. Furthermore, any object in view of the thrust chamber will be exposed to substantial radiative heating. A compact, vehicle-integrated engine installation such as might be used for a regeneratively or ablatively cooled engine is thus impossible. Fully radiatively cooled engines of more than a few thousand pounds thrust have not been demonstrated to date. This is due to the materials costs and systems integration difficulties of fabricating such engines. It may be noted, however, that some fairly large engines intended for upper-stage operation employ the extendable exit cones mentioned earlier, or fixed extensions, which are radiation-cooled. The nozzles of the Ariane launcher engines are radiation cooled.

Heat sink thrust chambers, where the chamber wall material simply accumulates the heat by bulk temperature increase during the burn, are fairly common as low-cost, short-duration ground test articles, which are required for injector performance characterization. They are rarely used in flight hardware, except as buried units subjected to brief, infrequent pulses. The ability of refractory metals such as niobium (columbium) to operate at very high temperatures allows use of a "hybrid" cooling scheme. The niobium thrust chamber/nozzle assembly acts as a refractory heat sink; however, the interior of the nozzle has a sufficiently good view of space that much of the heat energy can be radiated away, allowing long-duration operation. In most cases, boundary-layer cooling, i.e., excess fuel near the walls, helps minimize heat transfer. With adequate external insulation, this assembly can be buried in structure. The Space Shuttle attitude control thrusters use this approach.

An interesting concept, worthwhile only with hydrogen, is dump cooling. Hydrogen, if heated to a few hundred degrees and exhausted through a nozzle, has a specific impulse equal to many bipropellant combinations. In such an engine, the bipropellants would be burned in the conventional manner and exhausted, while hydrogen would pass through the chamber walls, being heated in the process and exhausted through its own nozzle. Previous studies have not shown the performance to be worth the complexity of a three-propellant system, but future applications may be possible.

Such concepts as spray cooling, in which the liquid coolant is sprayed against the combustion chamber wall rather than caused to flow over it, can accommodate very high heat fluxes but have not been required by propulsion systems used to date. Similarly, transpiration cooling, where the coolant is uniformly "sweated" through a porous wall, has not shown enough performance advantage over less expensive and more conventional boundary-layer cooling to justify its use. However, transpiration cooling has found use in some LO_2/LH_2 injectors, such as those for the J-2 and RL-10.

The F-1 engine used an interesting variant in which turbopump exhaust gas was dumped into a double-walled nozzle extension and then, via a series of holes in the inner wall, into the boundary layer of the main stream. This cooled the extension while getting rid of the often troublesome turbine exhaust by entraining it in the main flow.

Combustion Cycles

Rocket engine combustion cycles have grown steadily more complex over the years as designers have sought to obtain the maximum possible specific impulse and thrust from hardware of minimum weight. The current state of the art in this field is probably exemplified by the Space Shuttle main engine (SSME). However, basic designs remain in wide use, as exemplified by the fact that the simple pressure-feed system continues to be a method of choice where simplicity, reliability, and low cost are driving requirements.

As noted, a major driver in engine cycle development is the desire for higher performance. This translates to higher combustion chamber pressure, efficient use of propellant, and minimum structural mass. Since structural mass increases rapidly with tank size and pressure, the need for pump-fed, as opposed to pressure-fed, engines was recognized quite early. Dr. Robert H. Goddard began flying pump-fed engines in the 1920s to prove the concept, while the German rocket engineers at Peenemünde went immediately to pump-fed systems for the larger vehicles such as the V-2.

The early vehicles, of which the V-2 and the U.S. Army Redstone are classical examples, used a turbopump system in which the hot gas, which drove the turbine that in turn drove the pumps, was provided by a source completely separate from the rocket propellants. The hot gas was obtained by decomposing hydrogen peroxide into water and oxygen, a process actually accomplished by spraying the peroxide and a solution of potassium permanganate into a reaction chamber. These substances were stored in pressurized tanks, an approach with the virtue of simplicity in that operation of the turbine drive was decoupled from that of the main propulsion system. Also, the low-temperature turbine exhaust made turbine design relatively simple. On the negative side, there was a considerable mass penalty because of the low energy of the hydrogen peroxide and because of the extra tanks required to hold the peroxide and the permanganate. The basic concept worked quite well, however, and was applied in a variety of systems well into the late 1950s.

Next to be developed was the bootstrap concept, in which a small fraction of the main propellant is tapped off at the pump outlet and burned in a gas generator to provide turbine drive gas. This approach has several advantages. The use of existing propellants saves weight because the increase in tank size to accommodate the turbine requirements imposes a smaller penalty than the use of separate tanks. Also, in order to provide gas temperatures tolerable for turbine materials, it is necessary to operate well away from the stoichiometric fuel-to-oxidizer ratio. This is usually done by running substantially on the fuel-rich side, which provides a nonoxidizing atmosphere for the turbine.

A number of systems developed in the 1950s and 1960s used the bootstrap approach. The first such vehicles were the Navaho booster and the Atlas, Titan, Thor, and Jupiter missiles. The F-1 and J-2 engines for the Saturn series used similar cycles, as in fact have most of the vehicles in the U.S. inventory of launchers, excluding the Space Shuttle.

Individual engine systems vary in detail regarding implementation, particularly in the starting cycle. Some use small ground start tanks filled with propellant to get the engines started and up to steady-state speed. In other cases, the start tanks are mounted on the vehicle and refilled from the main propellant tanks to allow later use with vernier engines providing velocity trim after main engine shutdown. Still others, generally later versions, use solid-propellant charges that burn for about a second to spin up the pumps and provide an ignition source for the gas generator. The J-2 used hydrogen gas from an engine-mounted pressure bottle, which was repressurized during the burn to allow orbital restart.

The F-1, the 1.5-million-lb-thrust first-stage engine for the Saturn 5, used no auxiliary starting system at all. By the time this engine was designed, it was recognized that a bootstrap system could be self-starting. Simply opening the valves at the tank pressure used to provide inlet pump head and igniting the propellants would start the pumps, which would increase combustion chamber pressure, etc., in a positive-feedback process that

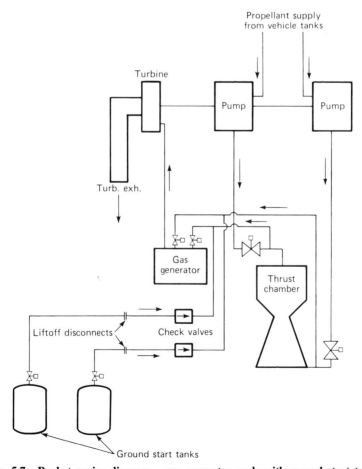

Fig. 5.7   Rocket engine diagram—gas generator cycle with ground start tanks.

continued until full thrust was obtained. On the F-1 this took some 8 or 9 s; however, because of the reduced structural loads associated with the slow start, this was an advantage. Figures 5.7–5.10 show some of the various engine cycles.

Of the workhorse engines of the past several decades, one in particular did not use the gas generator cycle. The Pratt & Whitney RL-10 LO_2/LH_2 series used in the Centaur and S4 upper stages takes advantage of the hydrogen fuel being heated in the thrust chamber cooling process and expands it through the turbine to provide energy to run the pumps. Essentially all the hydrogen is used for this purpose and then injected (as a gas) into the combustion chamber. In this cycle, none of the propellant is "wasted" as relatively low-energy turbine exhaust gas. Therefore, the overall performance tends to be better than that of the basic bootstrap cycle. The primary performance penalty for this system is the extra pumping energy required to counteract the pressure drop through the turbine. This engine allows a tank head start, as with the F-1, and seems more tolerant of throttling than most. The cycle is diagrammed in Fig. 5.11.

The SSME uses another more complex cycle. The liquid oxygen and liquid hydrogen are passed through completely separate turbopump packages driven by separate gas generators that maintain acceptable tempera-

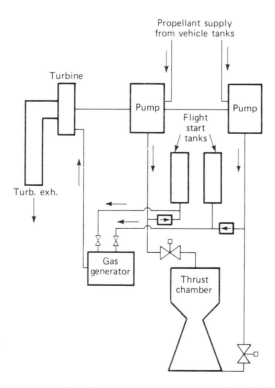

Fig. 5.8   Rocket engine diagram—gas generator cycle with flight start tanks.

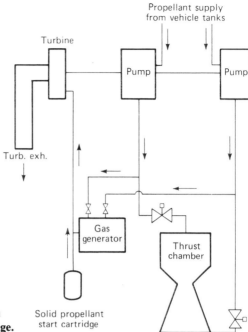

Fig. 5.9 Rocket engine diagram with solid propellant start cartridge.

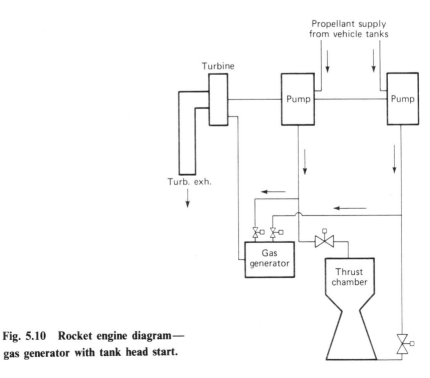

Fig. 5.10 Rocket engine diagram— gas generator with tank head start.

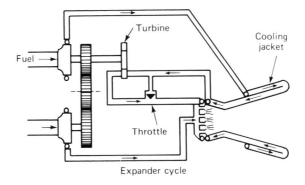

Fig. 5.11 Rocket engine diagram—expander cycle (RL-10).

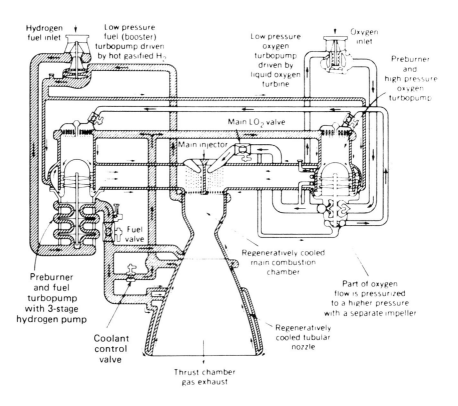

Fig. 5.12 Rocket engine diagram—preburner cycle (SSME).

tures by off-stoichiometric combustion. The turbine gas is generated by burning a portion of the total main engine flow at off-mixture-ratio conditions. This gas then flows through a turbine to power the pumps. It is then dumped into the main combustion chamber along with the flow from the other pump loop. Additional propellant is added to form the main thrust chamber flow. Figure 5.12 shows the cycle. The SSME is most notable for operating at a much higher chamber pressure than its predecessors and therefore yields very high performance. Even higher performance would be possible, except for the requirement in the Shuttle system for the engines to be operating at liftoff. The SSME is currently capable of throttling from 65 to 109% of rated thrust, which is 470,000 lbf in vacuum.

Combustion Chamber Pressure

Besides the obvious advantages indicated in Eq. (5.8) and discussed in Sec. 5.1 (Calculation of Specific Impulse), some additional benefits and problems accrue to the use of high chamber pressure. At a given thrust level, a higher chamber pressure engine is more compact. This may lend itself to easier packaging and integration. A high-pressure engine may allow more flexibility in the choice of expansion ratio and may be more amenable to throttling down within the atmosphere. On the negative side, all of the internal plumbing becomes substantially heavier. Sealing of joints, welds, and valve seats becomes progressively more difficult as pressure increases. Leakage, especially of hot gas or propellants such as hydrogen, becomes more dangerous and destructive. The amount of energy required to operate the fuel pumps that inject the propellants into the combustion chamber increases with chamber pressure. Inasmuch as this energy is derived from propellant combustion as part of the overall engine cycle, it represents a "tax" on the available propellant energy that mitigates the advantages of high-pressure operation. Finally, higher-pressure assemblies have historically demonstrated a greater tendency toward combustion instability than their lower-pressure counterparts.

5.2 ASCENT FLIGHT MECHANICS

Rocket-powered ascent vehicles bridge the gap between flight in the atmosphere, governed both by gravitational and aerodynamic forces, and space flight, shaped principally by gravitational forces punctuated occasionally by impulsive corrections. Purely astrodynamic considerations were discussed in Chapter 4; in this section we discuss the mechanics of the powered ascent phase. We first consider the basic equations of motion in terms of the physical parameters involved, followed by a discussion of special solutions of the equations.

Equations of Motion

In keeping with our approach, we present the simplest analysis that treats the issues of salient interest to the vehicle designer. To this end, we consider the planar trajectory of a vehicle over a nonrotating spherical planet. The geometric situation is as shown in Fig. 5.13. The equations of

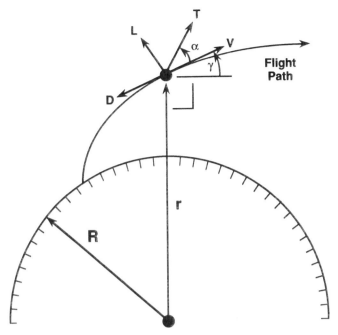

Fig. 5.13   Planar ascent from nonrotating planet.

motion are[3]

$$dV/dt = (T \cos\alpha - D)/m - g \sin\gamma \qquad (5.11a)$$

$$V\,d\gamma/dt = (T \sin\alpha + L)/m - (g - V^2/r) \cos\gamma \qquad (5.11b)$$

$$ds/dt = (R/r)V \cos\gamma \qquad (5.11c)$$

$$dr/dt = dh/dt = V \sin\gamma \qquad (5.11d)$$

$$dm/dt = -\dot{m}(t) \qquad (5.11e)$$

$$L = \tfrac{1}{2}\rho V'^2 SC_L \qquad (5.11f)$$

$$D = \tfrac{1}{2}\rho V'^2 SC_D \qquad (5.11g)$$

$$g = g_s[R/(R + h)]^2 \qquad (5.11h)$$

$$\alpha = \alpha(t) \qquad (5.11i)$$

where

V = inertial velocity magnitude
V' = speed relative to planetary atmosphere
R = planetary radius
h = height above surface
r = $R + h$ = radius from planetary center
s = down-range travel relative to nonrotating planet
γ = flight-path angle, positive above local horizon
T = thrust at time t
m = mass at time t
$\dot{m}$ = mass flow rate, a prescribed function
L = lift force, normal to flight path
D = drag force, parallel to flight path
C_L = lift coefficient
C_D = drag coefficient
ρ = atmospheric density
S = vehicle reference area for lift and drag
g = gravitational acceleration
g_s = surface gravitational acceleration
α = angle of thrust vector relative to flight path; i.e., vehicle pitch angle, a prescribed function

These equations are not solvable in closed form but may be integrated numerically, subject to appropriate initial conditions. Note that, in practice, this particular formulation would not necessarily be used for numerical calculations. If numerical integration is to be employed, it is often simplest to work in Cartesian coordinates directly with the vector equations,

$$\mathrm{d}\mathbf{r}/\mathrm{d}t = \mathbf{V} \qquad (5.12\mathrm{a})$$

$$\mathrm{d}\mathbf{V}/\mathrm{d}t = f(\mathbf{r},\mathbf{V},t) \qquad (5.12\mathrm{b})$$

where $f(\cdot)$ is the sum, per unit mass, of forces on the vehicle. When obtained, the results are easily transformed to a coordinate system that is more appropriate to ascent from a spherical planet. However, Eqs. (5.11) in the form given have the advantage of portraying the physical parameters of interest most directly and are used here for that reason.

The assumption of a nonrotating planet introduces three basic errors. First, Eqs. (5.11) are valid as written only in an inertial frame; neglect of planetary rotation involves neglect of the centrifugal and Coriolis forces generated by the transformation of the time derivatives to a rotating frame. Predictions of position and, to a smaller extent, velocity relative to the planetary surface will be in error if the rotational effect is omitted.

The atmosphere shares the planetary rotation, which tends to carry the vehicle along with it, thus altering the trajectory. Also, errors are introduced in the aerodynamic modeling of the flight vehicle if the atmosphere-relative velocity V' is not used.

Finally, planetary rotation aids the launch by providing an initial velocity in the direction of rotation. The extent to which this is helpful depends on the vehicle design and, as shown in Chapter 4, on the launch site latitude and launch azimuth. As an example, Space Shuttle payload capacity is approximately doubled for a due-east launch from Cape Canaveral, compared with a polar orbital mission from Vandenberg Air Force Base.

None of these factors is important in the present discussion. Provided lift and drag are computed using planetary-relative velocity, rotating atmosphere effects are usually ignored except for re-entry calculations, and often there as well. Coriolis and centrifugal terms are important at the preliminary design level when calculating ballistic missile trajectories, but usually not otherwise. Finally, the effect of planetary rotation on vehicle performance can be modeled simply by specifying the appropriate initial condition on the inertially referenced launch velocity.

The angle α, the vehicle pitch angle in conventional flight mechanics terminology, is a control variable. In general, it is desired that the vehicle adhere to some specific predetermined flight path (position and velocity history) as a function of time. This is accomplished by controlling the direction (and often the magnitude) of the thrust vector. It is the task of the ascent guidance system to provide the required commands to follow the chosen trajectory. The guidance commands ultimately translate to specification of a prescribed vehicle attitude, represented here by the pitch angle α.

The assumption that the vehicle pitch angle defines the thrust axis alignment ignores small, transient variations about this mean condition that are commonly used for implementation of vehicle steering commands via thrust vector control. For example, some or all of the engines may be gimballed slightly (a 2- to 5-deg range is typical) to generate a force perpendicular to the thrust axis and hence a moment about the center of mass to allow control of vehicle attitude. It is the task of the vehicle autopilot to translate attitude requests from the guidance logic into engine gimbal angles for steering. Since the gimbal angles are typically small, preliminary calculations often omit this effect; i.e., the autopilot is not modeled, and it is assumed that the vehicle points as required to shape the trajectory. Once a suitable family of ascent trajectories is found, higher-order models including guidance and autopilot functions are used to establish detailed performance.

As discussed previously, methods other than engine gimballing may be used to effect thrust vector control. These include nozzle injection, jet vanes or, when several engines are present, differential throttling. Finally, the vehicle may in some cases be steered, or at least stabilized, aerodynamically.

If the engines are throttled, as is the case with a Space Shuttle ascent, then the T is also a control variable. Calculation of thrust from basic engine parameters was discussed in the previous section. If thrust is constant, a state variable may be eliminated, as Eq. (5.11e) integrates to yield

$$m(t) = m_p - \dot{m}(t - t_0) \qquad (5.13)$$

The lift and drag coefficients contain the information on the vehicle aerodynamic behavior. For a specified body shape, C_L and C_D are functions of Mach number, Reynolds number, and the angle of attack. Except at very low speeds, which constitute an insignificant portion of the ascent flight, the dependence on Reynolds number is unimportant. For the flight regimes of interest in typical ascent performance calculations, and for a given Mach number, C_L is proportional to α and C_D is proportional to α^2.

As before, the preceding statement contains the implicit assumption that the vehicle thrust axis is aligned with the geometric centerline, to which the aerodynamic angle of attack is referenced, and that the center of mass lies along the centerline. These assumptions are usually appropriate at the preliminary design level, but will rarely be strictly true. If the vehicle center of mass is offset from whatever aerodynamic symmetry axes exist, as is the case with the Space Shuttle, the thrust axis cannot be aligned with the centerline and the pitch angle will not equal the aerodynamic angle of attack. And, as mentioned, if the vehicle is steered via thrust vector control, transient offsets of the thrust axis from the center of mass are used to generate attitude control moments. These effects may often be neglected for initial performance assessments; however, a complete six-degree-of-freedom simulation with guidance and autopilot models will include them.

For a particular vehicle shape, C_L and C_D are usually obtained as functions of Mach number and angle of attack from experimental data, taken either in wind tunnels or during flight tests. A wealth of such data[4,5] exists for various generic shapes of interest as well as for specific vehicles that have flown. For preliminary design purposes, data can usually be found that will be sufficiently representative of the actual vehicle. It has also become possible within the last decade to attempt the direct numerical solution of the governing fluid dynamic equations appropriate to many vehicle configurations of interest. These computational fluid dynamic methods can often provide data outside the envelope of wind-tunnel test capabilities. As in so many areas we have discussed, the Space Shuttle program again provides an excellent example. Substantial effort was expended during the 1970s in learning to compute high-speed flowfields over Space Shuttle configurations. In some regimes, the information obtained represented the only aerodynamic performance data available prior to the first flight. Subsequent comparisons with flight data have shown generally excellent agreement. Theoretical methods and results obtained are surveyed by Chapman[6] and Kutler.[7]

The Mach number is given by

$$M = V/a \qquad (5.14)$$

where a is the local speed of sound, which for perfect gases is a function of the temperature alone,

$$a^2 = kR_{\text{gas}}T \qquad (5.15)$$

Temperature in turn is a prescribed function of the altitude h, usually according to the dictates of a standard atmosphere model. Very detailed models exist for the Earth[8] and Mars[9] and to a lesser extent for other planets. Actual atmospheric probe data are necessary to obtain a temperature profile; planets for which these data have not yet been obtained are often idealized by very simple models based on what can be observed at the planet's cloud tops. However obtained, the temperature information is used with the hydrostatic equation,

$$dp = -\rho g \, dh \qquad (5.16)$$

and the perfect-gas equation of state,

$$p = \rho R_{gas} T \qquad (5.17)$$

to allow $\rho = \rho(h)$ to be computed. If the temperature profile is piecewise linear (the usual fitting procedure), the resulting density function has one of two forms:

$$\rho = \rho_1 \exp[-g_s(h - h_1)/R_{gas} T] \qquad (5.18a)$$

for isothermal layers, and

$$\rho = \rho_1 (T/T_1)^{-(1 + g_s/aR_{gas})} \qquad (5.18b)$$

for constant gradient layers where

$$T(h) = T_1 + a(h - h_1) \qquad (5.19a)$$

and

$$a = (T_2 - T_1)/(h_2 - h_1) \qquad (5.19b)$$

where $T_1 = T(h_1)$ and $T_2 = T(h_2)$ are constants from the measured temperature profile.

The preceding results are strictly true only for constant $g = g_s$, whereas, in fact, the gravitational acceleration varies according to Eq. (5.11h). Although the difference is rarely important, the preceding formulation applies exactly upon replacement of the altitude h in Eqs. (5.16–5.19) with the geopotential altitude,

$$h_G = [R/(R + h)]^2 h \qquad (5.20)$$

Rocket Performance and Staging

Let us consider Eqs. (5.11) under the simplest possible circumstances; i.e., neglect lift, drag, and gravitational forces and assume no steering, so that α and γ are zero. These assumptions are a poor approximation to planetary ascent but may faithfully represent operation in space far away from planetary fields. More importantly, these conditions are also appropriate to the case of acceleration applied in a near-circular orbit, where the terms $(g \sin\gamma)$ and $(V^2/r - g)$ are nearly zero and lift and drag are absent. This is often the situation for orbital maneuvers or for injection into an interplanetary trajectory from a parking orbit. In this case, Eqs. (5.11) reduce to

$$dV/dt = T/m = \dot{m}V_{eq}/m = -(gI_{sp}/m)\,dm/dt \qquad (5.21)$$

which integrates immediately to yield

$$\Delta V = gI_{sp}\,\ell n(m_i/m_f) = V_{eq}\,\ell n(m_i/m_f) \qquad (5.22)$$

where m_i and m_f are the initial and final masses, respectively. Defining the mass ratio MR as

$$MR \equiv m_i/m_f$$

we have, for the change in velocity during the burn,

$$\Delta V = V_{eq}\,\ell n MR = gI_{sp}\,\ell n MR \qquad (5.23)$$

If a burn to propellant exhaustion is assumed, this equation gives the maximum theoretically obtainable velocity increment from a single stage. Here we clearly see the desirability of high I_{sp} and a large mass ratio. This latter condition implies a vehicle consisting, as much as possible, of payload and propellant only.

It is often necessary to compute the propellant mass expended for a single ΔV maneuver; from Eq. (5.22) it is readily found that the propellant expenditure is

$$\delta m_p = \{1 - \exp[-(\Delta V/gI_{sp})]\}m_i \simeq (\Delta V/gI_{sp})m_i \qquad (5.23a)$$

or

$$\delta m_p = \{\exp[-(\Delta V/gI_{sp})] - \}m_f \simeq (\Delta V/gI_{sp})m_f \qquad (5.23b)$$

It is also useful to know the payload sensitivity to small changes in I_{sp}. Again, from Eq. (5.22), it is found that

$$\delta m_f/m_f \simeq (\Delta V/gI_{sp})\delta I_{sp}/I_{sp} \qquad (5.23c)$$

A variety of dimensionless quantities are used to describe the allocation of mass to various portions of the rocket vehicle. Note

$$m_i = m_p + m_s + m_{pl} \qquad (5.24)$$

where

m_p = total propellant mass
m_{pl} = payload mass
m_s = total structural mass (all other mass necessary to build and fly the vehicle, including tanks, engines, guidance, and other supporting structures)

If complete propellant depletion may be assumed, then

$$m_f = m_s + m_{pl} \tag{5.25}$$

Note that we do not require this assumption, and indeed it will never be satisfied exactly. Vehicles intended for multiple restarts will of course retain propellant for later use after each maneuver. Also, even if a given stage burns to depletion, there will remain some surplus fuel or oxidizer, since it is impossible to achieve exactly the required mixture ratio during the loading process. This excess fuel is termed ullage and is normally small. If significant propellant remains at engine cutoff, whether by accident or design, then the actual m_f must be used in performance calculations. When this is done intentionally, it will generally occur only with a single-stage vehicle or with the final stage of a multistage vehicle. If this is the case, the remaining propellant can be classed, for accounting purposes in what follows, with payload. Since we wish to consider the maximum attainable performance, we neglect any ullage in the analysis that follows.

In any case, we define the payload ratio λ as

$$\lambda = m_{pl}/(m_i - m_{pl}) \tag{5.26a}$$

or, with no ullage,

$$\lambda \simeq m_{pl}/(m_p + m_s) \tag{5.26b}$$

and the structural coefficient as

$$\varepsilon = m_s/(m_p + m_s) \tag{5.27a}$$

Again assuming no ullage at burnout,

$$\varepsilon \simeq (m_f - m_{pl})/(m_p + m_s) \tag{5.27b}$$

The mass fraction η is also used frequently in place of the structural coefficient ε:

$$\eta = m_p/(m_p + m_s) = 1 - \varepsilon \tag{5.28}$$

Assuming complete propellant depletion, the mass ratio becomes

$$MR = (m_p + m_s + m_{pl})/(m_s + m_{pl}) \tag{5.29a}$$

or, in terms of the previously defined nondimensional quantities,

$$MR = (1 + \lambda)/(\varepsilon + \lambda) \qquad (5.29b)$$

The advantage of a light structure (small ε) is clear. Since $(m_s + m_{pl})$ appears as a unit, structural mass trades directly for payload. The launch vehicle designer works to keep the structural coefficient as small (or propellant fraction as large) as possible.

There are limits on the minimum structural and control hardware required to contain and burn a given mass of propellant. We shall examine these limits in more detail later, but consider as an example the Shuttle external tank (ET), which carries no engines and very little other equipment. On STS-1 (the first Shuttle mission) the ET had an empty mass of approximately 35,100 kg and carried about 700,000 kg of propellant, yielding a structural coefficient of 0.0478 from Eq. (5.27). Subsequent modifications to the tank design produced a lightweight ET with a mass of approximately 30,200 kg and a structural coefficient of $\varepsilon = 0.0414$. Results such as these represent the currently practical limits for a vehicle that must ascend from Earth. Thus, improvement in overall performance must be sought in other areas, chiefly (at least for chemical propulsion systems) by means of vehicle staging.

Staging is useful in two ways. First and most obviously, expended booster elements are discarded when empty, so that their mass does not have to be accelerated further. A second consideration is that the engines needed for initial liftoff and acceleration of the fully loaded vehicle are usually too powerful to be used after considerable fuel has burned and the remaining mass is lower. Even in unmanned vehicles where crew stress limits are not a factor, the use of very high acceleration can cause much additional mass to be used to provide structural strength.

The analysis for a multistage vehicle is similar to that for a single stage. Assuming sequential operation of an N-stage vehicle, the convention is to define

m_{i_n} = nth stage initial mass, with upper stages and payload
m_{f_n} = nth stage final mass, with upper stages and payload
m_{s_n} = structural mass of nth stage alone
m_{p_n} = propellant mass for nth stage

The initial mass of the nth stage is then

$$m_{i_n} = m_{p_n} + m_{s_n} + m_{i_{n+1}} \qquad (5.30)$$

It is thus clear that the effective payload for the nth stage is the true payload plus any stages above the nth. The payload for stage N is the original m_{pl} from the single-stage analysis. By analogy with this earlier case, we define for the nth stage the ratios

$$\lambda_n = m_{i_{n+1}}/(m_{i_n} - m_{i_{n+1}}) \qquad (5.31)$$

$$\varepsilon_n = m_{s_n}/(m_{i_n} - m_{i_{n+1}}) \simeq (m_{f_n} - m_{i_{n+1}})/(m_{i_n} - m_{i_{n+1}}) \qquad (5.32)$$

$$MR_n = m_{i_n}/m_{f_n} \simeq (1 + \lambda_n)/(\varepsilon_n + \lambda_n) \qquad (5.33)$$

With these definitions, the basic result of Eqs. (5.23) still applies to each stage:

$$\Delta V_n = gI_{\text{sp}_n} \ell n MR_n \qquad (5.34)$$

The total ΔV is the sum of the stage ΔV_n:

$$\Delta V = \sum_{n=1}^{N} \Delta V_n \qquad (5.35)$$

and the mass ratio is the product of the stage mass ratios:

$$MR = \prod_{n=1}^{N} MR_n \qquad (5.36)$$

The approach outlined in the preceding equations must be applied with care to parallel-burn configurations such as the Atlas, Space Shuttle, Titan 3, etc. This is because fuel from more than one stage at a time may be used prior to a staging event, thus complicating the allocation of mass among the various stages. For example, m_{i_1} for the Shuttle consists of the Shuttle Orbiter, external tank, and two solid rocket boosters (SRBs). Following SRB separation, m_{i_2} consists of the Orbiter and the external tank, less the fuel burned by the Shuttle main engines prior to SRB separation. This complicates the definition of m_{s_n} and m_{p_n}; however, no fundamental difficulties are involved. Of greater concern is the fact that the SRBs and the Shuttle main engines have substantially different I_{sp}. In such cases, staging analysis as presented here may be of little utility.

Hill and Peterson[2] examine the optimization of preliminary multistage design configurations, subject to different assumptions regarding the nature of the various stages. In the simplest case, where ε and I_{sp} are constant throughout the stages ("similar stages"), it is shown that maximum final velocity for a given m_{pl} and initial mass m_{i_1} occurs when $\lambda_n = \lambda$, a constant for all stages, where

$$\lambda = (m_{pl}/m_{i_1})^{1/N}/[1 - (m_{pl}/m_{i_1})^{1/N}] \qquad (5.37)$$

The similar-stage approximation is unrealizable in practice; very often the last stage carries a variety of equipment used by the whole vehicle. Even if this is not the case, there are economies of scale that tend to allow large stages to be built with structural coefficients smaller than those for small stages. If we assume fixed I_{sp} for all stages but allow ε to vary, then for fixed m_{pl} and m_{i_1}, maximum final velocity occurs for

$$\lambda_n = \alpha \varepsilon_n/[1 - \varepsilon_n - \alpha] \qquad (5.38)$$

where α is a Lagrange multiplier obtained from the constraint on the ratio of payload to initial mass given by

$$m_{pl}/m_{i_1} = \prod_{n=1}^{N} \lambda_n/(1+\lambda_n) = \prod_{n=1}^{N} [\alpha\varepsilon_n/(1-\varepsilon_n-\alpha+\alpha\varepsilon_n)] \qquad (5.39)$$

If all stages have both varying ε_n and I_{sp}, then again for MR and number of stages N, it is found that the maximum velocity is obtained with stage payload ratios:

$$\lambda_n = \alpha\varepsilon_n/[gI_{sp_n}(1-\varepsilon_n)-\alpha] \qquad (5.40)$$

where α is again found from the constraint

$$m_{i_1}/m_{pl} = \prod_{n=1}^{N} (1+\lambda_n)/\lambda_n = \prod_{n=1}^{N} [(1-\varepsilon_n)(gI_{sp_n}-\alpha)/\alpha\varepsilon_n] \qquad (5.41)$$

With α known, λ_n may be found, and the mass ratio

$$MR_n = (1+\varepsilon_n)/(\varepsilon_n+\lambda_n) \qquad (5.42)$$

computed for each stage.

Finally, if it is desired to find the minimum gross mass for m_{pl}, final velocity V, and N, with both ε_n and I_{sp_n} known variables,

$$\lambda_n = (1-\varepsilon_n MR_n)/(MR_n-1) \qquad (5.43)$$

where

$$MR_n = (1+\alpha gI_{sp_n})/\alpha\varepsilon_n gI_{sp_n} \qquad (5.44)$$

and α is found from the constraint on final velocity,

$$V = \sum_{n=1}^{N} gI_{sp_n} \ell n[(\alpha gI_{sp_n}+1)/\alpha\varepsilon_n gI_{sp_n}] \qquad (5.45)$$

As stated previously, α in Eqs. (5.38–5.45) is a Lagrange multiplier resulting from the inclusion of a constraint equation. In general, it will be found necessary to obtain the roots of Eqs. (5.39), (5.41), and (5.45) numerically.

Again, we point out that the preceding results ignore the effects of drag and gravity. Essentially, these are free-space analyses and are thus of questionable validity for ascent through a gravity field with steering maneuvers and atmospheric drag. Furthermore, the results of this section are inapplicable to the case of parallel burn or other nonsequential staging configurations. Though for detailed performance analysis it will be necessary to resort to the direct numerical integration of Eqs. (5.11), the results given earlier are useful for preliminary assessment.

Ascent Trajectories

The objective of the powered ascent phase of a space mission is to put the payload, often desired to be as large as possible, into a specified orbit. The manner in which this is done is important because small changes in the overall ascent profile can have significant effects on the final payload that can be delivered, as well as on the design of the ascent vehicle itself. The usual desire in astronautics is to maximize payload subject to constraints imposed by structural stress limits, bending moments, aerodynamic heating, crew comfort, range safety requirements, mission-abort procedures, launch site location, etc.

During the ascent phase the rocket vehicle must in most cases satisfy two essentially incompatible requirements. It must climb vertically away from the Earth at least as far as necessary to escape the atmosphere and must execute a turn so that, at burnout, the flight-path angle has some desired value, usually near zero. Few if any missions are launched directly into an escape orbit; thus, a satisfactory closed orbit is practically a universally required burnout condition for the ascent phase of a mission, unless it is a sounding rocket or ICBM flight. Except on an airless planet, high altitude at orbit injection is needed to prevent immediate re-entry, and near-horizontal injection is usually necessary to prevent the orbit from intersecting the planet's surface.

A typical powered ascent into orbit will begin with an initially vertical liftoff for a few hundred feet, which is done to clear the launch pad prior to initiating further maneuvers. In general, the launch vehicle guidance system will be unable to execute pitch maneuvers about an arbitrary axis but will require such maneuvers to be done in a particular vehicle plane. This may also be a requirement due to the vehicle aerodynamic or structural configuration, as with the Titan 3 or Space Shuttle. In any case, if this plane does not lie along the desired launch azimuth, the rocket must roll to that azimuth prior to executing any further maneuvers. Following this roll, a pitch program is initiated to turn the vehicle from its initially vertical ascent to the generally required near-horizontal flight-path angle at burnout. The pitch program is often specified in terms of an initial pitch angle at some epoch (not necessarily liftoff) and a desired angular rate, $d\alpha/dt$, as a function of time. This closes Eqs. (5.11) and allows integration of the trajectory from the launch pad to burnout conditions.

Detailed examination of Eqs. (5.11) reveals a number of energy loss mechanisms that degrade ascent performance. These can be classified as thrust losses, drag losses, gravity losses, and steering losses. The selection of an ascent trajectory is governed by the desire to minimize these losses subject to the operational constraints mentioned earlier. Problems such as this are classically suited to the application of mathematical and computational optimization techniques, areas in which much theoretical work has been done. Recent examples with application to ascent trajectory optimization include the work of Bauer et al.,[10] Well and Tandon,[11] Brusch,[12] and Gottlieb and Fowler.[13]

Detailed discussion of these techniques is beyond the scope of this book and to some extent is also beyond the scope of the current state of the art in actual launch operations. In practice, ascent profiles are often optimized for given vehicles and orbital injection conditions through considerable reliance on trial and error and the experience of the trajectory designer. We will examine in this section some of the basic considerations in trajectory design and the tradeoffs involved in the selection of an ascent profile.

Thrust loss has already been discussed; here we are speaking of the degradation in specific impulse or thrust due to the pressure term in Eq. (5.1) when exit pressure is less than ambient pressure. If the engine is sized for sea level operation, it is then much less efficient for the high-altitude portions of its flight. It is thus advantageous, from the point of efficient utilization of the propulsion system, to operate the vehicle at high altitudes as much as possible.

The dependence of vehicle drag on atmospheric density, flight velocity, angle of attack, and body shape was discussed in Sec. 5.2 (Equations of Motion). From Eq. (5.14b), it is clearly desirable to operate at high altitudes, again as early as possible in the flight, since drag is proportional to atmospheric density. On the other hand, it is advantageous to ascend slowly to minimize the effect of the squared velocity in regions of higher density.

Gravity losses are those due to the effect of the term $(g \sin\gamma)$ in Eq. (5.11a). To minimize this term, it is desirable to attain horizontal flight as soon as possible. Also, careful consideration will show that, to minimize gravity losses, the ascent phase should be completed as quickly as possible, so that energy is not expended lifting fuel through a gravity field only to burn it later. Other factors being equal, a high thrust-to-weight ratio is a desirable factor; an impulsive launch, as with a cannon, is the limiting case here but is impractical on a planet with an atmosphere. However, electro-magnetic mass-drivers, which are essentially electric cannons, have been proposed for launching payloads from lunar or asteroid bases to the vicinity of Earth for use in orbital operations.[14] The opposite limiting case occurs when a vehicle has just sufficient thrust to balance its weight; it then hangs in the air, expending its fuel without benefit.

Finally, steering loss is that associated with modulating the thrust vector by the $\cos\alpha$ term in Eq. (5.11a). Clearly, any force applied normal to the instantaneous direction of travel is thrust that fails to add to the total vehicle velocity. Thus, any turning of the vehicle at all is undesirable. If done, it should be done early, at low speeds. This is seen in Eq. (5.11b), where, if we specify for example a constant flight-path angular rate (i.e., constant $d\gamma/dt$), the required angle of attack is

$$\alpha = \sin^{-1}\{[V \, d\gamma/dt + (g - V^2/r) \cos\gamma - L/M]/(T/M)\} \qquad (5.46)$$

It is seen that larger flight velocities imply larger angles of attack to achieve a fixed turning rate.

The preceding discussion shows the essential incompatibility of the operational techniques that individually reduce the various ascent losses. Early pitchover to near-horizontal flight, followed by a long, shallow climb to altitude, minimizes steering and gravity losses but dramatically increases drag and aggravates the problem by reducing the operating efficiency of the power plant. Similarly, a steep vertical climb can minimize drag losses while obtaining maximum engine performance, at the price of expending considerable fuel to go in a direction that is ultimately not desired. Experience reported by Fleming and Kemp[15] indicates that the various energy losses result in typical first-stage burnout velocities about 70% of the theoretical optimum as given in Eqs. (5.23) for the 0-g drag-free case.

There are a number of special cases in pitch rate specification that are worthy of more detailed discussion. The first of these is the gravity turn, which is defined by the specification of an initial flight-path angle γ and the requirement that the angle of attack be maintained at zero throughout the boost. In this way, no thrust is wasted in the sense of being applied normal to the flight path. All thrust is used to increase the magnitude of the current velocity, and α is controlled to align the vehicle (and hence the thrust vector) along the current velocity vector. Since the term $g \sin\gamma$ in Eq. (5.11b) produces a component normal to the current flight path, a gradual turn toward the horizontal will be executed for any case other than an initially vertical ascent. Setting the angle of attack to zero and solving Eq. (5.11b) for pitch rate, we find

$$d\gamma/dt = [L/m - (g - V^2/r) \cos\gamma]/V \qquad (5.47)$$

where we note that the lift L is generally small and is zero for rotationally symmetric vehicles at zero angle of attack. It is seen that low velocity or small flight-path angle increases the turn rate.

This approach would seem to be most efficient, as with zero angle of attack the acceleration V is maximized. However, this is strictly true only for launch from an airless planet. In the case of an Earth ascent, a rocket using a gravity turn would spend too much time at lower levels in the atmosphere, where other factors act to offset the lack of steering loss. Selection of higher initial values of γ, for more nearly vertical flight, does not generally allow the turn to horizontal to be completed within the burn time of the rocket. In general, gravity turns may comprise portions of an ascent profile but are unsuitable for a complete mission. An exception is powered ascent from an airless planet such as the moon; the trajectories used for Lunar Module ascent flight closely approximated gravity turns.

The case of constant flight-path angle is also of interest. Particularly with the final stage, the launch vehicle spends much of its time essentially above the atmosphere and accelerating horizontally to orbital velocity, with no need to turn the vehicle. In this case, Eq. (5.11b) is solved to yield for the pitch angle:

$$\alpha = \sin^{-1}\{[m(g - V^2/r) \cos\gamma - L]/T\} \qquad (5.48)$$

Most trajectories can be approximated by combinations of these two segments, plus the constant pitch rate turn noted earlier. In practice, once a desired trajectory is identified, implementation is often in the form of a series of piecewise constant steps in pitch rate, $d\alpha/dt$, chosen to approximate a more complicated curve. Such profiles tend in general to follow a decaying exponential of the form[15]

$$\mathrm{d}\alpha/\mathrm{d}t = A\, e^{-K(t-t_0)} \qquad\qquad (5.49)$$

where

A = amplitude factor
K = shape factor
t_0 = time bias

For such a case, Fleming and Kemp develop a convenient trajectory optimization method that allows substantial reductions in the time required to design representative two-stage ascent profiles. However, realistic ascent profiles can also be considerably more complex, as illustrated by the launch sequence[16] for STS-1, summarized in Table 5.2. Space Shuttle ascent guidance strategy and algorithms are reported by McHenry et al.,[17] Schleich,[18,19] Pearson,[20] and Olson and Sunkel.[21]

Rocket Vehicle Structures

As has been discussed, it is the sophistication of the electrical, mechanical, and structural design that produces low values of structural coefficients for each stage. Some general rules may be observed. Large stages tend to have lower values of ε than smaller stages. As mentioned, this is because

Table 5.2   STS-1 ascent timeline

Time	Altitude	Comments
8 s	400 ft	120-deg combined roll/pitch maneuver for head-down ascent at 90-deg launch azimuth.
32 s	8,000 ft	Throttle back to 65% thrust for maximum pressure at 429 kTorr.
52 s	24,000 ft	Mach 1, throttle up to 100%.
1 min 53 s	120,000 ft	Mach 3, upper limit for ejection seats.
2 min 12 s	27 n.mi.	Mach 4.5, SRB jettison.
4 min 30 s	63 n.mi.	Mach 6.5, limit of return to launch site (RTLS) abort. Pitch from $+19$ to -4 deg. Initially lofted trajectory for altitude in case of single engine failure.
6 min 30 s	70 n.mi.	Mach 15, peak of lofted trajectory.
7 min	68 n.mi.	Mach 17, 3-g acceleration limit reached. Throttle back to maintain 3 g maximum.
8 min 32 s	63 n.mi.	Main engine cutoff; 81×13 n.mi. orbit.
8 min 51 s	63 n.mi.	External tank jettison.
10 min 32 s	57 n.mi.	OMS[a] 1 burn, attain 130×57 n.mi. orbit.
44 min	130 n.mi.	OMS 2 burn, attain 130×130 n.mi. orbit.

[a]Orbital Maneuvering System.

Table 5.3 Structural coefficient vs mass

Stage	Liquid oxygen/liquid hydrogen stages Mass	Structural coefficient
Ariane 3rd stage	9,400 kg	0.127
Centaur G	16,635 kg	0.183
Centaur G'	22,861 kg	0.128
Saturn SIVB	105,000 kg	0.093
Saturn SII	437,727 kg	0.078

Stage	Earth storable or liquid oxygen/hydrocarbon Mass	Structural coefficient
Delta 2nd stage	6,499 kg	0.077
Titan III 2nd stage	33,152 kg	0.081
Titan III 3rd stage	124,399 kg	0.051
Ariane 2nd stage	36,600 kg	0.098
Ariane 3rd stage	160,000 kg	0.0875
Atlas	110,909 kg	0.036

some equipment required to construct a complete vehicle tends to be relatively independent of vehicle size. Also, the mass of propellant carried increases with the volume enclosed, but the mass of the tankage required to enclose it does not. Denser fuels allow more structurally efficient designs for given specific impulse, since a smaller structure can enclose a larger mass of propellant. This factor tends to remove some of the theoretical performance advantage of liquid hydrogen, particularly for first-stage operation and was a reason it was not selected for use on the first stage of the Saturn 5.

This area is the province of the structural design specialist, and its details are beyond the scope of this book. For preliminary design calculations and assessments, as well as to provide a "feel" for what is reasonable and possible, we include in Table 5.3 data on a wide range of vehicle stages and the structural coefficient for each.

5.3 LAUNCH VEHICLE SELECTION

Solid vs Liquid Propellant

The late 1950s and early 1960s were a time of strong debate between the proponents of solid propulsion and those of liquid propulsion. At stake was the direction of development of launch vehicles for the Apollo lunar mission and possibly even larger vehicles beyond that.

Solid propellants offer generally high reliability and high mass fraction resulting, respectively, from a relative lack of moving parts and high propellant density. Liquid-propellant engines generally achieve higher specific impulse and better thrust control, including throttling, restart capability, and accurate thrust termination. Development of liquid oxygen/liquid hydrogen stages with high specific impulse and good mass fraction has led to extensive use of this propellant combination for upper stages.

Considerable effort has gone into the development of solid rockets having some of the desirable liquid motor characteristics such as controlled thrust termination, multiple burns, and throttling. Various thrust termination schemes such as quenching and explosively activated vent ports have been successfully developed. Multiple burn and throttling concepts have been less productive, due in part to the fact that such features greatly increase the complexity of the motor and vitiate one of its main advantages, that of simplicity. (It should be noted that the simplicity of solid rocket motors refers to their operational characteristics. The design and fabrication of high performance solid boosters is a complex and demanding exercise.)

In some cases preprogrammed thrust variations can be used to accomplish for solid rocket motors what is done by throttling liquids. As solid propellant technology has matured, it has become possible to tailor the thrust vs time profile in a fairly complex manner. Except for the fact that the profile is set when the motor is cast and cannot be varied in response to commands, this thrust profile tailoring is almost as good as throttling for purposes such as moderating inertial or aerodynamic loads during ascent.

Thrust vector control has progressed substantially as well. Most early solid motor systems were spin stabilized. Although this is still common practice, it is not satisfactory for large vehicles or those requiring precise guidance. In such cases three-axis control is required. Early attempts to attain it included the use of jet vanes as in the liquid propellant V-2 or Redstone, or jetavator rings. These devices are swivel-mounted rings that surround the nozzle exit and are activated to dip into the flow, thus deflecting it. Multiple nozzles, each with a ring, can provide full three-axis control. Such an approach was used on the early Polaris missiles.

Vanes and rings are prone to failures through erosion and thermal shock effects and also reduce performance by introducing drag into the exhaust stream. Approaches circumventing these difficulties include the use of gimballed nozzles and nozzle fluid injection. The most notable use of fluid injection is found in the Titan 3 series of solid rocket boosters. Each booster carries a tank of N_2O_4 (nitrogen tetroxide) under pressure. Four banks of valves located orthogonally on the nozzle exit cone control the injection of the N_2O_4 through the nozzle wall into the supersonic flow. The intruding fluid produces a local shock wave that generates a downstream high-pressure region and a consequent flow deflection. Multiple valves provide various levels of sideloading and enhance reliability.

The fluid injection approach is simple and reliable, but again introduces performance losses due to the mass of the injection system and to the generation of shock waves used to turn the supersonic flow. The minimum performance degradation for a thrust vector control system is obtained through the use of a gimballed nozzle.

Providing a nozzle for a solid-propellant motor that can move freely under substantial thrust and aerodynamic loads while preventing leakage of very hot, high-pressure, and frequently corrosive gas is a major design challenge. Advances in mechanism and structural design as well as materials engineering have been required. Successful designs have evolved to meet requirements ranging from the small, multiple nozzle configurations of Minuteman and Polaris to the much larger single nozzle solid rocket boosters used on the Space Shuttle. A reminder of the difficulty involved in designing a gimballed nozzle for a solid rocket motor is provided by the in-flight failure of the IUS during the STS-6 mission in April 1983. The difficulty was traced to overheating and erosion of a fluid bearing seal in the nozzle gimbal mechanism.[22]

Solid-propellant rockets are most useful for applications requiring high thrust from a compact package in a single burn. First-stage propulsion, as on the various versions of Delta, Ariane, and Titan, as well as the Space Shuttle, is a prime example. Another major application is for apogee and perigee "kick motors" for Earth-orbiting spacecraft, typically communications satellites deployed in geostationary orbit. High thrust is not necessarily a virtue for these applications, but reliability, ease of integration, and simplified ground operational requirements are often crucial.

Liquid-propulsion systems are, in an operational sense, more complex than their solid counterparts. There must be some type of propellant flow control and some means of feeding propellant to the combustion chamber. In even the most simple systems, this requires the use of active components and introduces additional possibilities of failure. However, liquids offer flexibility in thrust levels, burn time, and number of burns, coupled with generally higher specific impulse.

Hybrid propulsion systems, normally consisting of a solid fuel and a liquid oxidizer, offer some of the advantages (and disadvantages) of both systems. Performance is better than solids but, as a rule, is inferior to liquid systems. Hybrid motors throttle readily, are easily restarted and lend themselves to low-cost production. These systems received considerable attention during the 1960s, but further development languished afterward. Interest in hybrids is again on the rise, in part because of their almost total freedom from the risk of detonation in even the most violent impact.

Of particular interest for liquid-propellant engines is their ability to be throttled. Although not trivial to develop, this capability is far easier to include in a liquid-propellant engine than in a solid rocket motor and is mandatory in many applications such as planetary landers (Surveyor, Apollo, Viking). Throttling capability is also highly desirable for some ascent propulsion systems, such as the Space Shuttle, to reduce structural loads and to obtain greater control over the ascent profile.

The primary difficulty in throttling is in maintaining an adequate injector pressure drop. For most types of combustion chamber injectors, the pressure drop across the injector is crucial to ensuring good atomization and mixing of the propellants. Adequate atomization and mixing are crucial to high performance and smooth operation. With a fixed injector area, pressure drop varies as the square of the flow rate according to

$$\Delta p = \dot{m}^2/2\rho C_d^2 A^2 \qquad (5.50)$$

where

$\dot{m}$ = mass flow rate
A = total injector orifice cross-sectional area
C_d = injector orifice discharge coefficient
ρ = propellant density
Δp = pressure drop across injector

For throttling ratios up to about 3:1 this can be tolerated by designing for an excessively high pressure drop at full thrust. This is probably acceptable only for relatively small systems, since the higher supply pressure will result in a substantially heavier overall system.

The lunar module descent engine required approximately a 10:1 ratio between full and minimum thrust using the liquid-propellant combination of nitrogen tetroxide and a 50/50 mixture of hydrazine and unsymmetric dimethylhydrazine. The propellant flow rate was controlled by variable-area cavitating venturi tubes (see Fig. 5.14). The contoured movable pintles were connected mechanically to a movable sleeve on the single element coaxial injector. As the venturi pintles moved to change the flow rate, the injector sleeve moved to adjust injector area to maintain a suitable pressure drop. This system could throttle over a range of nearly 20:1 while maintaining satisfactory performance.

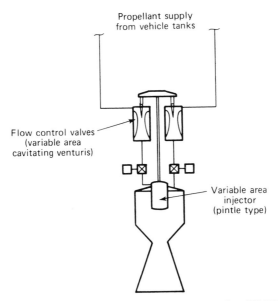

Fig. 5.14 Rocket engine diagram—throttling engine (LMDE type).

Alternative approaches such as injecting an inert gas into the injector to entrain the liquid and maintain the flow momentum during throttling have been tried with somewhat less success. Multielement concentric tube injectors (Fig. 5.15), which were used in the RL-10 and J-2, seem to offer the best overall performance with liquid hydrogen and are more tolerant of throttling than other fixed orifice injectors. This is probably due to the high velocity of the hydrogen and its tendency to flash quickly to a gas as the pressure drops. This tendency is enhanced by the temperature increase incurred in thrust chamber cooling. Micro-orifice sheet injectors (Fig. 5.16) are also less dependent upon pressure for atomization and mixing and thus are able to tolerate a wider range of injector pressure drops.

Design of engine throttling systems is complex, and the details are beyond the scope of this text and are only peripherally relevant to the overall system design problem. The spacecraft systems designer must be aware of the problems and potential solutions in order to allow evaluation of competing concepts. From a systems design point of view, the best approach to engine throttling may be to minimize the extent to which it is needed. The Surveyor spacecraft demonstrated a design approach that avoided the need for a difficult to attain deep-throttling capability. A solid-propellant motor was used to remove most of the lunar approach velocity. Small liquid-propellant engines that provided thrust vector control during the solid motor burn then took over control of the descent following termination of the solid rocket burn. A relatively small amount of velocity was removed during the final descent, with emphasis placed on controlling the maneuver along a predetermined velocity vs altitude profile until essentially zero velocity was reached just above the surface. For this descent maneuver sequence, a throttle ratio of 3:1 was adequate.

Selection of a liquid propellant, solid propellant, or some combination of these for ascent, upper stage, or spacecraft propulsion is another example

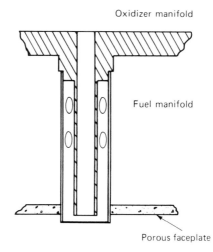

Fig. 5.15   Cross section—single element of concentric tube injector.

of a design issue with no single "correct" resolution. In general, if the problem can be solved at all, it can be solved in more than one way, selection of the final system will depend on such factors as cost, component availability, environmental considerations, etc., as well as traditional engineering criteria such as reliability and performance.

Available Vehicles and Stages

The number and variety of launch vehicles and upper stages, at least those built in the U.S., decreased substantially following the development of the Space Shuttle. Nonetheless, a significant range of launch vehicle options exists. This can be expected to increase as a result of difficulties in operating the Shuttle economically and at high flight rates. Accordingly, we include a discussion of currently available launch vehicles and upper stages. This is done primarily to convey to the reader a sense of typical vehicle capability, requirements, and constraints. Questions concerning the details of launch vehicle interface requirements, orbital performance, etc., should in all cases be referred to a current edition of the user's guide available from the manufacturer of the particular vehicle.

We do not consider launch vehicles produced by Communist bloc nations in this volume. However, with the changing political climate, U.S. payloads are being flown on Chinese launchers and may someday make use of Soviet vehicles as well. Other vehicles such as Japanese designs or vehicles developed commercially in the U.S. may also become available. At the moment we will confine our discussion to the larger U.S. vehicles and to the European Ariane.

Launch vehicles at present fall into two categories: the U.S. Space Transportation System (STS) and expendable vehicles. The expendable family comprises a variety of vehicles, many of them derived from military IRBMs and ICBMs, which can accommodate payloads ranging from a few kilograms in LEO to several thousand kilograms in interplanetary trajectories.

Space Shuttle payload accommodations. Even though basic aspects of the Space Shuttle system will be familiar to most readers, we will describe it briefly here for the sake of completeness. The major components of the system are shown in Fig. 5.17. The central component of the Shuttle "stack" is the Orbiter, a delta-winged aerospacecraft that contains the crew accommodations and cargo bay as well as main, auxiliary, and attitude propulsion systems and propellant for orbital maneuvers, along with power generation and control functions. The SRBs are used during the first 2 min of flight, after which they are jettisoned and recovered for refurbishment and reuse.

The external tank (ET) carries the entire supply of LO_2 and LH_2 for the main propulsion system. It is normally jettisoned just prior to orbital insertion and is essentially destroyed during re-entry (for a due-east launch from Cape Canaveral) over the Indian Ocean. It is worth noting that little extra energy is required to retain the ET through injection into orbit, where the tank material and the residual propellant it carries could in some

Fig. 5.16   Shuttle OMS engine injector. (Courtesy of Aerojet.)

circumstances be quite useful. Many scenarios have been advanced for the use of surplus ETs during heavy construction work in LEO.

Generally speaking, Shuttle payloads are carried in the cargo bay, which provides a clear space 15 ft in diameter and 60 ft long (4.57 × 18.3 m). Limited capacity for experiments also exists in the main cabin, which of course are restricted to those that do not require access to space and that pose no hazard to the flight crew and Orbiter systems. One of the first

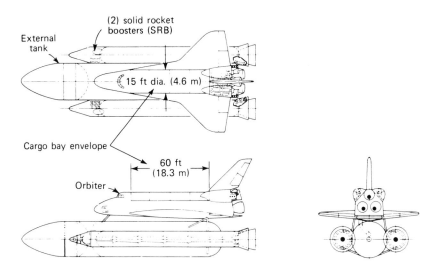

Fig. 5.17 Space shuttle flight system.

examples in this regard was the continuous-flow electrophoresis experiment by McDonnell Douglas and Ortho Pharmaceuticals, which flew aboard the Shuttle even during its initial test flights.

Various external means of carrying cargo have been suggested, most prominent among them being the so-called aft cargo container, which would fit behind the ET and would be especially useful for payloads for which the 15-ft-diam constraint of the payload bay poses a problem. A disadvantage is that the ET must be carried into orbit also and, as discussed, this does involve some performance penalty.

The payload bay is not pressurized and thus will see essentially ambient pressure during ascent, orbital flight, and descent. Access to space is obtained by opening double doors that expose the full length of the payload bay to space. Since the doors also hold radiators needed for thermal control of the Orbiter, they must be opened soon after orbit insertion and must remain so until shortly before re-entry. Payloads must therefore be designed to withstand the resulting environment, which may involve both extended cold soaking and lengthy periods of insolation. Space environmental effects are discussed in more detail in Chapter 3.

Provisions for mounting payloads in the bay are unique in launch vehicle practice and reflect the dual rocket/airplane nature of the vehicle, as well as the desire to accommodate a variety of payloads in a single launch. The support system shown in Fig. 5.18 consists of a series of support points along the two longerons that form the "doorsills" of the bay and along the keel located along the bottom (referenced to landing attitude) of the bay. Proper use of four of these attach points to apply restraint in selected directions as shown in Fig. 5.18a can provide a statically determinant attachment for even a large item such as an upper stage or a spacelab module.

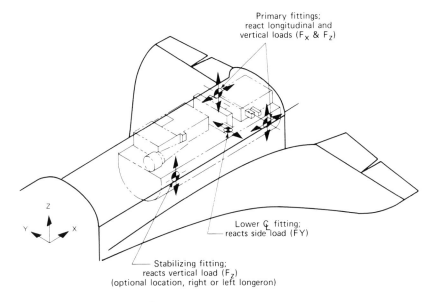

Fig. 5.18a   Statically determinant shuttle payload attach points.

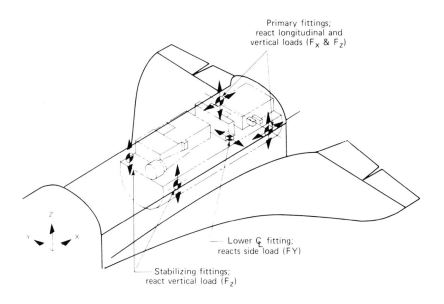

Fig. 5.18b   Five-point payload retention system (indeterminate).

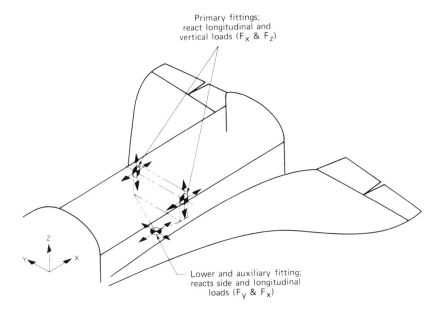

Fig. 5.18c   Three-point payload retention system (determinate).

Other payload attachment accommodations are also available (*STS User Handbook*, 1982). Spacecraft and upper stages are usually mounted in a cradle or adapter that provides an interface to the Shuttle attach points. The usual purpose of this procedure is to accommodate deployment mechanisms and to provide mounting locations for various auxiliary equipment. These structures impose some penalty on total Shuttle payload mass and volume and in theory at least could be eliminated by having the upper stage or payload interface directly to the Shuttle attach points. However, this may also impose a mass penalty or design constraint on the payload, since the interface structure often performs a load leveling or isolation function that, in its absence, would be required of the payload itself.

A variety of methods are used to deploy payloads from the Orbiter bay, depending on such factors as the payload stabilization mode and instrument requirements. The manipulator arm may also be used to deploy payloads or to grapple with and return them to the bay.

Payloads can be installed in the bay while the Shuttle is in the horizontal attitude in the Orbiter Processing Facility (OPF). This is typically done several weeks prior to launch, and the payload will thus remain with the Shuttle throughout erection, mating to the ET, and rollout to the pad. Environmental control may not be maintained throughout all these operations, and for some payloads this may be unacceptable. In such cases, the payload may be installed vertically on the pad using the Rotating Service Structure (RSS). This is probably most desirable for spacecraft on large upper stages such as IUS. Advantages may also exist in allowing later commitment to a particular spacecraft and in maintenance of environmental control.

The Shuttle is basically a low-orbit transportation system. Generally, the orbit achieved is nearly circular in the range of 300–400 km. The basic Shuttle design is intended to carry a 29,500-kg (65,000-lbm) payload into a 300-km circular orbit at the 28.5-deg inclination which results from a due-east launch from Cape Canaveral. The earliest Shuttle vehicles, Columbia and Challenger, had a payload capability in the 24,000- to 26,000-kg range and therefore did not fully meet the design requirement, whereas the later vehicles, Discovery and Atlantis, met or exceeded it. The replacement Orbiter for the ill-fated Challenger will incorporate all the reduced weight improvements and should meet specified performance. If higher orbits are required, lighter payloads are required. Some additional capability is available through the use of so-called OMS kits, which are extra propellant tanks for the Shuttle Orbital Maneuvering System (OMS) engines. These tanks are mounted in the payload bay, and their mass is charged against payload. Figure 5.19 shows circular orbit altitude vs payload capability for various OMS kit loads.

Use of OMS kits to move the entire Shuttle to a higher orbit in order to deploy a payload is not particularly efficient, since most of the propellant is used to move the Shuttle Orbiter itself, which has a dry weight of roughly 68,000 kg (150,000 lbm). Unless there is a requirement for actual Shuttle presence in the higher orbit, as, for example, when retrieving a satellite or when human intervention is needed, it is more efficient to employ an

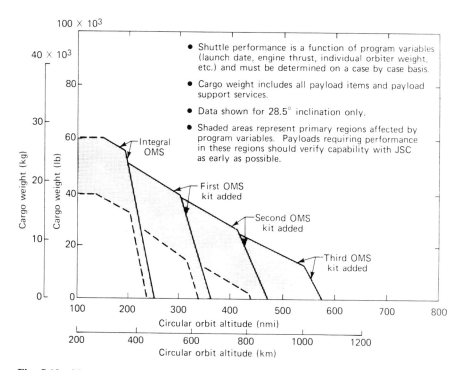

Fig. 5.19   Near-term cargo weight vs circular orbital altitude—KSC launch delivery only.

auxiliary propulsion system on the payload itself to effect the transfer. If, however, the addition of such a capability has an unacceptable payload impact, for example, by greatly increasing the cost and complexity of an otherwise simple device, then direct Shuttle deployment may be appropriate. Design tradeoffs, as always, are involved. No OMS kits have been built as yet, but the necessary preliminary design has been carried out and could be implemented if required.

Obviously, for missions beyond the maximum Shuttle altitude (about 1000 km), auxiliary propulsion is mandatory. Most traffic beyond LEO is destined for geosynchronous orbit, although particular scientific or military missions will require other orbits, as discussed in Chapter 2. A small but significant number of missions will be intended for lunar, planetary, or other deep space targets. A variety of upper stages have been developed or are under development to satisfy these requirements and will be discussed in some detail in later sections. The spacecraft designer is not restricted to the use of these stages, however, and may elect to design his own propulsion system as part of the spacecraft. As an example, Hughes Aircraft has elected to design its own propulsion stage for many of its geosynchronous orbit communications spacecraft.

Even where full orbital transfer capability is not included with the basic spacecraft design, some auxiliary propulsion is often required. Again using the GEO example, most upper stages provide only the capability for inserting the spacecraft into the so-called geosynchronous transfer orbit (GTO), i.e., the apogee raising maneuver from the initially circular orbit. This maneuver will require (see Chapter 4) on the order of 2.5 km/s. A second maneuver of about 1.8 km/s combining a plane change and a perigee raising burn must be done at the GTO apogee. The motor for this "apogee kick" is often designed as an integral part of the spacecraft.

The specialized stage approach, using mostly existing components, may become more popular in the future, since available stages are not often optimal for a given task. This choice will be influenced not only by the nature of the payload but by the number of missions to be flown, since a custom stage may be economically justifiable for use with a series of spacecraft, but impractical for a single application.

As this is written, the sole Shuttle launch capability exists at the Kennedy Space Center (KSC) in Florida. Launches from this facility can achieve orbital inclinations between approximately 28.5 and 57 deg. The lower limit is determined by the latitude of the launch site, as discussed in Chapter 4, and the upper limit results from safety constraints on SRB and ET impact zones. Inclinations outside these launch azimuth bounds can be attained from KSC by using a "dogleg" maneuver on ascent, or by executing a plane change maneuver once in orbit. Both of these procedures result in a reduction in net payload delivered to the desired orbit.

Orbital inclination for many missions is not a critical parameter, and when this is so, KSC is the launch site of choice. However, as discussed in Chapter 2, many missions (e.g., communications satellites) require equatorial orbits or (as with military reconnaissance spacecraft) near-polar orbits. The requirement for a high-altitude equatorial orbit is met rather easily from KSC by executing any required plane change at the apogee of the geosynchronous transfer orbit, which can be chosen to occur at the equator. As discussed in Chapter 4, such a plane change imposes a ΔV requirement of approximately 0.8 km/s when performed separately and less when combined with the usually desired circularization maneuver.

Since, due to safety constraints, polar orbits cannot be achieved from KSC, such requirements have been met by expendable launches from Vandenberg AFB near Lompoc, California. This site can accommodate orbits with inclinations from about 55 deg to slightly retrograde. It was planned to carry out Shuttle launches from this facility as well, even though performance penalties are substantial (Fig. 5.20). Because of facility, budget, and schedule concerns as well as increased concern for safety in the wake of the Challenger accident, these plans have been shelved.

Expendable launch vehicles. During the period of Space Shuttle conceptual design and early development, it was frequently stated that, when the STS became fully operational, expendable launch vehicles would become extinct. It now seems that this is unlikely; i.e., expendable vehicles will continue to meet requirements not addressed in a timely or economical fashion by the Shuttle. Most of the expendable vehicles that have been in use since the early 1970s or before still exist in some form. An additional option is the European Space Agency's Ariane launcher, designed concurrently with and as a competitor for the STS. These vehicles are considered in the following sections. Other launch vehicles are being considered for development, both by governmental agencies and private consortia. These potential additions to the available launch stable will be discussed as well.

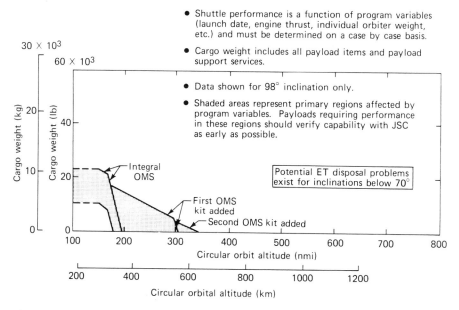

Fig. 5.20   Near-term cargo weight vs circular orbital altitude—VAFB launch delivery only.

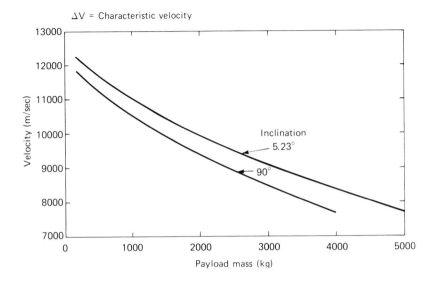

Fig. 5.21a   Characteristic velocity for Ariane.

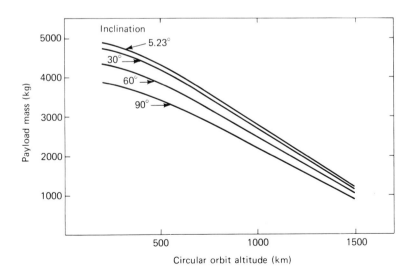

Fig. 5.21b   Performance in circular orbit—Ariane.

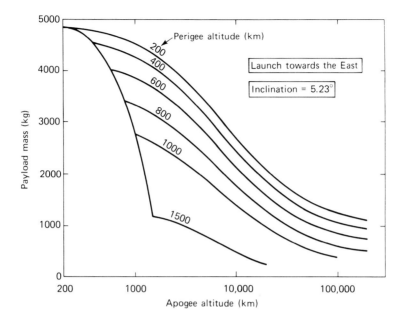

Fig. 5.21c   General performances (East) — Ariane.

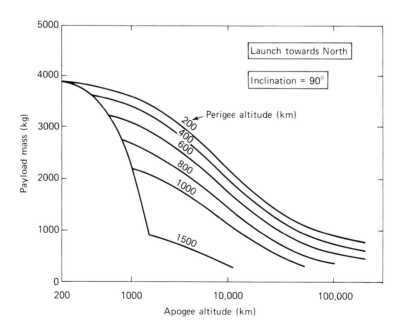

Fig. 5.21d   General performance (North) — Ariane.

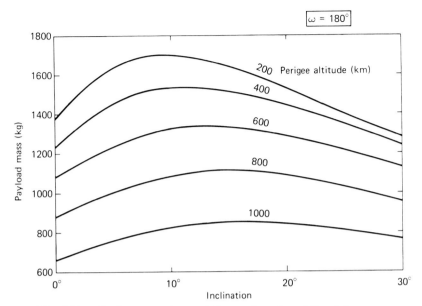

Fig. 5.21e   Performances in geostationary transfer orbit—Ariane.

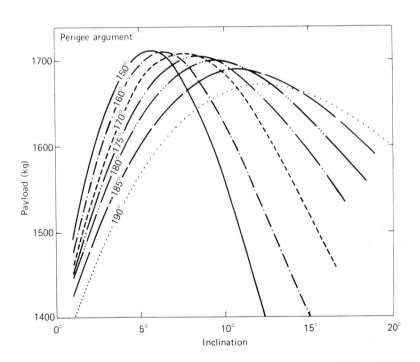

Fig. 5.21f   Performances in transfer orbit 20/35800—Ariane. Variation of inclination and of the argument of perigee.

(1) *Ariane.* This family of launch vehicles was developed by a French/ German consortium and marketed by a semiprivate organization called Arianespace. The Ariane vehicle was developed specifically as a competitor for the Shuttle, and in particular for the lucrative geosynchronous orbit market. Ariane's design is antithetical to that of the STS, consisting as it does of a very conventional three-stage expendable launcher (some consideration is being given to recovery and reuse of at least the first stage in later models).

The basic Ariane was optimized for delivery of payloads to GTO, an orbit with a perigee of 200 or 300 km and an apogee at the geosynchronous altitude of 36,000 km. Its third stage does not have an orbital restart capability, and there is thus no ability to coast in LEO prior to transfer orbit initiation. This significantly inhibits application of Ariane to planetary missions as well as others requiring multiple maneuvers, and it is probable that later versions will feature a restart capability in addition to an increased payload.

Figure 5.21 presents launch capability of the basic Ariane, or Ariane 1. Figure 5.22 shows advanced versions and indicates the concomitant enhancement of payload capability.

As noted, Ariane is a conventional three-stage vehicle with all stages using pump-fed liquid propellants. The lower two stages burn Earth-storable propellants, and the third uses cryogenic liquid oxygen and hydrogen. All launches are from Kourou, French Guiana, on the northeast coast of the South American continent. This site is only about 5° north of the equator and thus has significant performance advantages for low-inclination orbits and ample open sea areas to the east for down-range stage impact.

All three Ariane stages are three-axis stabilized for maximum injection accuracy. For spinning payloads, the entire third stage spins up prior to separation, with a maximum normal rate of 10 rpm provided. This is adequate for stability during separation; however, some spacecraft require a higher rate for later mission phases such as an apogee kick burn. The required additional rate must be supplied by the spacecraft after separation. Whether spinning or not, the third stage maneuvers after separation to remain clear of the spacecraft.

Umbilical electrical connectors for ground checkout, command, and power are provided. The Ariane vehicle does not normally provide power to the spacecraft after liftoff. The aft, or boattail portion of the fairing is radio transparent to allow for antenna checkout and inflight transmission prior to fairing jettison.

(2) *Atlas-Centaur.* The Atlas in various forms and combinations has been a major element of the U.S. space program since the late 1950s. Originally designed as an ICBM, the basic Atlas provided significant payload capability to LEO. This was first demonstrated in 1958 when an entire "bare Atlas" (i.e., no upper stages) was put in orbit as part of Project Score. Ostensibly a communications experiment, the mission probably had more significance as a counter to Soviet propaganda and as a

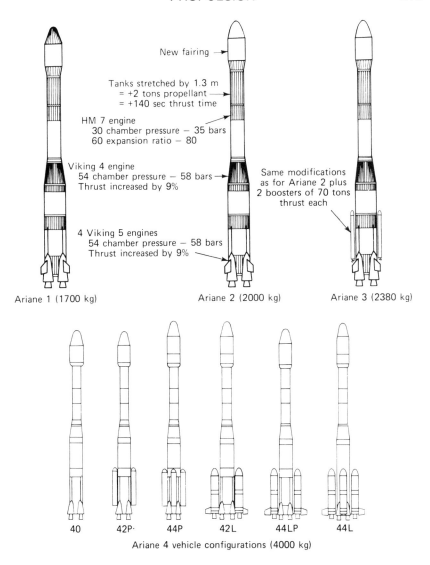

New fairing →

Tanks stretched by 1.3 m
= +2 tons propellant →
= +140 sec thrust time

HM 7 engine
30 chamber pressure − 35 bars
60 expansion ratio − 80

Viking 4 engine
54 chamber pressure − 58 bars →
Thrust increased by 9%

Same modifications
as for Ariane 2 plus
2 boosters of 70 tons
thrust each

4 Viking 5 engines
54 chamber pressure − 58 bars
Thrust increased by 9%

Ariane 1 (1700 kg)   Ariane 2 (2000 kg)   Ariane 3 (2380 kg)

40   42P·   44P   42L   44LP   44L

Ariane 4 vehicle configurations (4000 kg)

Fig. 5.22   Ariane configurations.

national morale booster. In any case, it was a portent of future developments.

A modified version of the operational Atlas D was used to launch the four manned orbital Mercury missions, beginning with John Glenn's three-orbit flight of Feb. 20, 1962. The Mercury flights employed no upper stages; most other Atlas applications have exploited the efficiency of additional staging to augment the basic vehicle. Possibly the most common Atlas upper stage was the Lockheed Agena, a storable liquid-propellant

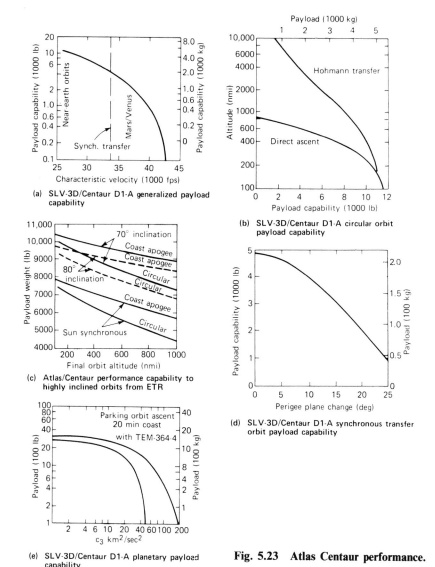

(a) SLV-3D/Centaur D1-A generalized payload capability

(b) SLV-3D/Centaur D1-A circular orbit payload capability

(c) Atlas/Centaur performance capability to highly inclined orbits from ETR

(d) SLV-3D/Centaur D1-A synchronous transfer orbit payload capability

(e) SLV-3D/Centaur D1-A planetary payload capability

Fig. 5.23   Atlas Centaur performance.

vehicle designed for use with both the Thor and Atlas boosters and later adapted for use with the Titan 3B. The Atlas-Agena launched a considerable variety of payloads, including most early planetary missions, and in modified form served as a docking target and orbital maneuvering stage for four two-man Gemini missions in 1966. However, it is no longer used.

The most capable Atlas derivative is the Atlas-Centaur. This vehicle uses a modified Atlas first stage and the LH_2/LO_2 Centaur as the second stage. This system has evolved into a highly reliable, adaptable vehicle that has launched many scientific and commercial spacecraft.

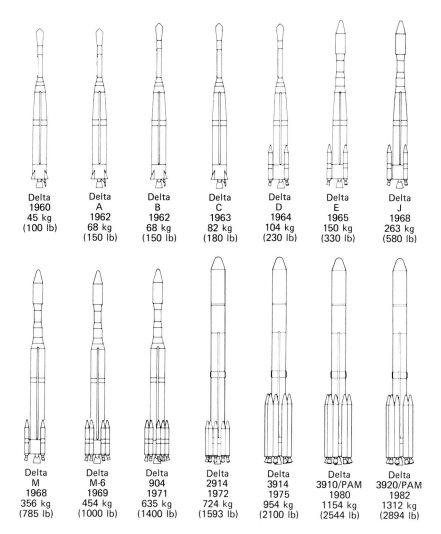

Delta 1960 45 kg (100 lb)	Delta A 1962 68 kg (150 lb)	Delta B 1962 68 kg (150 lb)	Delta C 1963 82 kg (180 lb)	Delta D 1964 104 kg (230 lb)	Delta E 1965 150 kg (330 lb)	Delta J 1968 263 kg (580 lb)

Delta M 1968 356 kg (785 lb)	Delta M-6 1969 454 kg (1000 lb)	Delta 904 1971 635 kg (1400 lb)	Delta 2914 1972 724 kg (1593 lb)	Delta 3914 1975 954 kg (2100 lb)	Delta 3910/PAM 1980 1154 kg (2544 lb)	Delta 3920/PAM 1982 1312 kg (2894 lb)

Fig. 5.24 Delta launch vehicle evolution.

The Atlas is unique among launch vehicles in its use of a "balloon" tank structure. The propellant tanks are used for the primary airframe structure itself and are constructed from welded, thin-gage stainless steel. Without substantial internal pressure, or tension supplied by external means, the tank structure cannot support itself or the payload. Although this design feature complicates ground-handling procedures, it allows a structural coefficient to be achieved that is still unmatched among liquid-propellant vehicles. The Centaur structure is the same.

Although a variety of Atlas variants have been in use, they are essentially gone now. The Atlas-Centaur very nearly became a casualty of the Shuttle-

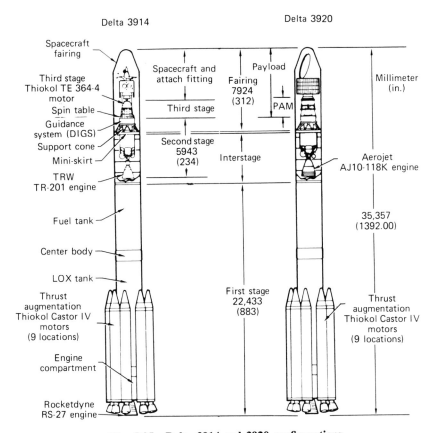

Fig. 5.25   Delta 3914 and 3920 configurations.

inspired purge of expendable launch vehicles. However, General Dynamics has elected to enter the commercial market with the Atlas-Centaur. The initial version, called simply Atlas I, is a strengthened and slightly upgraded version of the later model Atlas-Centaur. The primary change is an increase in nose-fairing diameter from 3.05 m (10 ft), the same as the tank diameter, to a choice of a 4.19 m diam × 12.2 m (13.75 ft × 40.1 ft) or a 3.3 m diam × 10.36 m (10.8 ft × 34 ft) fairing, depending on payload size.

General Dynamics has plans for an Atlas II with increased tank length and thrust in the first stage and a propellant mixture ratio change and length increase in Centaur. The Atlas IIA will incorporate these changes plus extendable nozzles on the Centaur engines and upgraded avionics. Further downstream, the Atlas IIAS will add two Castor II solid-propellant strap-ons to the Atlas IIA.

The Centaur is three-axis stabilized. Spinning payloads are mounted upon a spin table that is locked in place for launch. After final orbit insertion, the spin table is unlocked and spun up to the desired rate using small rockets. When proper spin rate and vehicle attitude are achieved, the payload is released. Following separation the Centaur can maneuver to a new attitude and apply a ΔV to allow a safe distance from the payload to be maintained. Typical separation rates are 0.5–1 m/s.

The Centaur has the capability for a large number of restarts and thus can fly complex mission profiles, including multiple payload deployment. Normal capability is for two-engine starts, with minor modifications required for additional burns. As many as seven restarts have been demonstrated on a single mission.

The Atlas-Centaur is launched from Cape Canaveral Air Force Station. As always, because of the 28.5° latitude of the launch site, performance suffers for delivery to an equatorial orbit. Figure 5.23 depicts Atlas I performance for various missions from low Earth orbit to planetary.

The Centaur is especially noted for high injection accuracy. This capability is summarized in Table 5.4.

(3) *Delta.* The Delta launch vehicle system, developed by McDonnell Douglas Astronautics, began as a derivative of the Thor IRBM. As was the case with its contemporary, the Atlas, a variety of upper stage systems have been used with the Thor, including Agena, Able, Abletar, and various solid motors. The Thor-Able was a mating of the Vanguard second and third stages (a storable liquid and a spinning solid, respectively) to the Thor in order to achieve better performance than offered by the small Vanguard first stage. The Thor-Able could deliver a few hundred pounds into low Earth orbit and a few tens of pounds into an escape trajectory. This vehicle, developed in the late 1950s, rapidly gave way to the Thor-Delta, a vehicle of similar appearance but improved performance and reliability.

Over the ensuing two decades the vehicle has evolved considerably, with the first stage gaining length and assuming a cylindrical rather than a combined conical/cylindrical form. The second stage has grown to equal

Table 5.4   Centaur injection accuracy

Mission	Flight data	Guidance (3σ)
Low Earth orbit		
Injection altitude, km	0.09–0.24	1.48
Apogee minus perigee, km	2–4	10.2
Injection, deg	0.004–0.017	0.035
Geosynchronous transfer		
Perigee accuracy, km	0.06–2.2	4.63
Apogee accuracy, km	6.8–48.7	143
Inclination, deg	N/A	0.024
Midcourse ΔV, m/s		
Inner planetary missions	1–7	7–17
Outer planetary missions (spinning solid 3rd stage)	14–20	108–117

the 8-ft diam of the first and has evolved through several engines. The first-stage engine has remained much the same in its design while being uprated from the original 150,000 lbf to over 200,000 lbf. Relatively early in the evolution of the vehicle, the concept of increasing liftoff thrust with strap-on solid-propellant motors was developed. The maximum number of such motors has now grown to nine (usually fired in a 6/3 sequence to avoid excessive peak loads), and the motors have increased substantially in size and thrust. Figure 5.24 shows the evolution of the configuration, and Fig. 5.25 displays profiles of the two most common current configurations.

A solid-propellant third stage is an option depending on the mission and payload. Three such options are available, the TE364-3, the TE364-4, and the Payload Assist Module (PAM).

It is clear that a substantial family of Delta launcher variants exists. In order to simplify identification, a four-digit code is used to identify a particular model (e.g., Delta 3914 or Delta 3920). Table 5.5 explains the code.

Essentially only 6000 and 7000 or Delta II series vehicles are available now, but this still provides a substantial range of launch options.

Delta launches are available from the Eastern Space and Missile Center (Cape Canaveral) and the Western Space and Missile Center (Vandenberg

Table 5.5 Delta identification

If the PAM third stage is flown, the Delta third-stage code is zero; e.g., the vehicle could be a Delta 3920/PAM.

Delta *WXYZ*

W: First-stage identification
    0: Long tank Thor, Rocketdyne MB-3 engine. (The "0" is usually omitted, in which case the configuration code is 3 digits.)
    1: Extended long tank Thor, Rocketdyne MB-3 engine, Castor II solids.
    2: Extended long tank Thor, Rocketdyne RS-27 engine, Castor II solids.
    3: Extended long tank Thor, Rocketdyne RS-27 engine, Castor IV solids.
    4: Extended long tank Thor, Rocketdyne MB-3 engine, Castor IVA solids.
    5: Extended long tank Thor, Rocketdyne RS-27 engine, Castor IVA solids.
    6: Extra extended long tank Thor, Rocketdyne RS-27 engine, Castor IVA solids.
    7: Extra extended long tank Thor, Rocketdyne RS-27 engine, GEM solids.
X: Number of solids
    3–9: Number of first-stage strap-on solid motors, as in first-stage identification.
Y: Second stage
    0: 65-in.-diam second stage and fairing, Aerojet AJ10-118F engine.
    1: 96-in.-diam second stage and fairing, TRW TR-201 engine.
    2: 96-in.-diam second stage and fairing, Aerojet AJ10-118K engine.
Z: Third stage
    0: No third stage.
    3: Thiokol TE364-3 engine.
    4: Thiokol TE364-4 engine.

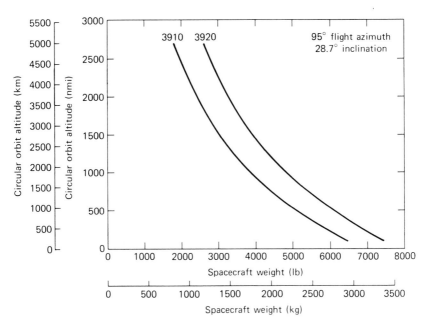

Fig. 5.26 Delta ETR performance, two-stage circular orbit capability.

Air Force Base). Figures 5.26–5.28 present performance capabilities of representative configurations from both sites.

If the third stage is used, it is mounted on a spin table. Prior to separation from the three-axis stabilized second stage, small rockets bring the spin rate up to the desired level. A typical value would be 50–60 rpm. The third stage is normally used to obtain geosynchronous transfer or interplanetary injection velocities.

Accuracy of low-orbit injection is quite good, with typical 3σ accuracy margins of 18.5 km in altitude and 0.05 deg in inclination. Addition of the spin-stabilized but unguided third-stage degrades naturally degrades these values. The 1–2.5 deg pointing errors typical of these stages result in 3σ inclination errors of 0.13–0.5 deg for the TE364 stages and 0.24–0.7 deg for the PAM and apogee errors as shown in Fig. 5.29.

(4) *Titan 3 payload accommodations.* The Titan 3 is derived from the Titan ICBM family, which is no longer a part of the U.S. strategic arsenal. The Titan 1 was a two-stage ICBM using cryogenic liquid propellants; it first flew in 1959. This vehicle saw little use and was rapidly replaced by the Titan 2, a substantially different system using storable hypergolic propellants (N_2O_4 oxidizer and 50/50 N_2H_4-UDMH fuel) in both stages. First flown in 1962, this launcher was subsequently man-rated for use in the Gemini program in the mid-1960s. Twelve missions, including 10 manned flights in 20 months, were conducted between 1964 and 1966.

Some of the Titan 2 vehicles that have been retired from service as missiles are being refurbished for use as orbital launchers by the military.

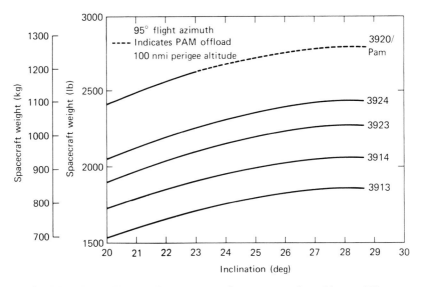

Fig. 5.27   Delta ETR performance, synchronous transfer orbit capability.

These vehicles do not incorporate all the upgrades used in the Gemini-Titan 2 and thus have lower performance. They are capable of placing over 2700 kg (6000 lb) in low Earth orbit. Various performance enhancers, e.g., solid strap-ons or third stages, are being studied.

The Titan 3 family was developed in an effort to provide a flexible, high-capability launch system using existing technology in a "building block" approach. The Titan 3A added two 120-in.-diam solid rocket boosters as a "zeroth stage" configuration to an uprated Titan 2 (also 120 in. in diam). The solids may be ignited in parallel with the basic Titan first stage, but this has not typically been done. A restartable upper stage, the so-called Transtage, was added to provide capability for complex orbital missions.

The Titan 3B used the improved core stages and an Agena upper stage, but no solid rocket boosters. An uprated version with stretched tanks has been developed.

The Titan 3C (Fig. 5.30) is similar to the "A" version, but uses five-segment solids. The Titan 3D omits the Transtage from the "C" vehicle and houses all avionics in the two core stages. All versions are inertially guided, with the "B" and "D" configurations allowing radio guidance as an option.

The Titan 3E is a Titan 3D modified to use a Centaur upper stage and a 14-ft-diam nose fairing. This vehicle, with inertial guidance supplied by the Centaur, was used to launch the Helios, Viking, and Voyager spacecraft in the mid-1970s.

The most recent operational version of the Titan 3 concept is the Titan 34D. This vehicle uses the stretched core stages of the B configuration and solid motors composed of five 1/2 segments. For launches from Cape

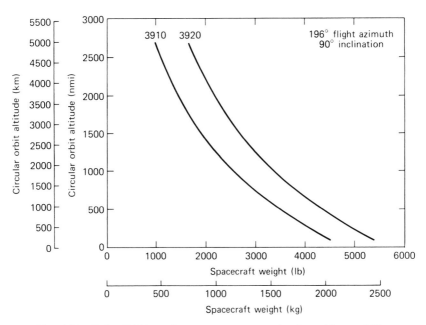

Fig. 5.28   Delta WTR performance, two-stage circular orbit capability.

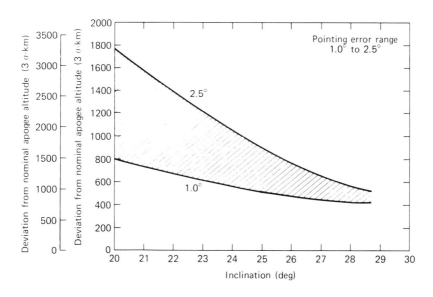

Fig. 5.29   Delta apogee error, typical synchronous transfer orbit deviations with PAM third stage.

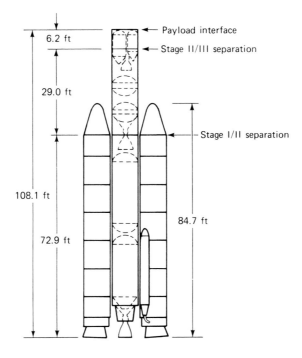

Fig. 5.30   Titan IIIC vehicle characteristics.

Canaveral Air Force Station, the booster would normally carry the two-stage Inertial Upper Stage (IUS) for maneuvering beyond low orbit and to provide inertial guidance for the vehicle. The version launched from Vandenberg has no IUS and uses radio guidance.

A new version of the Titan family, the Titan 4, is entering service under the USAF Complementary Expendable Launch Vehicle (CELV) program. This vehicle is similar in appearance to the Titan 3 family and uses seven-segment solid boosters, has a Centaur G or G′ upper stage, and can boost Shuttle-class payloads to geostationary orbit.

Performance of the Titan 3C is presented in Fig. 5.31 and that of the Titan 34D in Fig. 5.32. These data are for due-east launches from Cape Canaveral. As always, payload capability from Vandenberg Air Force Base is reduced because of inability to take advantage of the Earth's rotation.

Injection accuracy into a 185-km circular orbit for Titan 3C missions appears in Table 5.6.

Upper stages.   A variety of upper-stage vehicles has evolved over the last two decades. Much of this development was driven by the fact that the Space Shuttle is strictly a low-orbit vehicle and requires an upper stage of some type for delivery of payloads beyond LEO. Many of the upper stages derived to meet this need have been adapted for use on other launchers as

Table 5.6 Titan 3C injection accuracy (3σ), 185-km orbit

Component	Position, m	Velocity, m/s
Radial	1200	6
Normal	2400	9
Tangential	600	3

well. In some cases, e.g., the Centaur, the stages are derivatives of previously existing stage designs.

(1) *Centaur*.   It was recognized early in the development of the Shuttle that a complete space transportation system required orbital propulsion capabilities not designed to be met directly by the Shuttle. Concurrent development of a so-called interim upper stage (IUS) was planned to provide upper-stage propulsion prior to the deployment of an eventual high-capability vehicle, loosely called Space Tug. An adaptation of the basic 10-ft-diam Centaur used with Atlas and Titan was proposed (as were many other IUS concepts) but was rejected in favor of a multistage solid propellant vehicle that was expected to be simpler, less expensive, and, most importantly, safer. The latter property followed from the difficulties, thought at the time to be prohibitive, of off-loading potentially explosive liquid fuel from an upper-stage vehicle in the Shuttle bay in the event of a launch abort.

In later years it became obvious both that the Space Tug would be, for funding reasons, indefinitely delayed, and that the IUS would not be adequate for some missions. (As an indication of the staying power of a good acronym, note that by this time, IUS stood for "inertial upper stage.") This was due in part to cost increases resulting from escalation of the original requirements, which resulted in cancellation of some of the larger, high-performance IUS versions. As a result of these factors and because of the flexibility offered by Centaur, the concept of a Shuttle compatible version of Centaur was selected for development some 10 yr after the original proposal.

The basic Shuttle/Centaur concept evolved considerably and in fact became two distinct, closely related vehicles. The two versions reflect the two major mission classes that Centaur will accommodate, namely, payloads of modest mass and dimensions to very-high-energy interplanetary trajectories and larger payloads to high Earth orbit or low-energy interplanetary trajectories. Both versions use the same engines, avionics, and software, as well as having much structural and other hardware in common. Most of this equipment consists of upgraded versions of earlier Centaur subsystems, thus taking advantage of considerable prior experience. The two vehicles differ mostly in propellant capacity, as seen in Table 5.7. The smaller Centaur G has essentially the capacity of the original

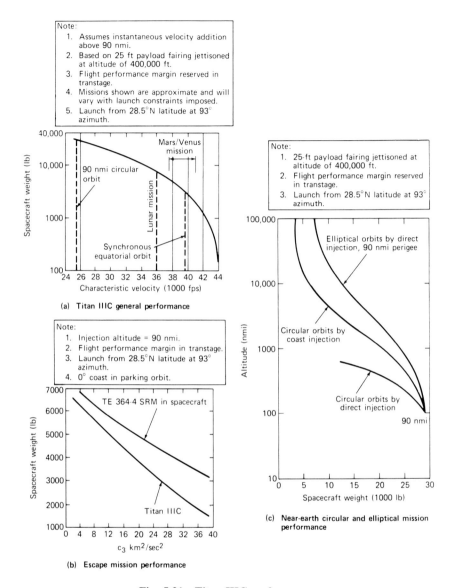

Note:
1. Assumes instantaneous velocity addition above 90 nmi.
2. Based on 25 ft payload fairing jettisoned at altitude of 400,000 ft.
3. Flight performance margin reserved in transtage.
4. Missions shown are approximate and will vary with launch constraints imposed.
5. Launch from 28.5°N latitude at 93° azimuth.

(a) Titan IIIC general performance

Note:
1. 25-ft payload fairing jettisoned at altitude of 400,000 ft.
2. Flight performance margin reserved in transtage.
3. Launch from 28.5°N latitude at 93° azimuth.

(c) Near-earth circular and elliptical mission performance

Note:
1. Injection altitude = 90 nmi.
2. Flight performance margin in transtage.
3. Launch from 28.5°N latitude at 93° azimuth.
4. 0° coast in parking orbit.

(b) Escape mission performance

Fig. 5.31   Titan IIIC performance.

vehicle, with a shorter and wider liquid hydrogen tank to take maximum advantage of the Shuttle payload bay dimensions. The Centaur G' has the approximate length of the original Centaur, but like the G version has grown to the maximum allowable diameter of about 4.3 m. Vehicle properties are given in Table 5.7; both versions may be flown with considerable propellant offload to allow optimization for a given mission. Performance capability of the two Centaur versions is given in further detail in Figs. 5.33 and 5.34.

Table 5.7   Shuttle/Centaur properties

	Centaur G	Centaur G'
Dry mass	2,795 kg	2,648 kg
		2,762 kg[a]
Propellant	13,872 kg	20,259 kg
Total vehicle	16,667 kg	22,907 kg
		23,021 kg[a]
Airborne support equipment	3,392 kg	3,650 kg
Total in Shuttle bay	20,059 kg	26,557 kg
		26,671 kg[a]
Length	5.9 m	8.8 m
Diameter	4.9 m	4.3 m
Available payload length	12.0 m	9.2 m
Engines (2) RL-10		
Thrust (each)	66,700 N	66,700 N
I_{sp}	446 s	446 s
Payload to GEO	4,800 kg	6,100 kg

[a]Dual-burn configuration.

Because of cost and schedule problems, but mostly because of safety concerns arising from the Challenger accident, the Shuttle/Centaur program was canceled in 1986. However, the Centaur G and G' stages will both be used as alternate upper stages for the Titan 4 vehicle. Since, in this application, the Centaur has to burn some propellant to achieve orbit, the high-orbit or escape performance is inferior to that of the Shuttle/Centaur situation in which a fully loaded Centaur was to be delivered to orbit.

(2) *Inertial Upper Stage (IUS).*   As noted earlier, the IUS was originally conceived as a low-cost interim upper stage for the Shuttle pending development of a high-performance Space Tug. A variety of IUS concepts, based mostly on existing liquid-propellant stages, was examined. The final selection, however, was a concept employing combinations of two basic solid-propellant motors, the 6000-lb motor and the 20,000-lb motor, producing approximately 25,000 and 62,000 lb of thrust, respectively. The basic two-stage vehicle would consist of a large motor first stage and small motor second stage, a combination optimized for delivery of payloads to geostationary orbit. Heavier payloads into other orbits would be handled by a twin-stage vehicle whose two stages both consisted of large motors. Both these combinations had substantial planetary capability for lower-energy missions such as those to the moon, Mars, or Venus. High-energy planetary missions were to be handled by a three-stage vehicle consisting of two large and one small motors.

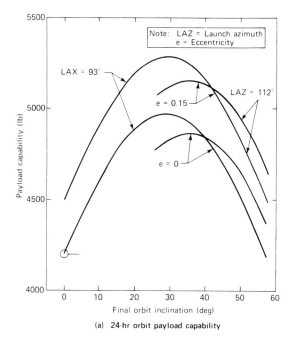

(a) 24-hr orbit payload capability

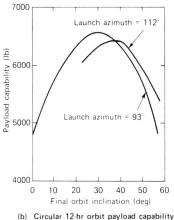

(b) Circular 12-hr orbit payload capability

Fig. 5.32   Titan 34D/IUS performance.

Complexity and rising cost forced cancellation of the two larger versions, leaving only the basic two-stage vehicle. The IUS is currently available in a Shuttle version, with full propellant capacity, and an off-loaded version for the Titan 34D. The payload interface is as for Titan launch vehicles.

(3) *Payload Assist Modules* (*PAM-A and PAM-D*).   The Centaur and IUS are excessively large for payloads of the class historically launched by the Atlas-Centaur or Delta vehicles. To meet the requirements for Shuttle deployment of these vehicles, McDonnell Douglas developed the PAM.

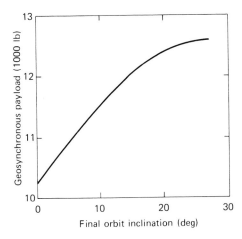

Fig. 5.33   Centaur G performance.

The vehicles are available in A and D models, the former denoting Atlas-Centaur class performance and the latter Delta class capability. Both are spin-stabilized vehicles using solid rocket motors. Avionics are provided for vehicle control for the period from Shuttle ejection (via springs) through postburn separation.

As noted previously, the PAM-D stage can be used as the third stage of the Delta expendable launch vehicle.

(4) *Transfer Orbit Stage.* A recent entry in the upper-stage market is the Transfer Orbit Stage (TOS). Built by Martin Marietta and marketed by Orbital Sciences Corporation, the TOS emphasizes delivery of relatively large (3100 kg from Shuttle, compared with 2270 kg for Shuttle/IUS and 5500 kg for Shuttle/Centaur) payloads to geostationary orbit at minimum cost. (It should be noted that TOS provides the large perigee burn only; in the fashion of the older expendable vehicles, apogee kick must be provided by the spacecraft. TOS performance figures assume optimal apogee motor performance.) This is accomplished with a design based on the IUS "large motor" and a simple avionics system relying heavily on "off-the-shelf" components.

TOS is a three-axis stabilized vehicle that is inactive during Shuttle ascent. One or more orbits before separation, power is turned on and vehicle checkout performed. TOS receives its attitude reference for perigee burn directly from the Shuttle prior to deployment by orienting the Shuttle payload toward the desired inertial perigee burn direction. The TOS gyros are then initialized and the vehicle deployed. During Shuttle separation the upper-stage reaction control system (RCS) thrusters are disabled for safety reasons, and the TOS is uncontrolled in attitude. When the RCS thrusters are enabled, TOS reacquires the original perigee burn attitude. Perigee injection accuracy is thus limited by Shuttle orientation errors, guaranteed to be less than 0.5 deg. This results in a relatively crude but often acceptable transfer orbit injection accuracy of 1100 km in apogee altitude

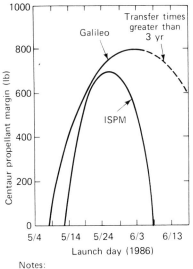

Notes:

A   For 1986 double launch
B   65,000 lb gross weight limit
C   Centaur contains fpr + 200 lb
    lvr in addition to propellant margin

(a) Centaur G' performance margins

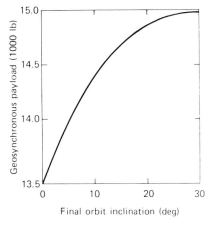

(b) Centaur G' geosynchronous orbit capability

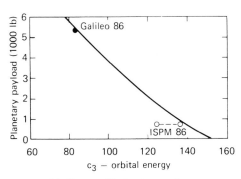

(c) Centaur G' planetary performance

Fig. 5.34   Centaur G' performance.

and 0.7 deg in orbital inclination. A liquid-propellant upper stage has been proposed for TOS to provide GEO circularization as well as more general maneuvering capability.

References

[1]Sutton, G. P., and Ross, D. M., *Rocket Propulsion Elements*, 4th ed., Wiley, New York, 1976.

[2]Hill, P. G., and Peterson C. R., *Mechanics and Thermodynamics of Propulsion*, Addison-Wesley, Reading, MA, 1965.

[3]Vinh, N. X., Busemann, A., and Culp, R. D., *Hypersonic and Planetary Entry Flight Mechanics*, Univ. of Michigan Press, Ann Arbor, MI, 1980.

[4]Horner, S. F., *Fluid Dynamic Lift*, Horner Fluid Dynamics, Bricktown, NJ, 1965.

[5]Horner, S. F., *Fluid Dynamic Lift*, Horner Fluid Dynamics, Bricktown, NJ, 1975.

[6]Chapman, D. R., "Computational Aerodynamics Development and Outlook," *AIAA Jorunal*, Vol. 17, Dec. 1979, pp. 1293–1313.

[7]Kutler, P., "Computation of Three-Dimensional Inviscid Supersonic Flows," *Progress in Numerical Fluid Dynamics*, Springer-Verlag Lecture Notes in Physics, Vol. 41, 1975.

[8]U.S. Standard Atmosphere, National Oceanic and Atmospheric Administration, NOAA S/T 76-1562, U.S. Government Printing Office, Washington, DC, 1976.

[9]Kliore, A. (ed.), "The Mars Reference Atmosphere," *Advances in Space Research*, Vol. 2, No. 2, Committee on Space Research (COSPAR), Pergamon, Elmsford, NY, 1982.

[10]Bauer, G. L., Cornick, D. E., Habeger, A. R., Peterson, F. M., and Stevenson, R., "Program to Optimize Simulated Trajectories (POST)," NASA CR-132689, 1975.

[11]Well, K. H., and Tandon, S. R., "Rocket Ascent Trajectory Optimization via Recursive Quadratic Programming," *Journal of the Astronautical Sciences*, Vol. 30, No. 2, April–June 1982, pp. 101–116.

[12]Brusch, R. G., "Trajectory Optimization for the Atlas-Centaur Launch Vehicle," *AIAA Journal of Spacecraft and Rockets*, Vol. 14, Sept. 1977, pp. 541–545.

[13]Gottlieb, R. G., and Fowler, W. T., "Improved Secant Method Applied to Boost Trajectory Optimization," *AIAA Journal of Spacecraft and Rockets*, Vol. 14, Feb. 1977, pp. 201–205.

[14]Chilton, F., Hibbs, B., Kolm, H., O'Neill, G. K., and Phillips, J., "Mass-Driver Applications," *Space Manufacturing from Nonterrestrial Materials*, Vol. 57, edited by G. K. O'Neill, Progress in Astronautics and Aeronautics, AIAA, New York, 1977.

[15]Fleming, F. W., and Kemp, V. E., "Computer Efficient Determination of Optimum Performance Ascent Trajectories," *Journal of the Astronautical Sciences*, Vol. 30, No. 1, Jan.–March 1982, pp. 85–92.

[16]Covault, C., "Launch Activity Intensifies as Liftoff Nears," *Aviation Week & Space Technology*, Vol. 114, April 1981, pp. 40–48.

[17]McHenry, R. L., Brand, T. J., Long, A. D., Cockrell, B. F., and Thibodeau, J. R. III, "Space Shuttle Ascent Guidance, Navigation, and Control," *Journal of the Astronautical Sciences*, Vol. 27, No. 1, Jan.–March 1979, pp. 1–38.

[18]Schleich, W. T., "The Space Shuttle Ascent Guidance and Control," AIAA Paper 82-1497, Aug. 1982.

[19]Schleich, W. T., "Shuttle Vehicle Configuration Impact on Ascent Guidance and Control," AIAA Paper 82-1552, Aug. 1982.

[20]Pearson, D. W., "Space Shuttle Vehicle Lift-Off Dynamics Occurring in a Transition from a Cantilever to a Free-Free Flight Phase," AIAA Paper 82-1553, Aug. 1982.

[21]Olson, L., and Sunkel, J. W., "Evaluation of the Shuttle GN&C During Powered Ascent Flight Phase," AIAA Paper 82-1554, Aug. 1982.

[22]Goodfellow, A. K., Anderson, T. R., and Oshima, M. T., "Inertial Upper Stage/Tracking Data Relay Satellite (IUS/TDRS) Mission Post-Flight Analysis," *Proceedings of the AAS Rocky Mountain Guidance and Control Conference*, AAS Paper 84-050, Feb. 1984.

Bibliography

The following are recommended as sources of up-to-date information on the various launch vehicles:

NASA, *Space Shuttle System Payload Accommodations*, Vol. XIV, JSC 07700.

NASA, *Space Transportation System User Handbook.*

Arianespace, *Ariane Users Manual.*

General Dynamics Commercial Launch Services, Inc., *Atlas Mission Planner's Guide.*

McDonnell Douglas Space Systems Co., *Delta II Spacecraft User's Manual.*

McDonnell Douglas Space Systems Co., *PAM-D/PAM-DII User's Requirements Document.*

Martin Marietta Corp. Denver Aerospace, *Payload User's Guide — Titan II Space Launch Vehicle.*

Martin Marietta Corp. Denver Aerospace, *Payloads User's Guide – Titan 34D/IUS.*

Note: No dates are specified on these documents since they are periodically updated. For example problems, any edition will suffice. For serious design work, the latest edition should be obtained from the manufacturer.

6
ATMOSPHERIC ENTRY

From the earliest days of space exploration, the problem of controlled atmospheric entry has been as difficult and constraining as that of rocket propulsion itself. Any Earth orbital mission for which the payload must be recovered and any interplanetary mission targeted for a planet with an atmosphere must address the issue of how to get down as well as how to get up. This obviously includes manned missions, and indeed some of the most challenging areas of manned spacecraft design are associated with systems and procedures for effective atmospheric entry.

The technology of atmospheric entry is a highly interdisciplinary area of space vehicle design. This is due to the many different functions that must be satisfied by the atmospheric entry system, and to the wide range of flight regimes and conditions encountered during a typical entry.

Basically, the atmospheric entry system must provide controlled dissipation of the combined kinetic and potential energy associated with the vehicle's speed and altitude at the entry interface. By "controlled dissipation," we imply that both dynamic loads and heat loads are maintained within acceptable limits during entry. This requires a carefully designed flight trajectory and possibly a precision guidance system to achieve the desired results. Control of the vehicle in response to guidance commands implies control of lift and drag throughout the flight. This is a nontrivial task for entry from Earth orbit, since it spans an aerodynamic flight range from subsonic speeds to Mach 25, and even higher speeds are encountered for hyperbolic entry. Finally, the entry system must provide suitable provisions for surface contact, usually with constraints on the landing location and vehicle attitude. These and other issues are addressed in this chapter.

6.1 FUNDAMENTALS OF ENTRY FLIGHT MECHANICS

Equations for Planar Flight

Figure 6.1 shows the geometry of atmospheric entry for planar flight over a spherical, nonrotating planet. Aside from the fact that thrust is typically zero for entry flight, the flight mechanics are the same as those discussed in Chapter 5 in connection with ascent vehicle performance. As in Sec. 5.2, we take this model to be the simplest one that allows presentation of the important phenomena, and Eqs. (5.11) will again apply. Assuming a

231

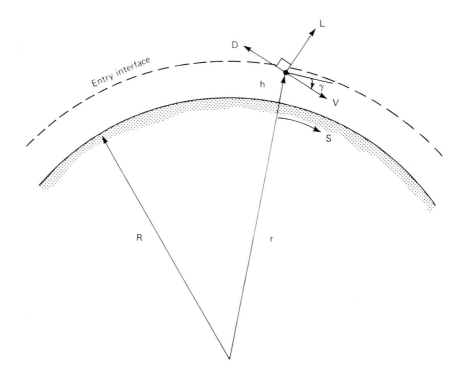

Fig. 6.1   Atmospheric entry geometry.

nonthrusting entry, we have

$$dV/dt = -D/m - g \sin\gamma \qquad (6.1a)$$

$$V \, d\gamma/dt = L/m - (g - V^2/r) \cos\gamma \qquad (6.1b)$$

$$ds/dt = (R/r)V \cos\gamma \qquad (6.1c)$$

$$dr/dt = dh/dt - V \sin\gamma \qquad (6.1d)$$

$$L = \tfrac{1}{2}\rho V'^2 S C_L \qquad (6.1e)$$

$$D = \tfrac{1}{2}\rho V'^2 S C_D \qquad (6.1f)$$

$$g = g_s[R/(R+h)]^2 \qquad (6.1g)$$

where

V = inertial velocity magnitude
V' = speed relative to planetary atmosphere
R = planetary radius

h = height above surface
r = $R + h$ = radius from planetary center
s = down-range travel relative to nonrotating planet
γ = flight-path angle, positive above local horizon
m = vehicle mass
L = lift force, normal to flight path
D = drag force, parallel to flight path
C_L = lift coefficient
C_D = drag coefficient
ρ = atmosphere density
S = vehicle reference area for lift and drag
g = gravitational acceleration
g_s = surface gravitational acceleration

These equations may be integrated forward in time subject to prescribed entry interface conditions (r_e, V_e, γ_e), a model for the atmosphere density $\rho(h)$, and specified values of the vehicle control parameters C_L and C_D. Indeed, this would be essential prior to specification of a flight vehicle configuration and entry trajectory. However, the comments in Sec. 5.2 in regard to direct numerical integration apply here as well. Such a procedure will produce more accurate results (which incidentally justifies the use of a more sophisticated mathematical and physical model), but at the cost of considerable loss of insight.

To obtain the broader perspective that is possible with an analytical solution, three simplifying assumptions are employed:

1) The atmosphere density is approximated by

$$\rho(h) = \rho_s\, e^{-\beta h} \qquad (6.2)$$

where

h = $r - R$
β^{-1} = scale height, assumed constant
ρ_s = surface density

2) The gravitational acceleration g is assumed constant.
3) In Eq. (6.1b) we employ the approximation

$$V^2/r \simeq V^2/R \simeq V^2/r_e \qquad (6.3)$$

Some comments on these assumptions are in order.

The use of atmosphere models has been discussed in Chapters 4 and 5 with respect to orbit decay and ascent vehicle flight. Although for numerical calculations a more detailed model such as the U.S. Standard Atmosphere[1] or the COSPAR International Reference Atmosphere 72[2] might be employed, such models are inappropriate for analytical work where closed-form results are desired.

Using the ideal gas law

$$p = \rho R_{gas} T \tag{6.4}$$

and the hydrostatic equation

$$dp = -\rho g \, dr \tag{6.5}$$

we obtain the differential relation

$$d\rho/\rho = -(g/R_{gas} + dT/dr) \, dr/T = -\beta \, dr \tag{6.6}$$

where

R_{gas} = specific gas constant
T = absolute temperature
P = pressure

Equation (6.2) is the integrated result of Eq. (6.6) subject to the assumption of constant scale height β^{-1}. Since by definition

$$\beta = -(g/R_{gas} + dT/dr)/T \tag{6.7}$$

it is seen that the assumption of constant scale height requires a locally isothermal atmosphere and fixed gravitational acceleration.

The Earth's atmosphere contains regions of strong temperature gradient, with resulting substantial variations in scale height. For entry analysis as given here, it is customary to select β^{-1} for the best fit according to some criteria. Chapman[3] recommends a weighted mean for β^{-1} of 7.165 km, and Regan[4] suggests 6.7 km as a better high-altitude approximation. Both Vinh et al.[5] and Regan[4] give extensive discussions of atmosphere models.

If flight in a particular altitude region is of primary interest, as may sometimes be the case, then ρ_s may be equated to the density near the altitude of interest. Careful selection of β^{-1} then allows a better fit of the density model to local conditions, at the expense of greater deviation elsewhere.

Gravitational acceleration varies according to Eq. (6.1g). For the Earth, with the entry interface altitude taken by convention as 122 km (400,000 ft), variations in g amount to no more than 4%, an acceptable error at this level of study. Even less error results if the reference altitude is chosen to lie between the surface and the entry interface.

The remaining assumption that variations in $(1/r)$ are negligible in Eq. (6.1b) contributes an error of about 2% over the entry altitude range of interest for the Earth. This is insignificant in comparison with other approximations thus far employed. In this chapter we will consistently use $(1/R)$ to replace $(1/r)$ in Eq. (6.1b) and derivations that follow from it. Other choices are possible, with $(1/r_e)$ being the most common alternative.

Two independent variable transformations are normally employed in conjunction with the assumptions discussed earlier. It is customary to eliminate time and altitude in favor of density through the kinematic relation

$$d/dt = (dr/dt)\,d/dr = V\sin\gamma\,d/dr \qquad (6.8)$$

and the density model

$$d\rho = -\rho_s\,e^{-\beta h}\beta\,dh \qquad (6.9)$$

or

$$d/dr = -\beta\rho\,d/d\rho \qquad (6.10)$$

With some additional manipulations,[6] Eqs. (6.1a) and (6.1b) are transformed to yield

$$d(V^2/gR)/d\rho = (SC_D/m)(1/\beta\,\sin\gamma)(V^2/gR) + 2/\rho\beta R \qquad (6.11a)$$

$$d(\cos\gamma)/d\rho = (1/2\beta)(SC_D/m)(L/D) - (gR/V^2 - 1)\cos\gamma/\rho\beta R \qquad (6.11b)$$

These are the reduced planar equations for flight over a nonrotating spherical plane. The reduced equations still cannot be integrated directly and thus require further approximations to obtain closed-form results. Nonetheless, they are worthy of some examination at this point.

In reduced form, the dependent variables are the nondimensional energy (V^2/gR) and γ, with ρ the independent variable. Since at the surface of the planet

$$gR = \mu/R = GM/R = V_{\text{circ}}^2 \qquad (6.12)$$

it is seen that the entry energy is normalized to the circular velocity at the planetary radius. If r_e is used instead of R, as discussed earlier, then the entry energy is referenced to the circular velocity at entry altitude.

Specification of entry interface conditions (V_e,γ_e,ρ_e) is sufficient to determine a particular trajectory subject to the fixed parameters in Eqs. (6.11). Trajectory solutions take the form of velocity and flight-path angle as a function of density, the independent variable. The location is obtained, if required, from Eqs. (6.1c) and (6.1d), with Eq. (6.2) relating atltitude to density.

Four parameters control the solution of Eqs. (6.11): two define the vehicle and two define the relevant planetary characteristics. The vehicle parameters are the L/D and the ballistic coefficient m/SC_D. The planetary entry environment is determined by the radius R and the atmosphere scale height β^{-1}.

In the following sections we will examine various solutions obtained by first simplifying and then integrating Eqs. (6.11). Such solutions implicitly assume the controlling parameters given earlier to be constant. Obviously they are not; we have devoted considerable discussion to this point both here and in Sec. 5.2. The sources of parameter variation can be both natural and artificial. That is, the L/D and the ballistic coefficient will vary considerably over the Mach 25 to 0 entry flight regime due to the differing flowfield dynamics. Additionally, however, the vehicle L/D is the primary control parameter available to the trajectory designer for tailoring the entry profile. Substantial mission flexibility can be gained with judicious L/D control. This cannot be modeled in the closed-form results derived from Eqs. (6.11) and offers another reason why detailed trajectory design requires a numerical approach.

The reader should note that it is common practice, particularly when English units are employed, to define the ballistic coefficient based on weight, i.e., mg/SC_D. When this is done, the ballistic coefficient C_B will have units of pounds per square feet or Newtons per square meters. We avoid this practice and will consistently express C_B in units of kilograms per square meters.

Some discussion of the physical significance of the terms in the reduced equations is in order, since in later sections we will obtain approximate solutions based on assuming a flat Earth, no gravity, small flight-path angle, etc. Equations (6.1) clearly show the influence of various terms such as lift, drag, and centrifugal force. The physical identity of the various terms is not as clear in the reduced form of Eqs. (6.11).

The first term on the right-hand side of Eq. (6.11a) is the reduced drag, and the second term $(2/\rho\beta R)$ is the reduced form of the tangential gravitational component, $g \sin\gamma$, in Eq. (6.1a). Depending on the vehicle configuration and flight conditions, one of these terms may be dominant.

The first term on the right-hand side of Eq. (6.11b) is the reduced lift force. The term $(g - V^2/r)$ in Eq. (6.1b) gives the net normal force contribution of gravity and centrifugal acceleration. The corresponding term on the left-hand side of Eq. (6.11b) is obvious; note, however, that gR/V^2 is the gravitational term, whereas "1" is the reduced centrifugal term.

The surface density ρ_s appears only through the auxiliary Eq. (6.2), which relates density to altitude. Care must be exercised in some cases to avoid the introduction of unrealistic density values (i.e., those corresponding to negative altitudes) when using integrated results from Eqs. (6.11). This will be seen in later sections where ballistic and skip entry are discussed.

As pointed out, Eqs. (6.11) are not directly integrable in the general case, and, if numerical integration is to be employed, there is little reason to use the reduced equations. If closed-form results are to be obtained, it is thus necessary to simplify the analysis even further. Such simplification yields several possible first-order entry trajectory solutions, classically denoted as ballistic, equilibrium glide, and skip entry. These solutions may be adequate within a restricted range of conditions, but most results are approximate only and are primarily suited to initial conceptual design. However, they are very useful in demonstrating the types of trajectories that can exist and the parameters that are important in determining them.

Ballistic Entry

First-order ballistic entry analysis involves two assumptions in addition to those thus far employed. By definition of ballistic entry, zero lift is assumed. We also employ the approximation

$$1/\beta R \simeq 0 \qquad (6.13)$$

which results in dropping terms in Eqs. (6.11) where βR is in the denominator. Some examination of these assumptions is in order.

The zero-lift approximation is often quite accurate and can be made more so when desired. Entry bodies possessing axial symmetry and flown at zero angle of attack will fly nominally ballistic trajectories. In a practical vehicle, small asymmetries will always produce an offset between the center of mass and the center of pressure. This causes the vehicle, unless it is spherical, to fly aerodynamically trimmed at some angle of attack, inducing a lift force. However, this may be dealt with by slowly rolling the vehicle during the entry to cancel out any forces normal to the velocity vector. For example, the Mercury spacecraft was rolled at a nominal 15-deg/s rate during re-entry.

Substantially higher roll rates are employed for ballistic entry of intercontinental ballistic missile (ICBM) warheads. The aerodynamic forces generated in this case can give rise to a significantly more complex entry trajectory. Such topics are considerably beyond the scope of this text. Platus[7] gives an excellent summary of the state of the art in ballistic entry analysis.

The second approximation is somewhat less valid. Although it is true that βR is typically large (approximately 900 for the Earth), this does not justify dropping all terms in Eqs. (6.11) where it appears in the denominator. In particular, $(2/\rho\beta R)$ represents the reduced gravitational force along the trajectory. By omitting it, we assume that the drag force dominates, which is not always true. The drag is always small and is usually comparable to the tangential gravitational force at the entry interface. Toward the end of the entry phase, when the velocity becomes small, $(2/\rho\beta R)$ will again dominate, and neglecting it will lead to inaccuracy.

From the preceding comments it is seen that our second ballistic entry assumption corresponds to neglecting gravity with respect to drag in Eq. (6.1a) and to neglecting the difference between gravity and centrifugal force in Eq. (6.1b). Thus, first-order ballistic entry may be viewed as a zero-g, flat-Earth solution.

In any case, if terms containing $(1/\beta R)$ are dropped and zero lift is assumed, Eq. (6.11b) integrates immediately to yield

$$\cos\gamma = \cos\gamma_e \qquad (6.14)$$

i.e., the flight-path angle remains constant at the entry value.

The validity of this result is obviously somewhat questionable. Intuition and experience suggest that, for shallow entry angles, such as those that are required for manned flight, the vehicle will undergo a lengthy high-altitude deceleration and then, its energy depleted, nose over into a nearly vertical trajectory. Also, the shallow entry angle produces a lengthier entry, with consequently more time for gravity to curve the flight path. As discussed earlier, this is the problem with neglecting $(1/\beta R)$ in Eqs. (6.11). Nonetheless, when entry occurs at a reasonably steep angle, as is typical for ICBMs, the flight path is indeed nearly straight. This is graphically illustrated in numerous time exposure photographs of entry body flight tests. Ashley[6] suggests that the surprisingly shallow value of $-\gamma_e > 5$ deg is sufficient to yield a good match to the ballistic entry assumptions.

Equation (6.11b) may be integrated subject to our assumptions to yield

$$V = V_e \exp[(1/2\beta)(\rho_s/\sin\gamma_e)(SC_D/m)\exp(-\beta h)] \qquad (6.15)$$

for the velocity as a function of altitude and entry flight-path angle. Of possibly greater interest is the derivative of velocity, the acceleration, which can be shown to have a peak value of

$$a_{\max} = -(\beta V_e^2/2e)\sin\gamma_e \qquad (6.16)$$

which occurs at altitude

$$h_{\mathrm{crit}} = (1/\beta)\ell n[(-1/\beta)(SC_D/m)(\rho_s/\sin\gamma_e)] \qquad (6.17)$$

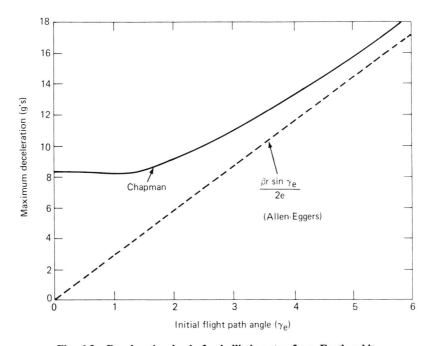

Fig. 6.2 Deceleration loads for ballistic entry from Earth orbit.

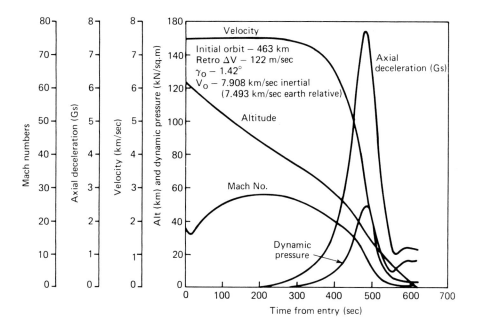

Fig. 6.3   Typical ballistic entry from Earth orbit.

and velocity

$$V_{\text{crit}} = V_e/e \qquad (6.18)$$

If the altitude of peak deceleration is to be real, the argument of the logarithm in Eq. (6.17) must be greater than unity. Assuming hypervelocity impact with the ground is to be avoided, the useful range of entry angles is defined by

$$0 < -\sin\gamma_e < (SC_D/m)(\rho_s/\beta) \qquad (6.19)$$

The ballistic entry results given above should be used with caution. For example, Eq. (6.16) predicts zero peak deceleration for $\gamma_e = 0$, a grazing entry. The first-order analysis provides a very poor model of the entry in such a case. The second-order analysis by Chapman,[3] summarized in Fig. 6.2, shows that entry from low circular Earth orbit has an irreducible deceleration load of about 8 g, which occurs for flight-path angles between 0 and -1 deg. For flight-path angles steeper than about -5 deg, the theories are in reasonable agreement as to trend, though the first-order theory underpredicts the deceleration by about 1 g.

Figure 6.3 shows a typical shallow angle ballistic entry solution, obtained by numerically integrating Eqs. (6.1). The sharp peak in entry deceleration and the rather high value of that peak are characteristic of ballistic entry. As indicated earlier, the peak is seriously underestimated in this case by Eq. (6.16), which predicts a 3.6-g maximum deceleration.

Note also in Fig. 6.3 the difference between the inertial and Earth-relative entry speed. The Earth-relative (hence atmosphere-relative) speed should be employed for V_e; however, the approximations inherent in a first-order solution are such that correcting for atmosphere-relative velocity may not be important. If used, the appropriate correction is

$$V'_e = [1 - (\omega r_e / V_e) \cos i] V_e \qquad (6.20)$$

where

ω = angular velocity of Earth, 7.292×10^{-5} rad/s
V_e = inertial entry velocity
V'_e = Earth-relative entry velocity
i = orbital inclination
r_e = entry interface radius

As mentioned, ballistic entry from Earth orbit will require loads of 8 g or more. Entry at hyperbolic speed is practical only at fairly steep angles and consequently very high deceleration if skip-out is to be avoided. It will be seen in later sections that a moderate value of L/D greatly eases the entry dynamic load.

Purely ballistic entry has a somewhat limited range of application; however, within this range it is quite useful and is widely employed. It is simple to mechanize, requiring little or no guidance beyond stabilization for the deorbit burn, if any. Entry and landing accuracy is determined primarily by the precision to which V_e and γ_e are controlled, knowledge of the ballistic coefficient, and variations in atmosphere properties. However, relatively large dispersions in the controlling parameters can usually be tolerated without disaster; the technique is quite robust when, again, one is within its range of applicability. Table 6.1[8] gives the ballistic entry vehicle dispersions used in the Mercury program. Actual flight performance was somewhat better than these estimates.

Ballistic entry has seen application to numerous vehicles, including the manned U.S. Mercury spacecraft and Russian Vostok/Voshkod series, as well as unmanned spacecraft, including Discoverer, Pioneer Venus, and Galileo. The Mercury procedure is typical of those used for entry from low Earth orbit, 200–500 km.

The Mercury entry sequence was initiated by performing a deorbit burn of approximately 150 m/s at a pitch attitude of 34 deg above the horizontal. This resulted in a nominal flight-path angle of −1.6 deg at the 122-km entry interface. At this point, indicated by a 0.05-g switch, a 15 deg/s roll rate was initiated, with pitch and yaw attitude rates controlled to zero. Attitude control was terminated at 12–15 km altitude upon deployment of a drogue stabilization parachute, followed by main parachute deployment at 3 km altitude. The flight time from retrofire to landing was about 20 min, and the range was approximately 5500 km.

Table 6.1 Mercury spacecraft entry dispersions

Error source	Tolerance	Dispersions, n.mi.		
		Overshoot	Undershoot	Cross range
Perigee altitude	± 0.05 n.mi.	11.0	11.0	
Eccentricity	± 0.0001	15.6	15.6	
Inclination	± 0.10 deg			6.0
Pitch angle	± 6.9 deg	65.0	10.0	
Yaw angle	± 8.1 deg			15.0
Retrofire velocity	± 2.4%	85.0	85.0	
Down-range position	± 5 n.mi.	5.0	5.0	
Cross-range position	± 5 n.mi.			5.0
Drag coefficient	± 10%	5.8	5.8	
Atmosphere density	± 50%	15.4	15.4	
Winds		2.5	2.5	4.0
rss Total		110.1	89.4	17.4

The overwhelming base of ballistic entry flight experience today lies with unmanned satellite reconnaissance vehicles, which in some cases utilize small entry vehicles for return of film canisters from orbit.[9] These vehicles perform shallow-angle, Mercury-type re-entries and routinely land within 20 km of their intended targets. Over 300 such flights have taken place successfully.

Steep entry can be much more accurate, and for this reason it has been favored for use with ICBMs. High dynamic loads, on the order of several hundred g, are possible, but unmanned vehicles can be designed to withstand this. Purely ballistic entry can yield a targeting accuracy on the order of a few hundred meters under such conditions.

The first-order ballistic entry solution given here follows the classical treatment of Allen and Eggers[10] and as previously emphasized is valid only at relatively steep flight-path angles. Higher-order theories that are more suited to shallow-angle entry are available.[3,5] However, none of these treatments produces closed-form results suited to rapid analysis of vehicle loads in preliminary design. It is our view that, if more accuracy is required, direct numerical integration of the basic equations is preferred to a numerical solution from a second-order theory.

Gliding Entry

In contrast to ballistic entry, first-order gliding entry analysis assumes that the vehicle generates sufficient lift to maintain a lengthy hypersonic glide at a small flight-path angle. Clearly this is an idealization. Substantial lift is readily obtained at hypersonic speeds, and it is possible to achieve shallow-angle gliding flight over major portions of the entry trajectory. However, a practical vehicle configuration for an extended hypersonic glide will be poorly suited to flight at low supersonic and subsonic speeds. Toward the end of its flight, such a vehicle must fly at a steeper angle to maintain adequate airspeed for approach and landing control.

This is readily illustrated by the Space Shuttle entry profile.[11,12] The entry guidance phase is initiated at the entry interface altitude of 122 km with the flight-path angle typically about -1.2 deg. It terminates when the Shuttle reaches an Earth-relative speed of about 750 m/s, at an altitude of approximately 24 km and a distance to the landing site of about 110 km. This phase of flight covers a total range of roughly 8000 km, with the average flight-path angle on the order of -1 deg.

Upon completion of entry guidance, the terminal area energy management (TAEM) phase is initiated. The goal of this procedure is to deliver the Orbiter to the runway threshold at the desired altitude and speed for approach and landing. This phase of flight covers a range of 110 km while descending through 24 km of altitude, at an average flight-path angle of about -12 deg, an order of magnitude steeper than that for the hypersonic phase.

The results of this section, although inadequate in the terminal flight regime, may well be appropriate for the major portion of a gliding entry. In keeping with the small angle assumption noted earlier, we assume $\sin\gamma \simeq \gamma$, $\cos\gamma \simeq 1$, and hence $d(\cos\gamma)/d\rho \simeq 0$. With these approximations

Eq. (6.11b) is reduced to an algebraic equation for energy as a function of density:

$$V^2/gR = 1/[1 + (R/2)(SC_D/m)(L/D)\rho] \qquad (6.21)$$

or, from Eq. (6.2),

$$V^2/gR = 1/[1 + (R/2)(SC_D/m)(L/D)\rho_s e^{-\beta h}] \qquad (6.22)$$

Equation (6.21) may be differentiated with respect to ρ and substituted into Eq. (6.11a) to solve for the flight-path angle. Consistent with the assumption of small γ, we neglect the tangential component of gravitational acceleration and obtain

$$\sin\gamma \simeq \gamma \simeq -2/[\beta R(L/D)(V^2/gR)] \qquad (6.23)$$

Note that, although the flight-path angle γ is assumed to be small and its cosine constant, γ is not itself assumed constant.

Equation (6.22) is an equilibrium glide result, where the gravitational force cancels the sum of the centrifugal and lift forces. This is readily seen by noting the physical identity of the various terms in Eq. (6.11b). Of course, the equilibrium is not exact because the derivative term in Eq. (6.11b) is not identically zero. For this reason the trajectory given by Eq. (6.22) is sometimes, and more correctly, referred to as a pseudoequilibrium glide.

To obtain the acceleration along the trajectory, note from Eq. (6.1a),

$$a = dV/dt \simeq -D/m = -(V^2/2)(SC_D/m)\rho \qquad (6.24)$$

where again we neglect the gravitational acceleration along the flight path. Solving Eq. (6.21) for ρ and substituting above gives the tangential acceleration,

$$a/g = (V^2/gR - 1)/(L/D) \qquad (6.25)$$

experienced by the vehicle during the equilibrium glide. Note that the maximum deceleration is encountered as the vehicle slows to minimum speed. Here we see the advantage of even small values of L/D in moderating entry deceleration loads. For a Gemini-class vehicle, with a hypersonic L/D of approximately 0.2,[8] the peak load is $5\,g$, substantially lower than for a Mercury-style ballistic entry at the same flight-path angle. For the Shuttle, which flies a major portion of its entry profile with a hypersonic L/D of about 1.1, Eq. (6.25) predicts essentially a 1-g re-entry. These results are consistent with flight experience.

By integrating the velocity along the entry trajectory, the total range of the equilibrium glide is found to be

$$s = (R/2)(L/D) \ell n[1/(1 - V_e^2/gR)] \qquad (6.26)$$

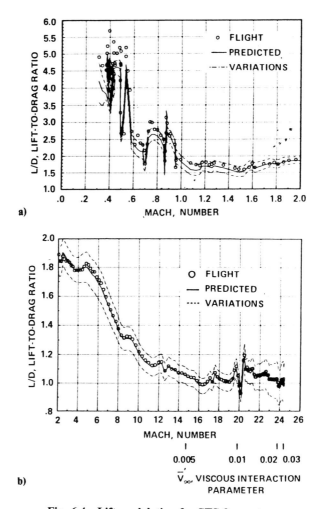

Fig. 6.4   Lift modulation for STS-2 re-entry.

Clearly, the greatest range is obtained when entry is performed at the maximum vehicle L/D.

As an example, consider a Space Shuttle entry at an atmosphere-relative speed of 7.5 km/s with a hypersonic L/D of 1.1 assumed. With these representative values, Eq. (6.26) yields a predicted range of about 8000 km, in good agreement with flight experience. This may be somewhat fortuitous, since the Shuttle in fact uses substantial lift modulation during entry to achieve landing point control. This is shown in Fig. 6.4[13] for the STS-2 re-entry. Nonetheless, an L/D of 1.1 closely approximates the high-altitude, high-speed portion of the entry, and it is this portion that obviously has the most effect on total range.

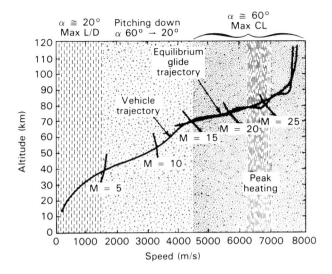

Fig. 6.5 HOTOL entry trajectory.

Equation (6.26) is of limited utility when the entry speed approaches the reference circular velocity. In this case the argument of the logarithm is almost singular and hence extremely sensitive to the value of the entry velocity. This reflects a limitation of the first-order theory rather than any real physical effect. Equation (6.26) is also invalid for the supercircular entry, since the logarithm becomes imaginary in this case.

Figure 6.5 shows an entry trajectory simulation result for the British horizontal takeoff and landing (HOTOL) concept. HOTOL is an unmanned, reusable, single-stage-to-orbit vehicle intended for runway launch and landing. Because it is quite light, it must fly a high, shallow-angle gliding entry to minimize peak dynamic and thermal loads. As shown, the result is an entry profile that closely approximates the pseudoequilibrium glide trajectory during the high-speed portion of the flight.

The first-order equilibrium glide solution given earlier is due to Eggers et al.[14] and is discussed by later authors, including Ashley[6] and Vinh et al.[5] As with ballistic entry, higher-order solutions are available but in our view are as annoying to implement as a complete numerical solution and yield little additional insight compared to the first-order theory.

Skip Entry

Gliding entry flight is not restricted to the equilibrium glide condition discussed in the previous section. Of particular interest is the case of supercircular entry with sufficient lift to dominate the gravitational and centrifugal forces. This is essentially the first-order ballistic entry model with lift added. With proper selection of parameters, the so-called skip or skip-glide entry may be obtained.

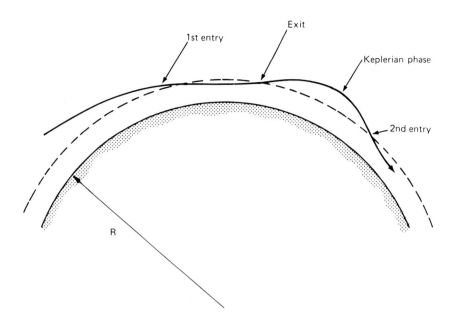

Fig. 6.6   Skip entry trajectory.

Consider the high-speed entry of a lifting vehicle at an initially negative flight-path angle. As always, the vehicle and atmosphere parameters are considered constant. With lift dominant over gravity, the flight path will be turned upward $(d\gamma/dt > 0)$ so that the vehicle enters, reaches a certain minimum altitude, pulls up, and eventually exits the atmosphere at reduced speed. Provided the exit velocity and flight-path angle are properly controlled, a brief Keplerian phase ensues, followed by a second entry that occurs somewhat downrange from the first. This is shown schematically in Fig. 6.6.

This type of trajectory offers considerable flexibility in the control of high-speed entry. For example, at lunar return speeds of about 11 km/s, the kinetic energy to be dissipated is about twice that of a typical low-Earth-orbit entry. This results in a very challenging thermal control problem, especially if the entry is required to occur in a single pass. With the skip entry, however, the vehicle can reduce its velocity sufficiently on its first pass to guarantee Earth capture. The brief suborbital lob allows radiative cooling and is followed by a second entry phase at lower speed.

Skip entry is also useful for range control and allows the vehicle to land in places that could not be reached via a single-entry phase. Again, lunar return provides an example, this time in connection with the Soviet Zond series of unmanned lunar probes. Astrodynamic constraints forced the initial entry point for the return vehicles to occur at latitudes well to the south of the USSR. The entry vehicles were brought in over the Indian Ocean and, following a pronounced skip, were targeted to land in the USSR.

With the previously mentioned assumptions (0 g, flat Earth) for a first-order model, Eqs. (6.11) are reduced to

$$d(V^2/gR)/d\rho = (SC_D/m)(1/\beta \sin\gamma)(V^2/gR) \qquad (6.27a)$$

$$d(\cos\gamma)/d\rho = (1/2\beta)(SC_D/m)(L/D) \qquad (6.27b)$$

Assuming, as always, constant ballistic coefficient and L/D, Eq. (6.27b) integrates immediately to yield the flight-path angle as a function of density (hence altitude):

$$\cos\gamma = \cos\gamma_e + (1/2\beta)(SC_D/m)(L/D)\rho \qquad (6.28)$$

with the approximation $\rho_e \simeq 0$. Since

$$dV/d\gamma = (dV/d\rho)[d\rho/d(\cos\gamma)][d(\cos\gamma)/d\gamma]$$

$$= -(V^2/gR)/(L/D) \qquad (6.29)$$

the velocity as a function of flight-path angle is found to be

$$V = V_e \exp[-(\gamma - \gamma_e)/(L/D)] \qquad (6.30)$$

Equations (6.28) and (6.30) constitute the first-order solution for gliding flight with lift large in comparison to other forces and range small with respect to the planetary radius. Though we are discussing skip entry, Vinh et al.[5] point out that these approximations are also appropriate to gliding entry at medium or large flight-path angle.

For the skip entry, however, we note that $\gamma = 0$ defines the pullup condition. Equation (6.28) then yields the density

$$\rho_{pullup} = \rho_{max} = 2\beta(1 - \cos\gamma_e)/[(SC_D/m)(L/D)] \qquad (6.31)$$

and Eq. (6.30) gives the corresponding velocity,

$$V_{pullup} = V_e \exp[\gamma_e/(L/D)] \qquad (6.32)$$

at pullup. Care must obviously be taken to ensure that the pullup density corresponds to a positive altitude. Though this is not typically a problem for the Earth, it can be a constraint when considering skip entry at a planet, such as Mars, with a tenuous atmosphere.

Observing that both entry and exit occur at the same defined altitude (and hence the same density, often assumed zero), the exit flight-path angle is simply

$$\gamma_{exit} = -\gamma_e \qquad (6.33)$$

From Eq. (6.30), the exit velocity is then

$$V_{\text{exit}} = V_e \exp[2\gamma_e/(L/D)] \tag{6.34}$$

The acceleration along the trajectory is found to be[5]

$$a = \tfrac{1}{2}[1 + (L/D)^2]^{\frac{1}{2}}(SC_D/m)\rho V^2 \tag{6.35}$$

Maximum deceleration occurs at a small negative flight-path angle, i.e., just prior to pullup. However, the value at pullup ($\gamma = 0$) is nearly the same and is much more easily obtained; hence,

$$a_{\text{max}} \simeq a_{\text{pullup}}$$

$$= [1 + 1/(L/D)^2]^{\frac{1}{2}}(1 - \cos\gamma_e)\beta V_e^2 \exp[2\gamma_e/(L/D)] \tag{6.36}$$

Taken together, Eqs. (6.34) and (6.36) imply the existence of an entry corridor, an acceptable range of flight-path angles, for supercircular skip entry. The lower (steep entry) bound on γ_e is determined for a given L/D by the acceptable deceleration load. For a manned vehicle, a reasonable maximum might be 12 g, the design limit for the Apollo missions. The upper (shallow entry) bound on γ_e for supercircular entry is determined by the requirement that the exit velocity be reduced to a sufficiently low level to allow the second phase of entry to occur within a reasonable time. The Apollo command module, for example, had battery power for only a few hours after the service module was jettisoned and could not tolerate a lengthy suborbital lob.

As an example, consider an Apollo-type entry with a vehicle L/D of 0.30, a lunar return speed of 11 km/s, and an atmospheric scale height of 7.1 km. Using the 12-g maximum acceleration design limit selected for Apollo, the steepest allowed entry angle is found from Eq. (6.36) to be -4.8 deg. Assuming circular exit velocity to be the maximum acceptable (the vehicle will not go into orbit because the flight-path angle is nonzero at the exit interface, which is itself too low for a stable orbit), we find from Eq. (6.34) that the shallowest possible entry is -2.9 deg.

An indication of the accuracy and limitations of the first-order skip entry analysis presented here is obtained by noting that the Apollo 11 entry was initiated at a velocity and flight-path angle of 11 km/s and -6.5 deg, respectively. The 12-g undershoot (steep entry) boundary was -7 deg, and the overshoot (shallow-angle) boundary was -5 deg. Figure 6.7 shows the predicted and actual L/D for the Apollo vehicle.[15]

The angular bounds on the entry corridor can be extrapolated backward along the entry hyperbola to yield the required B-plane targeting accuracy. This is done via the results of Sec. 4.1 (Motion in Hyperbolic Orbits).

The preceding discussion of entry corridor limits, though relevant, is oversimplified. In addition to errors introduced by the first-order model, other limitations must be considered. The total entry heating load is aggravated by an excessively shallow entry, whereas the heating rate (but not usually the total heat load) is increased by steepening the flight path. Either case may be prohibitive for a particular vehicle and may modify the entry corridor width determined solely from acceleration and exit velocity requirements.

The constant L/D assumption is an unnecessarily restrictive artifact of the analytical integration of the equations of motion. A more benign, and thus safer, entry can be obtained at an initially shallow angle with the lift vector negative, i.e., with the vehicle rolled on its back. Once the vehicle has been pulled into the atmosphere in this manner, it may be rolled over and flown with positive lift to effect the skip. This strategy was employed for the Apollo lunar return.[15]

As implied earlier, a skip entry sequence was possible with the Apollo command module and was initially selected as the nominal entry mode. Refinement of the entry guidance and targeting philosophy ultimately led to the use of a modulated-lift entry in which a full skip-out was avoided in favor of a trajectory that retained aerodynamic control throughout a nominal entry. However, the full skip phase was still available for trajectory control in the event of an off-nominal entry.[15] Because of the conservative aerothermodynamic design of the Apollo vehicle, heating loads were not a factor in entry corridor definition.

We will have additional comments on skip entry when aerocapture and aerobraking from supercircular entry conditions are discussed in a later section.

Cross-Range Maneuvers

Thus far we have assumed that the entry trajectory lies in the plane of the initial orbit. However, if a lifting entry vehicle is banked (lift vector rotated out of the vertical plane defined by r, V), then a force normal to the

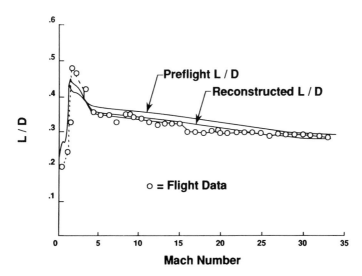

Fig. 6.7 Apollo spacecraft lift/drag ratio.

original orbit plane is generated and the vehicle flies a three-dimensional trajectory. This may be done with both gliding and skip entry profiles as discussed earlier.

The dynamics of three-dimensional flight within the atmosphere are beyond the intended scope of this text, and we shall not engage in a detailed study of lateral maneuvers. As with the planar trajectories discussed previously, however, first-order results are available[16] that yield considerable insight into the effect of banking maneuvers. Since cross-range control is often of interest even in the preliminary stages of entry vehicle and trajectory design, we will consider here some results of first-order three-dimensional entry analysis.

It is usually of interest to examine the maximum "footprint," or envelope of possible landing points, to which an entry vehicle can be steered. In-plane or down-range control for a lifting vehicle is attained through modulation of the lift-to-drag ratio. As seen in Eq. (6.26), maximum range is attained with flight at the highest available L/D. A landing at lesser range can be achieved by flying energy-dissipating maneuvers up and down in the entry plane, or back and forth across the initial plane. Cross-range maneuvers that do not cancel result in a lateral offset of the landing point, at some expense in downtrack range.

To effect a lateral maneuver, a lifting entry vehicle must bank to obtain a turning force normal to the initial plane and, upon attaining the desired heading change, reduce the bank angle again to zero. For maximum lateral range, the bank angle modulation must be performed in such a way that the downtrack range is not unduly reduced, or else the cross-range maneuver will not have time to achieve its full effect. There will thus be an optimum bank angle history that allows the maximum possible cross-range maneuver for a given down-range landing point.

For analytical purposes, the optimum, time-varying, bank angle history must be replaced by an equivalent constant value that provides similar results while allowing integration of the equations of motion. Although justified by the mean value theorem of integral calculus, this procedure renders invalid any consideration of the trajectory history, preserving only the maximum capability information. If in addition we assume an equilibrium glide with small changes in heading angle, first-order (Eggers)[16] and second-order (Vinh et al.)[5] results for maximum lateral range may be obtained. To first order, the angular cross range is

$$\phi = (\pi^2/48)(L/D)^2 \sin 2\sigma \qquad (6.37)$$

where

σ = optimum constant bank angle

ϕ = "latitude" angle attained relative to great-circle "equatorial" plane of initial entry trajectory

and, in terms of distance over the planetary surface, we have

$$s_\perp = R\phi \qquad (6.38)$$

Figure 6.8 shows the geometry for a cross-range maneuver.

The use of $(L/D)_{max}$ in Eq. (6.37) implies, as with the planar equilibrium glide, that maximum cross range is achieved with flight at maximum L/D. Note that the Eggers solution yields $\sigma = 45$ deg for the optimum constant bank angle. This is intuitively reasonable, since it implies that use of the vehicle lift vector is evenly divided between turning ($\sigma = 90$ deg) and staying in the air ($\sigma = 0$ deg) long enough to realize the result of the turn.

Vinh et al.[5] find that Eq. (6.37) overpredicts the cross-range travel that can be achieved with a given vehicle L/D. This is shown in Fig. 6.9, which compares the Eggers solution,[16] the second-order result of Vinh et al.,[5] and the cross range achieved with the true optimal bank angle history. It is seen that, for a vehicle L/D of 1.5 or less (small enough that the maximum achievable cross-range angle remains relatively small), the theories are in reasonable agreement.

As an example, consider the Gemini spacecraft with, as stated previously, an L/D of 0.2. Equations (6.37) and (6.38) yield a maximum cross-range travel of 52.4 km, or about 28.3 n.mi. This is in excellent agreement with the actual Gemini vehicle footprint data, shown in Fig. 6.10.[8]

Loh's Second-Order Solution

The first-order atmospheric entry results presented earlier adequately demonstrate the various entry profiles that can be obtained for particular initial conditions and vehicle parameters. However, these solutions have two important limitations that preclude their use for anything more sophisticated than initial conceptual design.

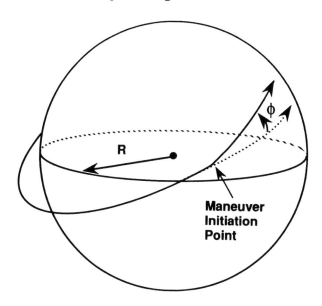

Fig. 6.8   Crossrange re-entry geometry.

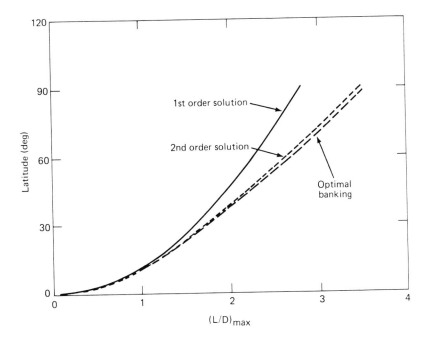

Fig. 6.9   Crossrange capability for varying lift-drag ratio.

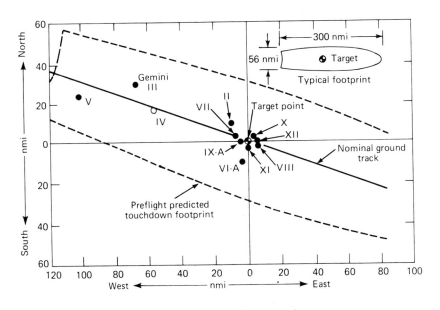

Fig. 6.10   Gemini landing footprint.

The obvious restriction of the first-order solutions is their accuracy. All results derived thus far incorporate the assumptions of constant gravitational force, strictly exponential atmosphere, and fixed reference altitude. These assumptions were discussed in Sec. 6.1 and contribute a total error on the order of 5%. An additional modeling error in the integrated results is the assumption of constant vehicle ballistic coefficient and L/D.

However, the more troublesome aspect of the first-order theory is the fact that one must know in advance what type of entry is under consideration. For example, the vehicle is assumed to be in either a steep ballistic or a shallow gliding trajectory. Shallow-angle ballistic entry and semiballistic entry at low but nonzero L/D do not readily fit into either category. A first-order theory that can address a wide range of entry interface conditions and vehicle parameters is not available.

As mentioned, second-order theories exist. We have also stated that in our view the implementation of higher-order solutions to approximate equations is more troublesome and has less benefit than direct numerical integration of the basic governing equations. There is one possible exception to this policy that is worthy of consideration here, and that is Loh's empirically derived second-order theory.[17] We include a brief discussion of the theory more for its pedagogical value than for any consideration of its practical utility.

Recall in Eq. (6.11b) that the term

$$\Delta \equiv (1/\rho\beta R)(gR/V^2 - 1)\cos\gamma \tag{6.39}$$

represents the difference between the gravitational and centrifugal force normal to the entry flight path. The various first-order solutions assume either that this term is zero (ballistic, skip) or that the rate of change of the flight-path angle is small (equilibrium glide). Another possible assumption, intuitively reasonable but discovered by Loh through a comprehensive numerical investigation, is that Δ is a nonzero constant throughout entry. Modeling errors are then due to departures from this value rather than to complete omission of the term and are of second order.

With Loh's assumption of constant Δ, Eq. (6.11b) becomes

$$d(\cos\gamma)/d\rho \simeq (1/2\beta)(SC_D/m)(L/D) - \Delta \tag{6.40}$$

and integrates immediately to yield

$$\cos\gamma = \cos\gamma_e + [(1/2\beta)(SC_D/m)(L/D) - \Delta](\rho - \rho_e) \tag{6.41}$$

To integrate the tangential equation, we introduce the additional assumption, used previously in the first-order analysis, that the tangential component of gravitational force is negligible in comparison with drag. This allows us to drop $(2/\rho\beta R)$ in Eq. (6.11a) and write

$$\sin\gamma \, d(V^2/gR)/d\rho \simeq (1/\beta)(SC_D/m)(V^2/gR) \tag{6.42}$$

As discussed previously, this assumption is quite reasonable everywhere except near the entry interface, where drag is small and comparable to the tangential gravitational force. And, indeed, it is found that the major area of difficulty with Loh's analysis lies in obtaining a match between the endo- and exoatmospheric trajectories at the entry interface.

From Eq. (6.40), note that

$$\Delta = -\mathrm{d}(\cos\gamma)/\mathrm{d}\rho + (1/2\beta)(SC_D/m)(L/D)$$

$$= \sin\gamma \; \mathrm{d}\gamma/\mathrm{d}\rho + (1/2\beta)(SC_D/m)(L/D) \qquad (6.43)$$

or

$$\sin\gamma \; \mathrm{d}\gamma/\mathrm{d}\rho = \Delta - (1/2\beta)(SC_D/m)(L/D)$$

$$= \Lambda = \mathrm{const} \qquad (6.44)$$

Dividing this result into Eq. (6.42) gives

$$\mathrm{d}(V^2/gR)/\mathrm{d}\gamma = (1/\Lambda\beta)(SC_D/m)(V^2/gR) \qquad (6.45)$$

which immediately integrates to

$$\ell n[(V^2/gR)/(V_e^2/gR)] = (1/\Lambda\beta)(SC_D/m)(\gamma - \gamma_e) \qquad (6.46)$$

As written, Eqs. (6.41) and (6.46) are not materially different from Eqs. (6.28) and (6.30). Taking $\Delta \equiv 0$ reproduces the first-order solution for skip or nonequilibrium gliding entry. However, substituting the definition of Δ into Eqs. (6.41), (6.43), and (6.46) allows the flight-path angle

$$\cos\gamma = \frac{\cos\gamma_e + (1/2\beta)(L/D)(SC_D/m)(\rho - \rho_e)}{1 + (1/\beta R)(gR/V^2 - 1)(1 - \rho_e/\rho)} \qquad (6.47)$$

to be determined. The density can be found as a function of flight-path angle and velocity as

$$\rho - \rho_e = [(gR/V^2 - 1)(m/SC_D)(\cos\gamma/R) \; \ell n(V^2/V_e^2)]$$

$$\div [(\gamma - \gamma_e) + \tfrac{1}{2}(L/D)\ell n(V^2/V_e^2)] \qquad (6.48)$$

Equations (6.47) and (6.48) constitute Loh's second-order solution for atmospheric entry. They are a coupled set of transcendental equations connecting the three variables V, γ, and ρ. Typically, we wish to specify the density (or altitude) and obtain velocity and flight-path angle as a result. This requires a numerical root-finding scheme that we consider to be as complex as directly integrating the differential equations of motion. The solution is slightly simpler, computationally, if we are able to make the usual assumption that $\rho_e \simeq 0$ in Eq. (6.47).

Loh[17] shows excellent results with this theory over a wide range of entry flight conditions and vehicle parameters. Speyer and Womble[18] have verified this conclusion numerically during their investigation of three-dimensional trajectories. The authors perform an interesting variation of Loh's analysis in which Δ is explicitly included as a constant in the differential equations of motion, which are then integrated numerically. Speyer and Womble show that, by periodically updating Δ with recent trajectory values, even better results than claimed by Loh can be obtained.

6.2 FUNDAMENTALS OF ENTRY HEATING

Up to this point we have considered only the particle dynamics of atmospheric entry, wherein the vehicle is completely characterized by its L/D and ballistic coefficient. This determines the flight trajectory and allows assessment of the vehicle acceleration and dynamic pressure loads, the down-range and cross-range travel, and the sensitivity of these quantities to the entry conditions and vehicle parameters.

Of equal importance are the thermal loads imposed on the vehicle during entry. These are of two types: the total heat load and the instantaneous heating rate. The total heat load is obviously a concern in that the average vehicle temperature will increase with the energy input. The allowed heating rate, either local or body-averaged, is a concern because of the thermal gradient induced from a heat flux according to Fourier's law:

$$q = -\kappa \nabla T \qquad (6.49)$$

where

q = power per unit area, W/m^2
κ = thermal conductivity, W/mK
∇T = gradient of temperature, K/m

In materials with a nonzero coefficient of thermal expansion, a temperature gradient causes differential expansion and mechanical stress in the vehicle wall material.

Tradeoffs between allowed heating rate and total heat load are often necessary. Sustained high-energy flight at high altitude (e.g., gliding entry) reduces the instantaneous heating rate but, by extending the duration of the flight, may unacceptably increase the total heat absorbed. A more rapid, high drag entry usually reduces the total energy input at the expense of incurring a very high local heating rate and may in addition result in unacceptable dynamic loads.

Entry vehicle heating results from the dissipation of the initial total (kinetic plus potential) energy through two heat-transfer mechanisms, convection and radiation. Convective heating occurs when the air, heated by passage through a strong bow shock in front of the vehicle, bathes the wall in a hot fluid stream. If the air is hot enough, significant thermal radiation will occur as well. Radiative heat transfer is important when the entry velocity is greater than about 10 km/s and may be significant at considerably lower speeds.

Peak aerodynamic heating will usually occur in stagnation point regions, such as on a blunt nose or wing leading edge. However, turbulent flow along the vehicle afterbody can under some conditions produce a comparable or greater heat flux. Conversely, delayed onset of turbulence (i.e., turbulent transition at a higher than expected Reynolds number) can produce a substantially cooler aft body flow than expected.

Thermal control is a major entry vehicle design challenge. As Regan[4] notes, the specific kinetic energy that is dissipated during entry from low Earth orbit is about 3×10^7 J/kg. This is sufficient to vaporize a heat shield composed of pure carbon ($h_v = 6 \times 10^7$ J/kg) equal to half the initial vehicle mass. If this is to be prevented, then the major portion of the entry kinetic and potential energy must be deflected to the atmosphere rather than the vehicle. A good aerothermodynamic design will allow only a few percent of this energy to reach the vehicle.

Thermal Protection Techniques

The design and analysis of an entry vehicle and flight profile to meet the thermal protection requirement is a multidiscipline task involving aerodynamics, chemistry, flight mechanics, structural analysis, and materials science. Three basic approaches to entry vehicle thermal control have evolved: heat sinking, radiative cooling, and ablative shielding.

The heat sinking technique, as the name implies, uses a large mass of material with a high melting point and heat capacity to absorb the entry heat load. The initial Mercury spacecraft design utilized this approach, employing a beryllium blunt body heat shield. This design was used on the Mercury-Redstone suborbital flights. However, the increased system weight for protection against the order-of-magnitude higher orbital entry heat load forced the use of an ablative shield on those missions. This is a typical and important limitation of the heat sink approach to entry thermal control.

The principle of radiative cooling is to allow the outer skin of the vehicle to become, literally, red-hot due to the convectively transferred heat from the flowfield around the vehicle. Blackbody radiation, primarily in the infrared portion of the spectrum, then transports energy from the vehicle to the surrounding atmosphere. Convective heating to the vehicle is proportional to the temperature difference between the fluid and the wall, whereas the energy radiated away is in proportion to the difference in the fourth powers of the fluid and wall temperatures. The net result is that thermal equilibrium can be reached at a relatively modest skin temperature provided that the rate of heating is kept low enough to maintain near-equilibrium conditions.

Radiative cooling obviously requires excellent insulation between the intense hot outer shell and the internal vehicle payload and structure. This is exactly the purpose of the Shuttle tiles, the main element of the Shuttle thermal protection system. Essentially a porous matrix of silica (quartz) fibers, these tiles have such low thermal conductivity that they can literally be held in the hand on one side and heated with a blowtorch on the other.

As stated, radiative cooling relies on equilibrium, or near-equilibrium, between the entry vehicle and its surroundings in order to shed the absorbed heat load. This is most easily achieved in a lengthy, high-altitude gliding entry where the instantaneous heating rate is minimized as the speed is slowly reduced. The vehicle aerodynamic design (ballistic coefficient and L/D), the entry flight trajectory, and the heat shield material selection are intimately related when radiative cooling is used. This complicates the design problem; however, significant mass savings are possible when a system-level approach is taken.[19]

A potential problem with an insulated, radiatively cooled vehicle having a lengthy flight time is that ultimately some heat will soak through to the underlying structure. Coolant fluid may thus need to be circulated through the vehicle so that this energy can be radiated away to a portion of the surroundings, such as the aft region, which is cooler. This can occur even if the flight time is sufficiently short that in-flight cooling is not required. Such is the case with the Shuttle, which must be connected to cooling lines from ground support equipment if postflight damage to the aluminum structure is to be prevented.

Although heat sinking is best suited to a brief, high drag entry and radiative cooling is more appropriate for a gliding trajectory, ablative cooling offers considerably more flexibility in the flight profile definition. Ablation cooling is also typically the least massive approach to entry heat protection. These advantages accrue at the expense of vehicle (or at least heat shield) reusability, which is a pronounced benefit of the other techniques.

Ablative cooling occurs when the heat shield material, commonly a fiberglass-resin matrix, sublimes under the entry heat load. When the sublimed material is swept away in the flowfield, the vehicle is cooled. This process can produce well over 10^7 J/kg of effective energy removal. Ablative cooling has been the method of choice for most entry vehicles, including the manned Mercury, Gemini, and Apollo vehicles.

Entry Heating Analysis

From the theory of viscous fluid flow[20] it is known that the flowfield about an atmospheric entry vehicle develops a thin boundary layer close to the body to which viscous effects, including skin friction and heat transfer, are confined. The heat flux to the wall is proportional to the local temperature gradient,

$$q_w = \kappa (\partial T / \partial y)_w = \varepsilon \sigma T_w^4 \qquad (6.50)$$

where y is the coordinate normal to the wall. The temperature gradient is obtained from the boundary-layer flowfield solution, determined from the boundary-layer edge properties and wall conditions. The edge conditions in turn follow from the inviscid solution for the flow over the entry vehicle. The vehicle heat-transfer analysis is thus dependent on knowledge of the flowfield. The right-hand equality in Eq. (6.50) implies that iteration between the convective and radiative heat flux equations will in general be necessary to fix the equilibrium wall temperature.

The difficulty of obtaining an accurate solution for the high-speed flowfield around an entry vehicle can hardly be overstated. The fluid is a chemically reacting gas, possibly not in equilibrium, probably ionized, and with potentially significant radiative energy transfer. Vehicle surface properties such as roughness and wall catalycity influence the flowfield and heat-transfer analysis.

The entry flight regime is equally demanding of an experimental approach. It is at present impossible to conduct a wind-tunnel experiment that simultaneously provides both Mach and Reynolds numbers appropriate to entry flight. Thus, the Space Shuttle received the first true test of its performance during its first flight, a potentially hazardous situation, since, unlike its predecessors, the Shuttle was not flight tested in an unmanned configuration.

A recurrent theme in this text is that recourse to all available analytical sophistication is desirable, even essential, prior to critical design and development. However, preliminary design and mission feasibility assessment would be virtually impossible without the use of simpler, less accurate, techniques. Accordingly, we rely on an approach to entry heating analysis first given by Allen and Eggers.[10] This approach assumes the primary source of energy input to be convective heating from the laminar boundary-layer flow over the entry vehicle. In this case, the local heating rate as given by Eq. (6.50) may be correlated with the total enthalpy difference across the boundary layer:[20]

$$q_w = \kappa(\partial T/\partial y)_w = (\kappa/C_p)(Nu_L/L)(H_{oe} - H_w)$$

$$= (Nu_L/Pr)(\mu/L)H_{oe}(1 - H_w/H_{oe}) \qquad (6.51)$$

The subscript oe denotes local flow conditions at the outer edge of the boundary layer and w the local wall values. Nu_L is the Nusselt number, based on an appropriate length scale L, for the particular boundary-layer flow in question. The Prandtl number, $Pr = \mu C_p/k$, ranges from 0.71 to 0.73 for air below 9000 K, but is often taken as unity for approximate calculations.

Total enthalpy is conserved for the inviscid flow across the normal shock portion of the bow wave; since it is this stream that wets the body:

$$H_{oe} = H = h + V^2/2 = C_p T + V^2/2 \cong V^2/2 \qquad (6.52)$$

The unsubscripted parameters denote, as usual, freestream or approach conditions. The right-hand approximate equality follows from the high-speed, low-temperature nature of the upstream flow. For example, assume an entry vehicle at 80 km altitude with $T = 200$ K, $V = 6000$ m/s, and $C_p = 1005$ J/kg K. Then $V^2/2C_p T = 90$, and the thermal energy content of the air provides a negligible contribution to the total enthalpy.

Multiplying and dividing Eq. (6.51) by $(\rho V)_{oe}$ yields an equivalent result,

$$q_w = (\rho V)_{oe}(Nu_L/PrRe_L)H_{oe}(1 - H_w/H_{oe}) = (\rho V)_{oe}StH_{oe}(1 - H_wH_{oe})$$
$$(6.53)$$

where $St = Nu/PrRe$ is the Stanton number, a local heat-transfer coefficient. The derivation of this result makes it clear that the Reynolds number Re is referenced to the same (as yet unspecified) length scale as the Nusselt number and to boundary-layer edge values of density and velocity.

Equation (6.53) offers no particular simplification; however, exploiting the Reynolds analogy for laminar boundary-layer flow,[20] we note that

$$St \cong C_f/2 \tag{6.54}$$

where C_f is the local skin friction coefficient. This approximation is typically valid to within about 20%. For example, White[20] shows that the Reynolds analogy factor $2St/C_f$ varies between 1.24 and 1.27 over the subsonic to Mach 16 range for laminar flow over a flat plate. With the Reynolds analogy and Eq. (6.52), Eq. (6.53) becomes

$$q_w = (\rho V)_{oe} V^2 (1 - H_w/H_{oe}) C_f/4 \tag{6.55}$$

Equation (6.55) shows that the heating rate to the body depends on the local wall temperature through the term $(1 - H_w/H_{oe})$. Since the flow is stagnant at the wall, $H_w \cong C_p T_w$, with the equality exact if T_w is low enough (below about 600 K) that the gas may be assumed calorically perfect. It is a conservative assumption, consistent with other approximations adopted here, to assume the wall sufficiently cool that H_w/H_{oe} is small. The heating rate is then

$$q_w \cong (\rho V)_{oe} V^2 C_f/4 \tag{6.56}$$

and we see that for a reasonably cool wall, the gross heat-transfer rate is independent of the body temperature.

Integration over the body wall area S_w gives the total heating rate (power input) to the body,

$$Q = \rho V^3 S_w C_F/4 \tag{6.57}$$

where C_F is the body-averaged skin friction coefficient defined as

$$C_F = (1/S_w) \int C_f [(\rho V)_{oe}/\rho V] \, ds \tag{6.58}$$

Again, the subscript oe denotes local boundary-layer outer edge values. As usual, the upstream or approach velocity V is found as a function of density ρ from trajectory solutions such as those obtained in Sec. 6.1. Skin friction coefficient calculations are discussed in more detail in a subsequent section.

Total Entry Heat Load

The total heat load (energy) into the vehicle can be obtained from

$$dE/dV = (dE/dt)(dt/dV) = Q \, dt/dV$$

$$= 2Q(m/SC_D)(1/\rho V^2)$$

$$= \tfrac{1}{2}(m/SC_D)S_w C_F V \qquad (6.59)$$

where as usual we have dropped the tangential gravitational force in Eq. (6.1a). Upon integrating from the entry velocity to the final velocity,

$$E = \tfrac{1}{4}m(V_e^2 - V_f^2)(S_w C_F/SC_D) \qquad (6.60)$$

If, as is usually the case, the final velocity is effectively zero, the total heat load has the particularly simple form

$$E/(\tfrac{1}{2}mV_e^2) = \tfrac{1}{2}S_w C_F/SC_D \qquad (6.61)$$

Equation (6.61) is valid with any entry profile (ballistic, glide, or skip) and for any vehicle sufficiently "light" that it slows before hitting the ground. This is the same requirement as for a deceleration peak with ballistic entry, and Eq. (6.19) may therefore be used to define a "light" vehicle. A dense vehicle on a steep ballistic trajectory may fail to meet this criterion. If this is the case, the first-order ballistic entry velocity profile given by Eq. (6.15) will be quite accurate and may be substituted in Eq. (6.60) to yield

$$E/(\tfrac{1}{2}mV_e^2) = \tfrac{1}{2}(S_w C_F/SC_D)\{1 - \exp[(\rho_s/\beta \, \sin\gamma_e)(SC_D/m)]\} \qquad (6.62)$$

By failing the light vehicle criterion of Eq. (6.19), the exponent has magnitude less than unity, and the expansion $e^\alpha \simeq 1 + \alpha$ may be employed inside the brackets to yield the "heavy body" result,

$$E/(\tfrac{1}{2}mV_e^2) \simeq -(\tfrac{1}{2}m)(S_w C_F \rho_s)/(\beta \, \sin\gamma_e) \qquad (6.63)$$

Equation (6.61) provides the rationale for the classical blunt body entry vehicle design. The total heat load is minimized when the skin friction drag C_F is small compared to the total drag C_D, and the wetted area S_w is as small as possible in comparison with the reference projected area S. Both of these conditions are met with an entry vehicle having a rounded or blunt shape.

Equation (6.63) shows that a dense ballistic entry vehicle should have a slender profile to minimize the total skin friction and hence the heat load.

Entry Heating Rate

The body-averaged heated rate is also of interest and is found from Eq. (6.57):

$$q_{avg} \equiv Q/S_w = \tfrac{1}{4}\rho V^3 C_F \tag{6.64}$$

The average heating rate can be found once a trajectory profile giving velocity as a function of atmosphere density (and hence altitude) is specified. This can be done as an adjunct to a numerical solution, or by substituting the previously obtained first-order results for ballistic, equilibrium glide, and skip trajectories. Using this latter approach, we obtain

$$q_{avg} = \tfrac{1}{4}C_F V_e^2 \rho \, \exp[(3/2)(SC_D/m)(\rho/\beta \sin\gamma_e)] \tag{6.65}$$

for the ballistic entry heating rate as a function of density. Similarly,

$$q_{avg} = \tfrac{1}{2}(m/SC_D)(1 - V^2/gR)gC_F V/(L/D) \tag{6.66}$$

gives the heating rate for gliding entry vs velocity. Finally,

$$q_{avg} = (m/SC_D)\beta V_e^3 \exp[-3(\gamma - \gamma_e)/(L/D)](\cos\gamma - \cos\gamma_e)/2(L/D) \tag{6.67}$$

is the heating rate for skip entry as a function of flight-path angle.

It is usually of greatest interest to find the value of the maximum body-averaged heating rate, as well as the altitude (or density) and velocity at which this rate occurs. This maximum heating rate will often constrain the entry trajectory. For ballistic entry, the maximum heating rate and critical trajectory conditions are

$$q_{avg_{max}} = (C_F/6e)(m/SC_D)\beta V_e^3 \sin\gamma_e \tag{6.68}$$

$$\rho_{crit} = -(2/3)(m/SC_D)\beta \sin\gamma_e \tag{6.69}$$

$$V_{crit} = V_e/e^{\frac{1}{3}} \tag{6.70}$$

For equilibrium gliding entry, we find

$$q_{avg_{max}} = (g^3 R/27)^{\frac{1}{2}}(m/SC_D)/(L/D) \tag{6.71}$$

$$\rho_{crit} = (4/R)(m/SC_D)/(L/D) \tag{6.72}$$

$$V_{crit} = (gR/3)^{\frac{1}{2}} \tag{6.73}$$

The corresponding parameters are slightly more difficult to obtain for skip entry. To obtain explicit algebraic results it is necessary to assume small γ_e.[5] This assumption is nearly always satisfied, and the results are

$$q_{avg_{max}} \cong (\beta/2)(m/SC_D)\gamma_e^2 V_e^3 \exp[3\gamma_e/(L/D)]/(L/D) \tag{6.74}$$

$$\rho_{\text{crit}} \simeq 2\beta\gamma_e (m/SC_D)/(L/D) \tag{6.75}$$

$$V_{\text{crit}} \simeq V_e \exp[\gamma_e/(L/D)] \tag{6.76}$$

$$\gamma_{\text{crit}} \simeq -3\gamma_e^2/(L/D) \tag{6.77}$$

We urge caution in the application of the results given here. The heat load calculations of this section implicitly incorporate the approximations in the trajectory solutions for ballistic, gliding, or skip entry. For example, we have seen that the first-order ballistic entry analysis underpredicts the acceleration load for shallow entry angles. Since this result is incorporated in Eqs. (6.60) and (6.65), it is expected that at shallow entry angles the ballistic entry heating rate would be underpredicted and the total heat load overpredicted. This is the case, as shown in Fig. 6.11, which compares the first-order theory with that of Chapman.[3]

Moreover, the entry heating analysis is itself approximate, since it assumes laminar boundary-layer heating, invokes the Reynolds analogy to eliminate the Stanton number, ignores radiant energy input, and neglects vibrational and chemical excitation ("real gas effects") in the gas. Collectively, these assumptions are quite valid at low speeds, below about 2 km/s, and become progressively less so as typically atmospheric entry speeds are approached. There are some mitigating effects; for example, neglect of radiant heating partially offsets the calorically perfect gas assumption.

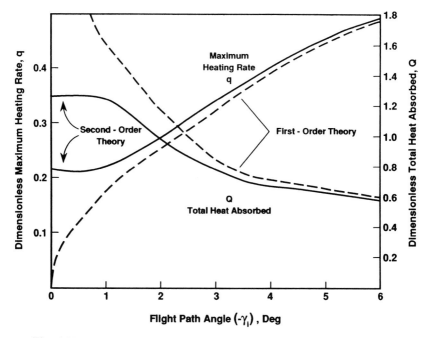

Fig. 6.11   Comparison of first- and second-order entry heating analysis.

High-altitude entry flight, with its attendant low Reynolds number, tends to favor laminar flow, particularly for short, blunt vehicles.

Furthermore, we must point out that all too frequently even the most sophisticated calculations yield poor accuracy. Prabhu and Tannehill[21] compared Shuttle flight data with theoretical heat-transfer results using a state-of-the-art flowfield code together with both equilibrium air and calorically perfect gas models. It was found that, provided a proper value of k (the ratio of specific heats) is chosen (the authors recommend $k = 1.2$), the calorically perfect model does as well as the equilibrium air model. In some cases substantially better agreement with flight data was obtained with the simpler model!

Space Shuttle flight experience further illustrates the points given earlier. For STS missions 1-5, Williams and Curry[22] show generally excellent agreement between preflight analysis and flight data, particularly in the higher-temperature regions. Heating in the cooler, leeside areas (where the flowfield is typically quite complex) has been significantly lower than preflight predictions, even those based on wind-tunnel data. Throckmorton and Zoby[23] attribute this to delayed onset of turbulent flow as compared with subscale wind-tunnel test results. Scott[24] discusses the effects of wall catalysis on Orbiter heat transfer and notes that the heat flux has increased from flight to flight as the Shuttle tile properties change with age and use.

Cumulative uncertainties as to model validity argue for due caution in interpreting the results of all heat-transfer analysis. We regard entry heating analysis as presented here to be an order-of-magnitude theory, useful in preliminary design but unsuited for detailed work. Even detailed calculations are not generally regarded as accurate to better than 10%.

Skin Friction Coefficient

The body-averaged skin friction coefficient C_F is seen to be a key parameter in determining both the heating rate and the total heat load for an entry vehicle. As shown by Eq. (6.58), C_F is determined by integration of the local skin friction coefficient C_f over the body. C_f is defined by

$$C_f = 2\tau_w/(\rho V^2)_{oe} \qquad (6.78)$$

where

$$\tau_w = \mu(\partial V/\partial y)_w \qquad (6.79)$$

is the boundary-layer shear stress at the wall.

Clearly, the boundary-layer flowfield solution must be known to evaluate the wall shear stress and skin friction coefficient. Since the skin friction coefficient was introduced in Sec. 6.2 (Entry Heating Analysis) to avoid precisely this difficulty, further approximation is required. To this end, we include some results from laminar boundary-layer theory, which, when used with judgment, allow estimation of C_F for preliminary vehicle design.

From low-speed boundary-layer theory we have the classical result for incompressible laminar flow over a flat plate[25] that

$$C_f = 0.664/Re_x^{\frac{1}{2}} \qquad (6.80)$$

where Re_x is the Reynolds number referenced to boundary-layer edge conditions and to the x or streamwise coordinate as measured from the leading edge of the plate:

$$Re_x = (\rho V)_{oe} x / \mu \qquad (6.81)$$

The streamlines that wet the outer edge of the boundary layer obey the steady flow continuity result

$$(\rho V)_{oe} = \rho V \qquad (6.82)$$

Combining Eqs. (6.82) and (6.58) and integrating over a plate of unit width and length L yields the low-speed result

$$C_F = 1.328 / Re_L^{\frac{1}{2}} \qquad (6.83)$$

Flat-plate theory is useful in aerodynamics because most portions of a flight vehicle are of a scale such that the local body radius of curvature dwarfs the boundary-layer thickness. Thus, most of the body appears locally as a flat plate, and good approximate results for skin friction can be obtained by ignoring those portions, small by definition, which do not. This assumption can be invalid for flight at very high altitude, where the reduced density lowers the Reynolds number and produces a thicker boundary layer.

Equation (6.83) can be extended to high-speed, hence compressible, flow through the reference-temperature approach.[20] It is found that, in the worst case (adiabatic wall), $C_F / \sqrt{(Re_L)}$ varies from 1.328 at low speed to approximately 0.65 at Mach 20. Compressibility thus has an important but not overwhelming effect on skin friction coefficient and, for entry heating calculations such as presented here, may with some justification be ignored or included in an ad hoc fashion. In any case, the use of the low-speed value is conservative from an entry heating viewpoint.

Stagnation Point Heating

Both the total heat load and the body-averaged heating rate are important in entry analysis, since either may constrain the trajectory. Their relative importance will depend on the entry profile and vehicle parameters, and, as we have mentioned, relief from one is usually obtained by aggravating the other.

Of equal importance is the maximum local heating rate imposed on any part of the entry vehicle, which determines the most severe local thermal protection requirement. With the possible exception of local afterbody hot spots due to turbulent effects and shock–boundary-layer interactions, the body heating rate is maximized at the stagnation point. Any realistic vehicle will have a blunt nose or wing leading edge, and this will be a region of stagnation flow, shown schematically in Fig. 6.12.

The stagnation region behind a strong normal shock is one of particularly intense heating. For example, at an entry speed of Mach 25 the perfect

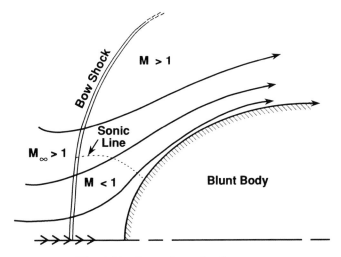

Fig. 6.12 Stagnation point flow.

gas shock tables[26] yield $T_{t_2}/T_1 = 126$, where T_1 is the freestream static temperature and T_{t_2} is the stagnation temperature behind the shock. Assuming $T_1 = 166$ K for the standard atmosphere at 80 km, the total temperature behind the shock is 20,900 K! For comparison, the surface temperature of the sun is approximately 5780 K.

This temperature is of course not attained. Its calculation assumes the atmosphere to be a calorically perfect gas for which the enthalpy and temperature are related by

$$h = C_p T \qquad (6.84)$$

where the heat capacity C_p is a constant, 1005 J/kg K for air. In fact, a major fraction of the available thermal energy is used to dissociate and ionize the air molecules, effectively increasing the heat capacity of the gas and lowering the stagnation temperature. For a Shuttle entry, the nose cap stagnation region reaches a peak temperature of approximately 2500°F (1644 K).[22]

The preceding example is interesting and informative regarding attempts to predict individual flowfield properties for high-speed and therefore high-energy flows. A cautionary note should be added, however. The wall heat flux q_w is the parameter of importance in entry vehicle design and is driven by the total enthalpy difference $(H_{oe} - H_w)$ between the wall and the outer edge of the boundary layer. The temperature difference is not the relevant parameter, despite what Eq. (6.50) would imply. For a calorically perfect gas, where Eq. (6.84) applies, no distinction between temperature and enthalpy need be made. In a chemically reacting gas, dissociation and ionization will alter the balance between effective heat capacity and temperature and thus significantly affect the flowfield. However, the net effect on the boundary-layer flowfield total enthalpy difference $(H_{oe} - H_w)$ and hence the wall heat flux may be small.

The implication is that the neglect of real gas effects, although horrifying to a physical gas dynamicist, may be fairly reasonable for our purposes. This is especially true when chemical equilibrium exists in the boundary-layer flowfield, an approximation that is reasonable in the stagnation region. The assumption of a nonequilibrium boundary-layer flow with a fully catalytic wall, so that surface equilibrium exists by definition, yields similar results.

Our approximate analysis follows that of Sec. 6.2 (Entry Heating Analysis), with Eq. (6.51) again applicable:

$$q_w = (Nu_L/Pr)(\mu/L)H_{oe}(1 - H_w/H_{oe}) \tag{6.85}$$

Previously we rearranged this equation to employ the Stanton number instead of the Nusselt number, then used Reynolds' analogy to cast the results in terms of the skin friction coefficient. This was done because skin friction data are more easily obtained (if only empirically) and generalized than are heat-transfer data when the complete flowfield solution is not available. However, the boundary-layer flow in the stagnation region shown in Fig. 6.12 is sufficiently well understood that a more direct approach is possible.

In the low-speed stagnation region behind a strong bow shock, incompressible flow theory applies. For such a flow over a rounded nose or wing leading edge, the Nusselt number is found to be[27]

$$Nu_L = \eta Pr^{\frac{2}{5}}(K\rho/\mu)^{\frac{1}{2}}L \tag{6.86}$$

where K is the stagnation point velocity gradient in the x, or streamwise, direction at the edge of the boundary layer:

$$K = (dV_{oe}/dx)_{sp} \tag{6.87}$$

and the subscript sp denotes stagnation point conditions. For axisymmetric flow, $\eta = 0.763,$[27] whereas for two-dimensional flow, as over a wing leading edge, $\eta = 0.570.$[20] Employing Eq. (6.52), Eq. (6.85) now becomes

$$q_w = (\eta/2)Pr^{-0.6}(\rho_{oe}\mu_{oe})^{\frac{1}{2}}_{sp}V^2(1 - H_w/H_{oe})(dV_{oe}\,dx)^{\frac{1}{2}}_{sp} \tag{6.88}$$

The stagnation point velocity gradient $(dV_{oe}/dx)_{sp}$ is evaluated for high-speed flow by combining the Newtonian wall pressure distribution with the boundary-layer momentum equation and the inviscid flow solution at the stagnation point. This yields[20]

$$K = (dV_{oe}/dx)_{sp} = (V/R_n)(2\rho/\rho_{oe})^{\frac{1}{2}} \tag{6.89}$$

where R_n is the nose radius of curvature. The term (ρ/ρ_{oe}) is the density ratio for the inviscid flow across a normal shock at upstream Mach number M:

$$\rho/\rho_{oe} = [(k-1)M^2 + 2]/(k+1)M^2 \qquad (6.90)$$

This ratio varies from unity at Mach 1 to $(k-1)/(k+1)$ at Mach infinity.

Equation (6.89) provides the well-known result that stagnation point heating varies inversely with the square root of the nose radius. This does not indicate that a flat nose eliminates stagnation point heating; the various approximations employed invalidate the model in this limiting case. It remains true, however, that stagnation point heating scales with leading-edge radius of curvature as given earlier.

Equation (6.88) is a perfect gas result and omits the effects of vibrational and chemical excitation. The landmark analysis of stagnation point heating including these effects was given by Fay and Riddell[28] and later extended by Hoshizaki[29] and by Fay and Kemp[30] to include the effects of ionization. Experimental work in support of these theories includes that of Rose and Stark[31] and Kemp et al.[32] We summarize here the important conclusions from this work.

Fay and Riddell[28] found the stagnation point heat flux for a nonradiating "binary gas" consisting of atoms (either O or N) and molecules (N_2 or O_2) to be

$$q_w = (\eta/2)Pr^{-0.6}(\rho_{oe}\mu_{oe})sp^{0.4}(\rho_w\mu_w)sp^{0.1}V^2(1 - H_w/H_{oe})$$
$$\times (dV_{oe}/dx)^{\frac{1}{2}}_{sp}[1 + (Le^\varepsilon - 1)h_d/H_{oe}] \qquad (6.91)$$

where

ε	$= 0.52$ for equilibrium boundary-layer flow
	$= 0.63$ for frozen flow with fully catalytic wall
	$= -\infty$ for frozen flow with noncatalytic wall
Le	$=$ Lewis number $= 1.4$ for air below 9000 K
h_d	$= \Sigma c_i(\Delta h_f^o)_i =$ average dissociation energy
c_i	$= i$th species concentration
$(\Delta h_f^o)_i$	$= i$th species heat of formation[33]

The Fay and Riddell analysis,[28] which agrees quite well with experimental data for typical Earth orbital speeds, essentially modifies Eq. (6.88) by a factor

$$D = [1 + (Le^\varepsilon - 1)h_d/H_{oe}](\rho_w\mu_w/\rho_{oe}\mu_{oe})^{0.1}_{sp} \qquad (6.92)$$

due to dissociation. Kemp and Riddell[34] show this factor to increase the stagnation heat flux by about 20% over the calorically perfect gas result for entry from low Earth orbit.

A few comments on the use of Eq. (6.91) are in order. The equation is evaluated in the forward direction; i.e., the wall temperature is specified and the heat flux computed. If q_w rather than T_w is known, the wall temperature must be found by iteration.

The freestream density ρ and velocity V are known from the trajectory solution. Specification of density fixes, through the standard atmosphere model, freestream pressure and temperature. Given the wall temperature and state relations for the gas comprising the chemically reacting boundary layer, the quantities ρ_w, μ_w, h_d, and H_w may be computed. The gas properties may be determined from first principles[33] or, with somewhat less effort, found in tables.[35,36] Reasonably accurate empirical relationships such as Sutherland's viscosity law[20] are also useful.

Although hand calculator evaluation of Eq. (6.91) is feasible, it is somewhat tedious, and therefore is to be avoided when possible. Kemp and Riddell[34] have used the Fay and Riddell result to correlate stagnation point heating for entry from Earth orbit as a function of freestream density and velocity and obtain

$$q_w = 20,800 \ \text{Btu/ft}^{1.5}/\text{s} \ (1/R_n)^{\frac{1}{2}}(\rho/\rho_s)^{\frac{1}{2}}$$

$$\times (V/V_{\text{circ}})^{3.25}(1 - H_w/H_{oe}) \qquad (6.93)$$

$V_{\text{circ}} = 26,000$ ft/s and $\rho_s = 0.002378$ slug/ft^3 were employed; the correlation is claimed accurate to within 5%. Combining the various constants and converting the SI units yield for the stagnation heating rate in air

$$q_w = 8.225 \times 10^{-5} \text{kg}^{\frac{1}{2}} \ \text{s}^{\frac{1}{4}} \ \text{m}^{-5/4}(\rho/R_n)^{\frac{1}{2}}V^{3.25}(1 - H_w/H_{oe}) = \qquad (6.94)$$

Thus far we have discussed only continuum flow results; stagnation heating in rarefied flow may be important when considering satellites that orbit, or at least have periapsis, at very low altitudes. Kemp and Riddell give the free molecular heating rate as

$$q_w = 2.69 \times 10^7 \sigma(\rho/\rho_s)(V/V_{\text{circ}})^3 \ \text{Btu/ft}^2/\text{s} \qquad (6.95)$$

where σ is an accommodation coefficient, upper bounded by unity, accounting for the energy transfer efficiency into the vehicle. With the constants combined, the free molecular stagnation point heating rate becomes

$$q_w = 0.5005\sigma\rho V^3 \qquad (6.96)$$

6.3 ENTRY VEHICLE DESIGNS

In previous sections we have seen that the key entry vehicle parameters are the ballistic coefficient, the L/D, and the body radius of curvature at the nose or wing leading edge. The topic of entry vehicle aeroshell design to achieve suitable combinations of these parameters is somewhat beyond

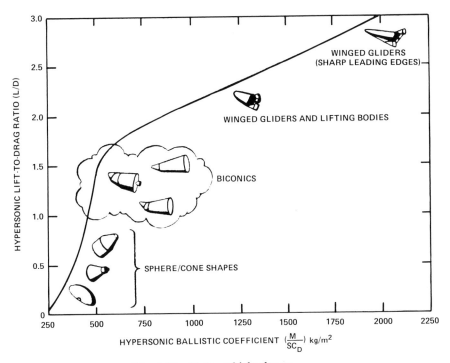

Fig. 6.13   Entry vehicle shapes.

the scope of this book. Consequently, our discussion in this area will be of a qualitative nature only.

Figure 6.13[37] shows vehicle L/D vs C_B for a range of typical entry vehicle aerodynamic designs.

6.4   AEROASSISTED ORBIT TRANSFER

A technique of great promise and extensive current interest for advanced space operations is that of aeroassisted orbit transfer. Many analyses[38] have demonstrated that propulsive requirements for both interplanetary and orbital operations can be significantly reduced with maneuvers that utilize the atmosphere of a nearby planet for braking or plane change ΔV.

Figure 6.14 shows this concept as applied to aerocapture of an interplanetary spacecraft into a low orbit at a target planet. The concept is also applicable for transfer from high orbit to low orbit around a given planet.

As discussed earlier, aerocapture is a technique requiring a fairly sophisticated, high L/D aeroshell design that imposes significant configuration constraints on the internal payload. In return, it offers the maximum flexibility in the entry flight trajectory design and control. Where the entry

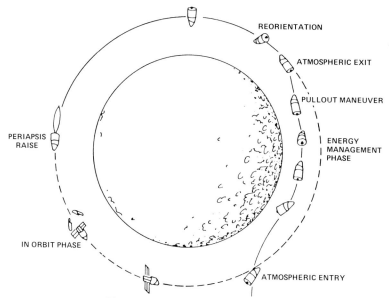

Fig. 6.14   Aerocapture flight plan.

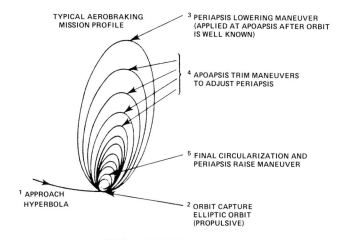

Fig. 6.15   Aerobraking scenario.

TYPICAL AEROBRAKING PARAMETERS

PLANET	VENUS	EARTH	MARS
VEHICLE MASS, kg	1630	1630	1630
DRAG BRAKE DIAMETER, m	9	9	9
INITIAL ORBIT PERIOD, hr	12	12	12
PERIAPSIS ALTITUDE, km	132	86	123
INITIAL APOAPSIS ALTITUDE, km	37,218	18,247	40,033
PERIAPSIS VELOCITY, m/s	9594	4604	10,363
SHIELD TEMP, k	664	571	592
FINAL PERIAPSIS ALTITUDE, km	691	556	692
NUMBER OF ORBITS	244	53	344
ELAPSED TIME, DAYS	44	10	61
ΔV TO CIRCULARIZE, m/s (at 300 km)	205	142	225

requirements are less severe, lower L/D or even ballistic designs may be suitable and usually lead to more advantageous packaging arrangements. Aeroassisted orbit transfer with low or zero L/D is commonly denoted "aerobraking." An application to low-orbit planetary capture is shown in Fig. 6.15. Again, the same scenario can be employed for transfer from high orbit to low orbit about a planet.

References

[1]U.S. Standard Atmosphere, National Oceanic and Atmospheric Administration, NOAA S/T 76-1562, 1976.

[2]*COSPAR International Reference Atmosphere*, Akademie-Verlag, Berlin, 1972.

[3]Chapman, D. R., "An Approximate Analytical Method for Studying Entry into Planetary Atmospheres," NACA TN-4276, 1958.

[4]Regan, F. J., *Re-Entry Vehicle Dynamics*, Education Series, AIAA, New York, 1984.

[5]Vinh, N. X., Busemann, A., and Culp, R. D., *Hypersonic and Planetary Entry Flight Mechanics*, Univ. of Michigan Press, Ann Arbor, MI, 1980.

[6]Ashley, H., *Engineering Analysis of Flight Vehicles*, Addison-Wesley, Reading, MA, 1974.

[7]Platus, D. H., "Ballistic Re-entry Vehicle Flight Dynamics," *Journal of Guidance, Control, and Dynamics*, Vol. 5, Jan.–Feb. 1982, pp. 4–16.

[8]"Guidance and Navigation for Entry Vehicles," NASA SP-8015, Nov. 1968.

[9]Mayer, R. T., "MOSES (Manned Orbital Space Escape System)," *Journal of Spacecraft and Rockets*, Vol. 20, March–April 1983, pp. 158–163.

[10]Allen, H. J., and Eggers, A. J., "A Study of the Motion and Aerodynamic Heating of Missiles Entering the Earth's Atmosphere at High Supersonic Speeds," NACA TR-1381, 1958.

[11]Harpold, J. C., and Graves, C. A., Jr., "Shuttle Entry Guidance," *Journal of the Astronautical Sciences*, Vol. 27, July–Sept. 1979, pp. 239–268.

[12]Harpold, J. C., and Gavert, D. E., "Space Shuttle Entry Guidance Performance Results," *Journal of Guidance, Control, and Dynamics*, Vol. 6, Nov.–Dec. 1983, pp. 442–447.

[13]Romere, P. O., and Young, J. C., "Space Shuttle Entry Longitudinal Aerodynamic Comparisons of Flight 2 with Preflight Predictions," *Journal of Spacecraft and Rockets*, Vol. 20, Nov.–Dec. 1983, pp. 518–523.

[14]Eggers, A. J., Allen, H. J., and Neice, S. E., "A Comparative Analysis of the Performance of Long-Range Hypervelocity Vehicles," NACA TN-4046, 1957.

[15]Graves, C. A., and Harpold, J. C., "Re-Entry Targeting Philosophy and Flight Results from Apollo 10 and 11," AIAA Paper 70-28, Jan. 1970.

[16]Eggers, A. J., "The Possibility of a Safe Landing," *Space Technology*, edited by H. S. Seifert, Wiley, New York, 1959, Chap. 13.

[17]Loh, W. H. T., "Entry Mechanics," *Re-Entry and Planetary Entry Physics and Technology*, edited by W. H. T. Loh, Springer-Verlag, Berlin, 1968.

[18]Speyer, J. L., and Womble, M. E., "Approximate Optimal Atmospheric Entry Trajectories," *Journal of Spacecraft and Rockets*, Vol. 8, Nov. 1971, pp. 1120–1125.

[19]Wurster, K. E., "Lifting Entry Vehicle Mass Reduction Through Integrated Thermostructural/Trajectory Design," *Journal of Spacecraft and Rockets*, Vol. 20, Nov.–Dec. 1983, pp. 589–596.

[20]White, F. M., *Viscous Fluid Flow*, McGraw-Hill, New York, 1974.

[21]Prabhu, D. K., and Tannehill, J. C., "Numerical Solution of Space Shuttle Orbiter Flowfields Including Real-Gas Effects," *Journal of Spacecraft and Rockets*, Vol. 23, May–June 1986, pp. 264–272.

[22]Williams, S. D., and Curry, D. M., "Assessing the Orbiter Thermal Environment Using Flight Data," *Journal of Spacecraft and Rockets*, Vol. 21, Nov.–Dec. 1984, pp. 534–541.

[23]Throckmorton, D. A., and Zoby, E. V., "Orbiter Entry Leeside Heat Transfer Data Analysis," *Journal of Spacecraft and Rockets*, Vol. 20, Nov.–Dec. 1983, pp. 524–530.

[24]Scott, C. D., "Effects of Nonequilibrium and Wall Catalysis on Shuttle Heat Transfer," *Journal of Spacecraft and Rockets*, Vol. 22, Sept.–Oct. 1985, pp. 489–499.

[25]Liepmann, H. W., and Roshko, A., *Elements of Gasdynamics*, Wiley, New York, 1957.

[26]Ames Research Staff, "Equations, Tables, and Charts for Compressible Flow," NACA Rept. 1135, 1953.

[27]Sibulkin, M., "Heat Transfer Near the Forward Stagnation Point of a Body of Revolution," *Journal of the Aeronautical Sciences*, Vol. 19, Aug. 1952, pp. 570–571.

[28]Fay, J. A., and Riddell, F. R., "Theory of Stagnation Point Heat Transfer in Dissociated Air," *Journal of the Aeronautical Sciences*, Vol. 25, Feb. 1958, pp. 73–85.

[29]Hoshizaki, H., "Heat Transfer in Planetary Atmospheres at Super-Satellite Speeds," *ARS Journal*, Oct. 1962, pp. 1544–1552.

[30]Fay, J. A., and Kemp, N. H., "Theory of Stagnation Point Heat Transfer in a Partially Ionized Diatomic Gas," *AIAA Journal*, Vol. 1, Dec. 1963, pp. 2741–2751.

[31]Rose, P. H., and Stark, W. I., "Stagnation Point Heat Transfer Measurements in Dissociated Air," *Journal of the Aeronautical Sciences*, Vol. 25, Feb. 1958, pp. 86–97.

[32]Kemp, N. H., Rose, P. H., and Detra, R. W., "Laminar Heat Transfer Around Blunt Bodies in Dissociated Air," *Journal of the Aerospace Sciences*, July 1959, pp. 421–430.

[33]Anderson, J. D., Jr., *Modern Compressible Flow*, McGraw-Hill, New York, 1982.

[34]Kemp, N. H., and Riddell, F. R., "Heat Transfer to Satellite Vehicles Re-entering the Atmosphere," *Jet Propulsion*, 1957, pp. 132–147.

[35]Hilsenrath, J., and Klein, M., "Tables of Thermodynamic Properties of Air in Chemical Equilibrium Including Second Virial Corrections from 1500 K to 15,000 K," Arnold Engineering Development Center Rept. AEDC-TR-65-68, 1965.

[36]Stull, D. R., *JANAF Thermochemical Tables*, National Bureau of Standards NSRDS-NBS 37, 1971.

[37]Cruz, M. I., "The Aerocapture Vehicle Mission Design Concept—Aerodynamically Controlled Capture of Payload into Mars Orbit," AIAA Paper 79-0893, May 1979.

[38]Walberg, G. D., "A Survey of Aeroassisted Orbit Transfer," *Journal of Spacecraft and Rockets*, Vol. 22, Jan.–Feb. 1985, pp. 3–18.

Bibliography

Cohen, C. B., and Reshotko, E., "Similar Solutions for the Compressible Laminar Boundary Layer with Heat Transfer and Pressure Gradient," NACA TN-3325, 1955.

Florence, D. E., "Aerothermodynamic Design Feasibility of a Mars Aerocapture Vehicle," *Journal of Spacecraft and Rockets*, Vol. 22, Jan.–Feb. 1985, pp. 74–79.

Hansen, C. F., "Approximations for the Thermodynamic and Transport Properties of High Temperature Air," NASA TR-R-50, 1959.

Miller, C. G., Gnoffo, P. A., and Wilder, S. E., "Measured and Predicted Heating Distributions for Biconics at Mach 10," *Journal of Spacecraft and Rockets*, Vol. 23, May–June 1986, pp. 251–258.

Vinh, N. X., Johannesen, J. R., Mease, K. D., and Hanson, J. M., "Explicit Guidance of Drag-Modulated Aeroassisted Transfer Between Elliptical Orbits," *Journal of Guidance, Control, and Dynamics*, Vol. 9, May–June 1986, pp. 274–280.

7

ATTITUDE DETERMINATION AND CONTROL

7.1 INTRODUCTION

In this chapter we discuss what is generally considered the most complex and least intuitive of the space vehicle design disciplines, that of attitude determination and control. We agree but would add that the more complex topics are of primarily theoretical interest, having limited connection with practical spacecraft design and performance analysis. Exceptions exist, of course, and will be discussed here because of their instructional value. But we believe that the most significant features of attitude determination and control system (ADCS) design can be understood in terms of rigid body rotational mechanics modified by the effects of flexibility and internal energy dissipation. At this level, the subject is quite accessible at the advanced undergraduate or beginning graduate level.

Even so, we recognize that the required mathematical sophistication will be considered excessive by many readers. Attitude dynamics analysis is necessarily complex due to three factors. Attitude information is inherently vectorial, requiring three coordinates for its complete specification. Attitude analysis deals inherently with rotating, hence noninertial, frames. Finally, rotations are inherently order-dependent in their description; the mathematics that describes them therefore lacks the multiplicative commutativity found in basic algebra.

We will try to alleviate this by appealing to the many analogies between rotational and translational dynamics and, as always, by stressing applications rather than derivations of results. Those requiring more detail are urged to consult one of the many excellent references in the field. Hughes[1] provides an especially good analytical development of attitude dynamics analysis, and includes extensive applications to practical spacecraft design. Wertz[2] is a definitive text on operational practices in attitude determination, as well as including brief but cogent summaries of many other topics of interest in space vehicle design.

7.2 BASIC CONCEPTS AND TERMINOLOGY

Definition of Attitude

Attitude determination and control is typically a major vehicle subsystem, with requirements that quite often drive the overall spacecraft design. Components tend to be relatively massive, power consuming, and demanding of specific orientation, alignment tolerance, field of view, structural frequency response, and structural damping. Effective attitude control system design is unusually demanding of a true systems orientation.

Spacecraft *attitude* refers to the angular orientation of a defined body-fixed coordinate frame with respect to a separately defined external frame. The spacecraft body frame may be arbitrarily chosen; however, some ways of defining it offer more utility than others, as we will see. The external frame may be one of the "inertial" systems discussed in Chapter 4 (GCI or HCI), or it may be a noninertial system such as the local vertical, local horizontal (LVLH) frame, which is used to define the flight path angle (Fig. 4.9).

Astute readers will note that we have mentioned only the angular orientation between a spacecraft and an external frame, whereas in general some translational offset will also exist between the two. This is illustrated in Fig. 7.1, and leads one to question the influence of parallax in performing spacecraft attitude measurements with respect to the "fixed" stars, which serve as the basis for inertial frames.

The concept of parallax is shown in Fig. 7.2. As seen, measurements of angles with respect to a given star will differ for frames whose origins are located apart. However, in almost all cases of practical interest parallax effects are insignificant for spacecraft. The nearest star system, α-Centauri, is approximately 4.3 lightyears (LY) from Earth. Using the Earth's orbital diameter as a baseline, and making measurements six months apart, an object will show a parallax of 1 arcsecond at a distance of 3.26 LY, a quantity defined for obvious reasons as a *parsec*. Thus, even α-Centauri has

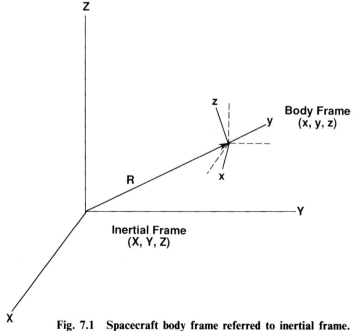

Fig. 7.1 Spacecraft body frame referred to inertial frame.

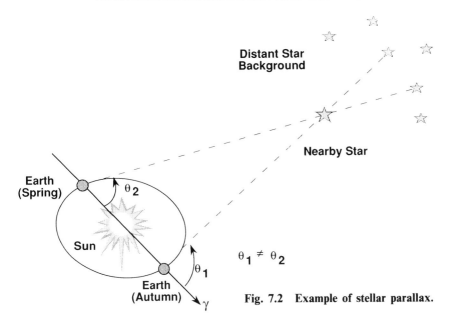

Fig. 7.2   Example of stellar parallax.

a parallax of only about 0.75 arcsecond; all other stars have less. For practical purposes, then, the location of a spacecraft will not influence measurements made to determine its attitude.

As always, exceptions exist. The Hubble Space Telescope (HST) is expected to make astrometric measurements of the parallax of many of the nearer stars so that their distances may be more accurately determined. Obviously, the "error" due to parallax is the measurement sought. Of greater relevance is the requirement of HST to track and observe moving objects within the solar system to within 0.01 arcsecond. At this level, parallax errors induced by HST movement across its Earth-orbital baseline diameter of 13,500 km are significant. Mars, for example, periodically approaches to within approximately 75 million km of Earth. During a half-orbit of HST it would then appear to shift its position by about 180 μrad, or roughly 36 arcseconds, against the background of fixed guide stars. The tracking accuracy requirement would be grossly violated if this apparent motion were not compensated for in the HST pointing algorithm.

Attitude *determination* refers to the process (to which we have already alluded) of measuring spacecraft orientation. Attitude *control* implies a process, usually occurring more or less continuously, of returning the spacecraft to a desired orientation, given that the measurement reveals a discrepancy. In practice, errors of both measurement and actuation will always exist, so both these processes take place within some tolerance.

Errors may result from inexact execution of reorientation maneuvers that are themselves based on inexact measurements, and will in addition arise from disturbances both internally and externally generated. The spacecraft is not capable of responding instantly to these disturbances; some time is always consumed in the process of measuring an error and computing and

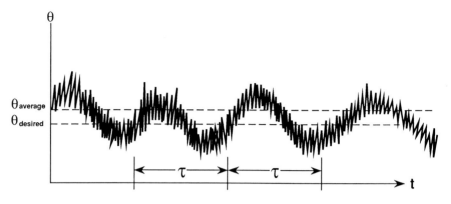

Fig. 7.3 Typical spacecraft pointing history.

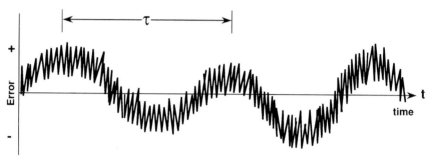

Fig. 7.4 Zero-mean attitude history.

applying a correction. This leads to a typical pointing history such as shown in Fig. 7.3. Close examination of this figure reveals several features of interest.

The low-frequency, cyclic departure from and restoration to an average value is the result of the error detection and correction process implemented by the ADCS. It is roughly periodic, an artifact of the finite interval required to sense an error and implement a correction. This fundamental period, τ, implies a limit to the frequency response of the spacecraft, called the *bandwidth* or *passband*, of about $1/\tau$ Hz. A disturbance (such as an internal vibration or external impulsive torque), which has a frequency content higher than this, is simply not sensed by the spacecraft ADCS. Only the longer term integrated effect, if such exists, is correctable.

This inability to sense and respond to high-frequency disturbances produces the *jitter* on the signal shown in Fig. 7.3. Jitter then refers to the high-frequency (meaning above the spacecraft passband) discrepancy between the actual and desired attitude. Attitude error, as we shall use it henceforth, implies the low-frequency (within the passband) misalignment that is capable of being sensed and acted upon.

We shall return shortly to the discussion of attitude jitter. For the moment, note that a long-term integration (several τ periods) of the data in Fig. 7.3 would clearly yield an average value θ_a displaced from the desired value θ_d. This *bias* in the attitude could be due to sensor or actuator misalignment, to the effects of certain types of disturbances, or to more subtle properties of the control algorithm. Note further that θ_d is not always (and maybe not even very often) a constant. If not, we are said to have a *tracking* problem, as opposed to the much simpler constant-angle *pointing* problem. Tracking at higher rates or nonconstant rates generally yields poorer average performances than does pointing or low-rate tracking, or requires more complex engineering to achieve comparable performance.

Attitude Jitter

Returning to our discussion of jitter, if we subtract the average value from the data, we produce by definition a zero-mean history such as shown in Fig. 7.4. The smooth central curve results from filtering or smoothing the data so as to remove the jitter, i.e., the components above the spacecraft passband. This curve is what we have denoted the attitude error.

If we subtract the smooth central curve as well, we retain only the jitter, as shown in Fig. 7.5. This jitter can have sources both deterministic and random. An example of the former could be the vibration of an attitude sensor due to an internal source at a structural frequency above the control system passband. Random jitter may be due to many causes, including electronic and mechanical noise in the sensors and actuators. Our use of the term "noise" in this sense somewhat begs the question; perception of noise often depends on who is using the data. The spacecraft structural

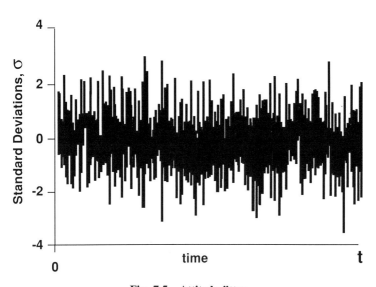

Fig. 7.5   Attitude jitter.

engineer will regard only the electronic effects as noise; the structural vibrations are, if included in the data, "signal" to him. The sensor designer may, given the data with mechanical effects removed, find much value in isolating the electronic disturbances. To the ADCS engineer, it is all noise, but understanding its source characteristics may be instrumental in removing or coping with it.

In ADCS performance analysis, it is profitable to assume that the jitter is random and that at any instant in time its amplitude has a zero-mean Gaussian or normal probability distribution (which is guaranteed in our example by the way we have constructed Fig. 7.5). This seemingly restrictive (but enormously convenient) assumption is usually quite well satisfied. This results from the application of the central limit theorem of statistics,[3] which loosely states that the sum of many independent zero-mean probability distributions converges in the limit to a Gaussian distribution. In practice (and this is a forever surprising result) "many" may be as few as four or five, and rarely more than ten, unless we are at the extremes of the normal curve shown in Fig 7.6. Since we usually have many independent noise sources in a system, we in most cases rely quite comfortably on the assumption of zero-mean Gaussian noise.

If the jitter amplitude data are squared, we obtain the instantaneous *power* in the signal, usually called noise power, $N(t)$. If the amplitude data are Gaussian distributed, $N(t)$ is Gaussian as well. If $N(t)$ is averaged over

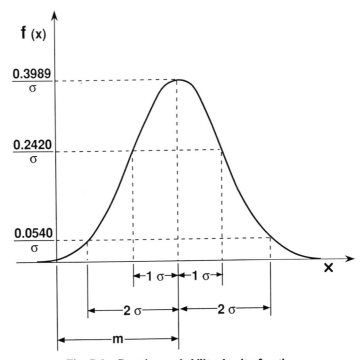

Fig. 7.6   Gaussian probability density function.

the system time constant τ and found to yield the same average value, N_0, at any epoch, the noise is said to be *white*. This term derives from the fact that the noise power is a constant at all frequencies ("colors"), hence is "white" by analogy to optics. White noise cannot truly exist, as it possesses infinite total signal power; however, in usual applications the assumption of white Gaussian noise (WGN) is universal. It is also reasonable, in that the system passband may be quite narrow with respect to the variations in the noise spectrum. Thus, in any such narrow segment, the noise power may indeed be approximately constant.

Under the zero-mean WGN jitter model, we note that the maximum amplitude excursion to be expected in the data can be loosely said to fall at the 2σ point on the normal curve. (Strictly, 99.74% of the data fall within $\pm 3\sigma$ of the mean, which in this case is zero.) This defines the corresponding 2σ and 1σ levels, at approximately 95.4% and 68.3%, respectively. To discuss the "average" value requires a little more care; as mentioned repeatedly, the average jitter amplitude is zero. This is not a useful concept in characterizing the system performance. But if we square the data, average over time interval τ, and then square-root the result (the so-called root-mean-square, or rms, operation) we obtain a useful "average" system jitter. For Gaussian processes, the rms and 1σ levels are synonymous.

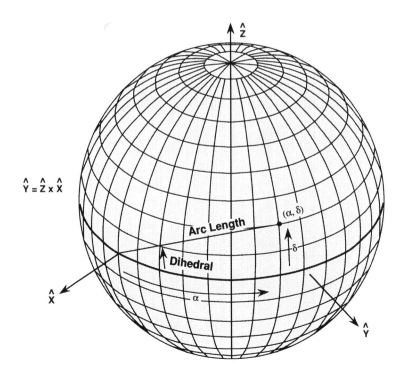

Fig. 7.7   Attitude measurements on the unit celestial sphere.

Jitter in a spacecraft must be accepted; by its definition, it is the error for which we do not compensate. It may, however, be reduced or controlled through proper mechanical, configuration, and structural design, as well as through attention to use of low-noise subsystems in the vehicle. If this proves insufficient, more sophisticated control system design is required to compensate for disturbances at a finer level.

Rotational Kinematics and Celestial Sphere

Figure 7.7 depicts a *celestial sphere* centered in the origin of a coordinate frame. As we have discussed, length scales do not influence attitude determination and control, so we may consider the sphere to be of unit radius. Directions may be specified in several ways on the celestial sphere. Possibly the most obvious is to use the Cartesian (x,y,z) coordinates of a particular point. Since

$$x^2 + y^2 + z^2 = 1 \qquad (7.1)$$

only two of the three coordinates are independent. It is common in astronomy to use the right ascension, α, and the declination, δ, defined as shown in Fig. 7.7, to indicate direction.[4] Note

$$x = \cos\alpha \, \cos\delta \qquad (7.2a)$$

$$y = \sin\alpha \, \cos\delta \qquad (7.2b)$$

$$z = \sin\delta \qquad (7.2c)$$

The use of Euler angles to describe body orientation is common in rotational kinematics. An Euler angle set is a sequence of three angles and a prescription for rotating a coordinate frame through these angles to bring it into alignment with another frame. Figure 7.8 shows a typical Euler rotation sequence; other choices are often encountered as well. In Fig. 7.9 we define an Euler angle set (φ,θ,ψ) corresponding to roll, pitch, and yaw

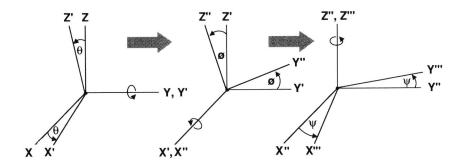

Fig. 7.8   Euler angle rotation sequence.

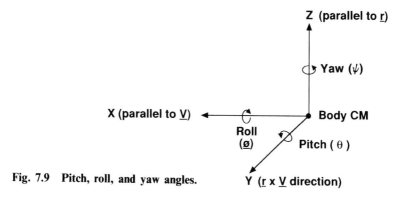

Fig. 7.9   Pitch, roll, and yaw angles.

angles of the spacecraft body frame relative to a rotating local vertical frame. Note that this frame is referenced to the spacecraft velocity vector, and not necessarily to the local horizontal. Using $C\theta$ and $S\theta$ to represent $\sin\theta$ and $\cos\theta$, the transformation matrix which rotates the body frame into the "inertial" frame via rotations in roll, pitch, and yaw is

$$
T_{B/I} = \begin{bmatrix} C\psi & S\psi & 0 \\ -S\psi & C\psi & 0 \\ 0 & 0 & 1 \\ & (\text{yaw}) & \end{bmatrix} \begin{bmatrix} 1 & 0 & 0 \\ 0 & C\varphi & S\varphi \\ 0 & -S\varphi & C\varphi \\ & (\text{roll}) & \end{bmatrix} \begin{bmatrix} C\theta & 0 & -S\theta \\ 0 & 1 & 0 \\ S\theta & 0 & C\theta \\ & (\text{pitch}) & \end{bmatrix} \quad (7.3)
$$

or, in combined form,

$$
T_{B/I} = \begin{bmatrix} C\psi C\theta + S\psi S\theta S\phi & S\psi C\phi & -C\psi S\theta + S\psi C\theta S\phi \\ -S\psi C\theta + C\psi S\theta S\phi & C\psi C\phi & S\psi S\theta + C\psi C\theta S\phi \\ S\theta C\phi & -S\phi & C\theta C\phi \end{bmatrix} \quad (7.4)
$$

Transformation matrices possess a number of useful properties. They are orthonormal, so the inverse transformation (in this case, from inertial to body coordinates) is easily found by transposing the original:

$$
T_{I/B} = T_{B/I}^{-1} = T_{B/I}^{t} \quad (7.5)
$$

Recall that matrix multiplication is not commutative; thus, altering the order of the rotation sequence produces a different transformation matrix. This is reflective of the fact that an Euler angle set implies a defined sequence of rotations, and altering this sequence alters the final orientation of the body if the angles are of finite size. It is readily shown for small angles that the required matrix multiplications are commutative, corresponding to the physical result that rotation through infinitesimal angles is independent of order.

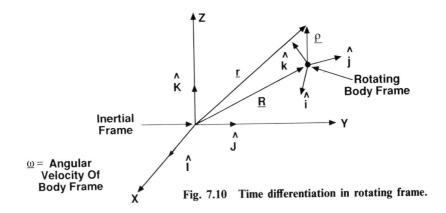

Fig. 7.10   Time differentiation in rotating frame.

Euler angle representations of spacecraft rotation are important in attitude analysis because they are easily visualized; they are suited to the way in which most humans think. They can be computationally inconvenient because all such formulations implicitly contain a singularity corresponding exactly to the mechanical engineer's "gimbal lock" problem in multiple-gimbal systems. The Euler angle set chosen here (from among 12 possible sets) is among the more convenient, in that the singularity can often be kept out of the working range of rotations. However, it cannot be eliminated altogether in any three-parameter attitude representation.

Relief is possible, however. Euler's theorem in rotational kinematics states that the orientation of a body may be uniquely specified by a vector giving the direction of a body axis and a scalar parameter specifying a rotation angle about that axis. A redundant fourth parameter is now part of the attitude representation. As a fourth gimbal allows a mechanical engineer to eliminate the possibility of gimbal lock, so too this analytical redundancy avoids coordinate singularities. From this result is derived the concept of quaternion, or Euler parameter, representation of attitude.[2] Hughes[1] considers the Euler parameter formulation to be, on balance, the most suitable choice for practical work.

The overview of attitude kinematics given here is sufficient only to acquaint the reader with the nature of the problem. More detailed discussion of attitude representations and rotational kinematics may be found in Wertz,[2] Kaplan,[5] or Hughes.[1]

7.3   REVIEW OF ROTATIONAL DYNAMICS

A goal of attitude determination and control analysis is to describe the rotational behavior of a spacecraft body frame subject to the forces imposed upon it. This requires the use of Newton's laws of motion and the tools of calculus for the formulation and solution of such problems. From sophomore physics we recall that time-differentiation in a rotating (hence noninertial) coordinate system produces extra terms, and so we are prepared for some additional complication in attitude analysis.

Figure 7.10 shows the essential geometry. We have a vector ρ given in a rotating body frame, whereas Newton's laws describe motion in an inertial frame, and require the use of second derivatives. Recalling the basic rule for time differentiation in a rotating frame, we write

$$(d\rho/dt)_i = (d\rho/dt)_b + \omega \times \rho \qquad (7.6)$$

where ω is the angular velocity vector of the rotating frame *in body coordinates*.

Newtonian dynamics problems involve the position vector r and its derivatives velocity v and acceleration a. If ρ is a position vector in a body frame having angular velocity ω, it is given in the inertial frame as

$$r = R + \rho \qquad (7.7)$$

hence

$$V = (dr/dt)_i = dR/dt + (d\rho/dt)_b + \omega \times \rho \qquad (7.8)$$

and

$$a = (d^2r/dt^2)_i = d^2R/dt^2 + (d^2\rho/dt^2)_b + 2\omega \times (d\rho/dt)_b$$
$$+ d\omega/dt \times \rho + \omega \times (\omega \times \rho) \qquad (7.9)$$

The third term on the right is commonly called the Coriolis force, while the last term on the right-hand side is the centrifugal force.

The fundamental quantities of interest in Newtonian translational dynamics are mass, momentum, and kinetic energy. Conservation laws for these quantities provide the basis for the description of dynamical systems in classical physics. In rotational dynamics, the analogous quantities are the moment of inertia, angular momentum, and rotational kinetic energy.

The angular momentum of a mass is the moment of its linear momentum about a defined origin. From Fig. 7.10, the angular momentum of mass m_i about the origin in the inertial frame is

$$H = r_i \times m_i v_i \qquad (7.10)$$

and for a collection of point masses, the total angular momentum is

$$H_t = \sum r_i \times m_i v_i \qquad (7.11)$$

If we apply Eqs. (7.7) and (7.8) with $V = dR/dt$, and if we assume that 1) the origin of the rotating frame lies at the body center of mass ($\sum m_i \rho_i = 0$), and 2) the position vectors ρ_i are fixed in the body frame (i.e., we have a rigid body, with $d\rho/dt = 0$), we obtain

$$H_t = \left(\sum m_i\right) R \times V + \sum m_i \rho_i \times d\rho_i/dt = H_{\text{orb}} + H_b \qquad (7.12)$$

The first term on the right is the angular momentum of the rigid body due to its translational velocity V in the inertial frame. The second term is the body angular momentum due to its rotational velocity about its own center of mass. If we consider the body to be an orbiting spacecraft, the first term is the orbital angular momentum introduced in Chapter 4, while the second is the angular momentum in the local center-of-mass frame, which is of interest for attitude dynamics analysis.

Equation (7.12) gives the important result that, for a rigid body, it is possible to choose a coordinate frame which decouples the spin angular momentum from the orbit angular momentum. Clearly, this is not always possible, and so-called "spin-orbit coupling" can at times be an important consideration in attitude control. However, unless stated otherwise, we employ the rigid body assumption in the discussion to follow, and will be concerned only with the body angular momentum.

Subject to the rigid body assumption, Eq. (7.6) yields

$$d\boldsymbol{\rho}_i / dt = \boldsymbol{\omega} \times \boldsymbol{\rho}_i \tag{7.13}$$

and from Eq. (7.12) the body angular momentum is

$$\boldsymbol{H} = \sum m_i \boldsymbol{\rho}_i \times d\boldsymbol{\rho}_i / dt = \sum m_i \boldsymbol{\rho}_i \times (\boldsymbol{\omega} \times \boldsymbol{\rho}_i) = \boldsymbol{I}\boldsymbol{\omega} \tag{7.14}$$

where I is a real, symmetric matrix called the *inertia matrix*, with components

$$I_{11} = \sum m_i (\rho_{i2}^2 + \rho_{i3}^2) \tag{7.15a}$$

$$I_{22} = \sum m_i (\rho_{i1}^2 + \rho_{i3}^2) \tag{7.15b}$$

$$I_{33} = \sum m_i (\rho_{i1}^2 + \rho_{i2}^2) \tag{7.15c}$$

$$I_{12} = I_{21} = -\sum m_1 \rho_{i1} \rho_{i2} \tag{7.15d}$$

$$I_{13} = I_{31} = -\sum m_i \rho_{i1} \rho_{i3} \tag{7.15e}$$

$$I_{23} = I_{32} = -\sum m_i \rho_{i2} \rho_{i3} \tag{7.15f}$$

The diagonal components of the inertia matrix are called the moments of inertia, while the off-diagonal terms are referred to as the products of inertia. Since I is a real, symmetric matrix, it is always possible[5] to find a coordinate system in which the inertia products are zero; i.e., the matrix is diagonal. The elements of the inertia matrix may then be abbreviated I_1, I_2, and I_3, and are referred to as the principal moments of inertia, while the corresponding coordinates are called principal axes. These are the "natural" coordinate axes for the body, in that a symmetry axis in the body, if it exists, will be one of the principal axes.

Because of the generality of this result, it is customary to assume the use of a principal axis set in most attitude analysis. Unless otherwise stated, we assume such in this text. This convenient analytical assumption is often violated in the real world. Spacecraft designers will normally select a principal axis coordinate frame for attitude reference purposes. However, minor asymmetries and misalignments can be expected to develop during vehicle integration, leading to differences between the intended and actual principal axes. When this occurs, attitude error measurements and control corrections intended about one axis will couple into others.

A force F_i applied to a body at position ρ_i in center-of-mass coordinates produces a *torque* defined by

$$T_i = \rho_i \times F_i \qquad (7.16)$$

about the center of mass. The net torque from all such forces is then

$$T = \sum \rho_i \times F_i = \sum \rho_i \times m_i \, d^2 r_i/dt^2 \qquad (7.17)$$

After expanding $d^2 R_i/dt^2$ as before, we obtain

$$T = dH/dt = (dH/dt)_{body} + \omega \times H \qquad (7.18)$$

The total kinetic energy of a body consisting of a collection of lumped masses is given by

$$E = \tfrac{1}{2} \sum m_i (dr_i/dt)^2 = \tfrac{1}{2} \sum m_i (dR_i/dt + d\rho_i/dt)^2 \qquad (7.19)$$

If center-of-mass coordinates are chosen for the ρ_i, then the cross terms arising in Eq. (7.19) vanish, and the kinetic energy, like the angular momentum, separates into translational and rotational components,

$$E = \tfrac{1}{2} \sum m_i (dR_i/dt)^2 + \tfrac{1}{2} \sum m_i (d\rho_i/dt)^2 \qquad (7.20)$$

If the body assumption is included, such that $H = I\omega$, we may, after expanding Eq. (7.20), write

$$E_{rot} = \tfrac{1}{2} \omega' I \omega \qquad (7.21)$$

Equations (7.14) and (7.21) define momentum and energy for rotational dynamics, and are seen for rigid bodies to be completely analogous to translational dynamics, with ω substituted for v and I replacing m, the body mass. Equation (7.18) is Newton's second law for rotating rigid bodies.

A particularly useful formulation of Eq. (7.18) is obtained by assuming a body-fixed principal axis center-of-mass frame in which to express $\boldsymbol{H}$, $\boldsymbol{T}$, and $\boldsymbol{\omega}$. In this case, we have

$$(\mathrm{d}\boldsymbol{H}/\mathrm{d}t)_{\mathrm{body}} = \boldsymbol{T} - \boldsymbol{\omega} \times \boldsymbol{I}\boldsymbol{\omega} \qquad (7.22)$$

which becomes, on expansion into components,

$$\dot{H}_1 = I_1\dot{\omega}_1 = T_1 + (I_2 - I_3)\omega_2\omega_3 \qquad (7.23\mathrm{a})$$

$$\dot{H}_2 = I_2\dot{\omega}_2 = T_2 + (I_3 - I_1)\omega_3\omega_1 \qquad (7.23\mathrm{b})$$

$$\dot{H}_3 = I_3\dot{\omega}_3 = T_3 + (I_1 - I_2)\omega_1\omega_2 \qquad (7.23\mathrm{c})$$

These are the Euler equations for the motion of a rigid body under the influence of an external torque. No general solution exists for the case of an arbitrarily specified torque. Particular solutions for simple external torques do exist; however, computer simulation is usually required to examine cases of practical interest.

7.4 RIGID BODY DYNAMICS

An understanding of the basic dynamics of rigid bodies is crucial to an understanding of spacecraft attitude dynamics and control. Although in practice few if any spacecraft can be accurately modeled as rigid bodies, such an approximation is nonetheless the proper reference point for understanding the true behavior. The Euler equations derived in the previous section can in several simple but interesting cases be solved in closed form, yielding insight not obtained through numerical analysis of more realistic models.

The most important special case for which a solution to the Euler equations is available is that for the torque-free motion of an approximately axisymmetric body spinning primarily about its symmetry axis; i.e., a spinning top in free fall. Mathematically, the problem is summarized as

$$\omega_x, \omega_y \ll \omega_z = \Omega \qquad (7.24)$$

$$I_x \cong I_y \qquad (7.25)$$

With these simplifications, the Euler equations become

$$\dot{\omega}_x = -K_x\Omega\omega_y \qquad (7.26\mathrm{a})$$

$$\dot{\omega}_y = K_y\Omega\omega_x \qquad (7.26\mathrm{b})$$

$$\dot{\omega}_z \cong 0 \qquad (7.26\mathrm{c})$$

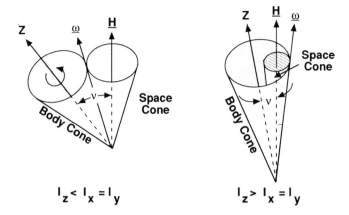

Fig. 7.11   Possible cases for torque-free motion of a symmetric rigid body.

where

$$K_x = (I_z - I_y)/I_x \qquad (7.27a)$$

$$K_y = (I_z - I_x)/I_y \qquad (7.27b)$$

The solution for the angular velocity components is

$$\omega_x(t) = \omega_{x0}\cos\omega_n t \qquad (7.28)$$

$$\omega_y(t) = \omega_{y0}\sin\omega_n t \qquad (7.29)$$

where the natural frequency ω_n is defined by

$$\omega_n^2 = K_x K_y \Omega^2 \qquad (7.30)$$

The conceptual picture represented by this solution is that of a body with an essentially symmetrical mass distribution spinning rapidly about the axis of symmetry, which we have defined to be the z axis. This rapid rotation is at essentially constant speed $\omega_z = \Omega$. However, a smaller x-y plane component of angular velocity, time-varying in its orientation, also exists. This component rotates periodically around the body z axis at a natural or "nutation" frequency ω_n determined by the body's inertia ratios. This results in a circular motion of the body z axis around the angular momentum vector H at the nutation frequency. (Recall H is fixed in inertial space since no torques are present.) The motion can have one of two general patterns, depending on the ratio of I_z to I_x or I_y. Figure 7.11 shows the two cases. The angle, v, between the body z-axis and the inertially fixed H vector is called the nutation angle.

The "space cone" in Fig. 7.11 refers to the fact that H is fixed in inertial space. Conversely, Eqs. (7.27) are expressed in body coordinates, so the "body cone" is defined relative to the body principal axis frame. The space cone may lie inside or outside the body cone, depending on whether the spinning body is "pencil shaped" ($I_z < I_x \cong I_y$) or "saucer shaped" ($I_z > I_x \cong I_y$).

The preceding discussion can be generalized[1,5] to include the case where $I_x \neq I_y$. We then have the possibility that the spin axis inertia I_z is intermediate between I_x and I_y. In such a case, Eqs. (7.27) and (7.30) show that $\omega_n^2 < 0$; i.e., the nutation frequency is imaginary. The previous sinusoidal solution for ω_x and ω_y becomes an exponential solution, expressible if desired in terms of hyperbolic sines. Thus, the body cannot have a fixed nutation angle when spinning primarily about an axis of intermediate inertia moment. Elementary stability analysis shows that if a rigid body is initially spinning perfectly about such an axis, any perturbing torque will result in the growth of the nutation angle until the body is spinning about either the maximum or minimum inertia axis, depending on initial conditions. We thus have the important result that a rigid body can rotate about its extreme inertia axes, but not the intermediate axis.

This important conclusion is further modified if flexibility is considered. If the body is not rigid, energy dissipation must occur as it flexes. Since total system energy must be conserved, this energy, dissipated as heat, must be derived from the rotational kinetic energy, Eq. (7.21). Thus, a flexible spinning body causes E_{rot} to decrease. At the same time, however, the angular momentum for the (torque-free) system must be constant. Ignoring for the moment the vector nature of these quantities, we have then the constraints

$$E_{\text{rot}} = I\omega^2/2 \qquad (7.31)$$

$$H = I\omega \qquad (7.32)$$

hence

$$E_{\text{rot}} = H^2/2I \qquad (7.33)$$

Clearly, if energy is to be dissipated (and the second law of thermodynamics guarantees that it will), the moment of inertia must increase. If the body is not to be permanently deformed, the spin axis must shift (in body coordinates, of course; the spin axis is fixed in space by the requirement that H be constant) to the maximum-inertia axis. To visualize this, we imagine an energy-dissipating body with its angular momentum vector initially aligned with the minimum-inertia axis. As energy is lost, the nutation angle grows so as to satisfy Eq. (7.33). Eventually, the nutation has grown to 90 deg, the maximum possible. The body will be spinning at a slower rate about the maximum-inertia principal axis. This is colloquially referred to as a "flat spin," a name instantly evocative of the condition.

Thus, in the absence of external torques a real body can spin stably only about the axis of maximum moment of inertia. This so-called "major axis rule" was discovered, empirically and embarrassingly, following the launch of Explorer 1 as the first U.S. satellite on February 1, 1958. The cylindrically shaped satellite was initially spin stabilized about its long axis, and had four flexible wire antennas for communication with ground tracking stations. Within hours, the energy dissipation inherent in the antennas had caused the satellite to decay into a flat spin, a condition revealed by its effect on the air-to-ground communications link. This initially puzzling behavior was quickly explained by Ron Bracewell and Owen Garriott, then of Stanford University (Garriott later became a Skylab and Shuttle astronaut).[6]

This result can be illustrated most graphically by considering the homely example of a spinning egg. It is well known that a hard-boiled egg can be readily distinguished from a raw egg by attempting to spin it about its longitudinal axis. Due to internal energy dissipation by the viscous fluid, the raw egg will almost immediately fall into a flat spin, while the boiled egg will rotate at some length.

It should be carefully understood that the arguments made earlier, while true in the general terms in which they are expressed, are heuristic in nature. Equations (7.31) and (7.32) must hold, leading to the behavior described. However, it is equally true that Newton's laws of motion must be satisfied; physical objects do not move without the imposition of forces. The energy-momentum analysis outlined earlier is incomplete, in that the origin of forces causing motion of the body is not included. Nonetheless, "energy sink" analyses based on the arguments outlined earlier can be quite successful in predicting spacecraft nutation angle with time.

Further refinements of these conclusions exist. For example, the major axis rule strictly applies only to simple spinners; complex bodies with some parts spinning and others stationary may exhibit more sophisticated behavior. In particular, a so-called "dual spin" satellite can be stable with its angular momentum vector oriented parallel to the minor principal axis. We will address the properties of dual spin satellites in a later section.

By assuming a flexible (e.g., nonrigid) body, we have violated a basic constraint under which the simplified results of Eq. (7.12) and those following were derived. Specifically, the spin motion and orbital motion are no longer strictly decoupled. Much more subtle behavior may follow from this condition.

Even if rigid, an orbiting spacecraft is not in torque-free motion. An extended body will be subject to a number of external torques to be discussed in the following section, including aerodynamic, magnetic, and tidal or gravity-gradient torques. The existence of the gravity-gradient effect, discussed in Chapter 4, renders a spinning spacecraft asymptotically stable only when its body angular momentum vector is aligned with the orbital angular momentum vector, i.e., the orbit normal.

The topic of stability analysis is a key element of spacecraft attitude dynamics. Even when the equations of motion cannot be solved in closed form, it may be determined that equilibria exist over some useful parametric range. If stable, such equilibria can be used as the basis for passive stabilization schemes, or for reducing the workload upon an active control system.

7.5 SPACE VEHICLE DISTURBANCE TORQUES

As mentioned, operating spacecraft are subject to numerous disturbance forces which, if not acting through the center of mass, result in a net torque being imparted to the vehicle. Assessment of these influences in terms of both absolute and relative magnitude is an essential part of the ADCS designer's task.

Aerodynamic Torque

The role of the upper atmosphere in producing satellite drag was discussed in Chapter 4 in connection with orbit decay. The same drag force will, in general, produce a disturbance torque on the spacecraft due to any offset between the aerodynamic center of pressure and the center of mass. Assuming r_{cp} to be the center-of-pressure (CP) vector in body coordinates, the aerodynamic torque is

$$T = r_{cp} \times F_a \tag{7.36}$$

where, as in Chapter 4, the aerodynamic force vector is

$$F_a = (\tfrac{1}{2})\rho V^2 S C_D V/V \tag{7.37}$$

and

ρ = atmosphere density
V = spacecraft velocity
S = spacecraft projected area $\perp V$
C_D = drag coefficient, usually between 1 and 2 for free-molecular flow

It is important to note that r_{cp} varies with spacecraft attitude and, normally, with the operational state of the spacecraft (solar panel position, fuel on board, etc.). As we discussed in Chapter 4, major uncertainties exist with respect to the evaluation of Eq. (7.37). Drag coefficient uncertainties can easily be of order 50%, while upper atmosphere density variations approaching an order of magnitude relative to the standard model are not uncommon. Thus, if aerodynamic torques are large enough to be a design factor for the attitude control system, they need to be treated with appropriate conservatism.

As an example, assume a satellite with a frontal area $S = 5 \text{ m}^2$ and $C_D = 2$ orbiting at 400 km (with atmosphere density $\rho = 4 \times 10^{-12} \text{ kg/m}^3$). Assuming circular velocity at this altitude, the magnitude of the disturbance torque is $T/r_{cp} = 1.2 \times 10^{-3} N$. This seems small; to put it in perspective, let us assume it to be the only torque acting on the spacecraft, through a moment arm of $r_{cp} = 1$ cm, and that the spacecraft moment of inertia about the torque axis is $I = 1000 \text{ kg m}^2$. Equations (7.23) simplify to

$$T = dH/dt = I \, d\omega/dt = I \, d^2\theta/dt^2 \tag{7.38}$$

with initial conditions on angular position and velocity

$$\theta(0) = 0 \qquad (7.39)$$

$$\omega(0) = d\theta/dt = 0 \qquad (7.40)$$

We find for this example

$$\theta(t) = (T/I)t^2 = (1.2 \times 10^{-5} \text{ Nm}/1000 \text{ kg m}^2) \, t^2$$
$$= 1.2 \times 10^{-8} \text{s}^{-2} t^2 \qquad (7.41)$$

The angular displacement predicted by Eq. (7.41) certainly seems small. However, left uncorrected, the spacecraft will drift about 0.012 rad, or 0.7 deg, after 1000 s. This is in most cases a large error for an attitude control system. Even worse, the error growth is quadratic, so this aerodynamic torque applied over 10,000 s, about 3 h or 2 orbits, would produce a 1 rad pointing error! This is unacceptable in most conceivable applications.

Gravity-Gradient Torque

Planetary gravitational fields decrease with distance R from the center of the planet according to the Newtonian $1/R^2$ law, provided higher order harmonics as discussed in Chapter 4 are neglected. Thus, an object in orbit will experience a stronger attraction on its "lower" side than its "upper" side. This differential attraction, if applied to a body having unequal principal moments of inertia, results in a torque tending to rotate the object to align its long (minimum inertia) axis with the local vertical. Perturbations from this equilibrium produce a restoring torque toward the stable vertical position, causing a periodic oscillatory or "librational" motion. Energy dissipation in the spacecraft will ultimately damp this motion.

The gravity-gradient torque for a satellite in a near-circular orbit is

$$\boldsymbol{T} = 3n^2 \hat{\boldsymbol{r}} \times I \cdot \hat{\boldsymbol{r}} \qquad (7.42)$$

where

$\hat{\boldsymbol{r}} = \boldsymbol{R}/R =$ unit vector from planet to spacecraft
$n^2 = \mu/a^3 \cong \mu/R^3 =$ orbital rate
$\mu =$ gravitational constant (398,600 km^3/s^2 for Earth)
$I =$ spacecraft inertia matrix

In the spacecraft body frame with pitch, roll, and yaw angles given as in Eqs. (7.4), the unit vector to the spacecraft from the planet is, from Fig. 7.12,

$$\hat{\boldsymbol{r}} = (-\sin\theta, \ \sin\varphi, \ 1 - \sin^2\theta - \sin^2\varphi)^t \cong (-\theta, \ \varphi, \ 1)^t \qquad (7.43)$$

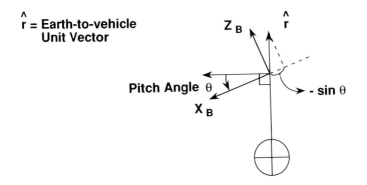

Fig. 7.12   Pitch plane geometry for gravity gradient torque.

with the approximation holding for small angular displacements. Then the gravity gradient torque vector may be expressed in the body frame as

$$T = 3n^2[(I_z - I_y)\varphi,(I_z - I_x)\theta,0]^t \qquad (7.44)$$

It is seen that the spacecraft yaw angle ψ does not influence the gravity-gradient torque; this is intuitively reasonable, since yaw represents rotation around the local vertical. We also note from Eq. (7.44) that the gravity-gradient influence is proportional to $1/R^3$. The torque magnitude clearly depends upon the *difference* between principal moments; thus, spacecraft that are long and thin are more affected than those that are short and fat.

To get an idea of a typical gravity-gradient torque magnitude, consider a low-orbiting spacecraft with $n \cong 0.001$ rad/s and an inertia moment difference in the relevant axis of 1000 kg m². Then $T = 5 \times 10^{-5}$ Nm/deg.

Solar Radiation Pressure Torque

Solar radiation pressure and its effect upon spacecraft orbital dynamics was discussed in Chapter 4. As with aerodynamic drag, solar radiation pressure can produce disturbance torques as well as forces, which may require compensation from the ACS. The solar radiation pressure torque is, in body coordinates,

$$T = r \times F_s \qquad (7.45)$$

where

$r =$ vector from body center of mass to spacecraft optical center of pressure

$F_s = (1 + K)p_s A_\perp$

$K =$ spacecraft surface reflectivity $(0 < K < 1)$

$A_\perp$ = spacecraft projected area normal to sun vector
$p_s = I_s/c$
$I_s \cong 1400$ W/m^2 @ 1 A.U.
$c = 2.9979 \times 10^8$ m/s

Solar radiation torque is independent of spacecraft position or velocity, as long as the vehicle is in sunlight, and is always perpendicular to the sun line. It will, in many cases, thus have no easily visualized relationship with the previously considered aerodynamic and gravity-gradient disturbance torques. For an order of magnitude comparison, consider typical values to be $A_\perp = 5$ m^2, $K = 0.5$, $r = 0.1$ m, and the spacecraft to be in Earth orbit. Then the torque magnitude is $T = 3.5 \times 10^{-6}$ Nm. This would be about two orders of magnitude below the representative aerodynamic torque computed earlier for a satellite orbiting at 400 km altitude. As noted, however, the solar torque is independent of position, while the aerodynamic torque is proportional to atmospheric density. Above 1000 km altitude, solar radiation pressure usually dominates the spacecraft disturbance torque environment.

At geostationary orbit altitude, solar radiation pressure can be the primary source of disturbance torque, and designers must take care to balance the geometrical configuration to avoid center-of-mass to center-of-pressure offsets. The useful lifetime of a geostationary satellite is often controlled by the mass budget available for stationkeeping and attitude control fuel. Poor estimates of the long-term effect of disturbance torques and forces can and do result in premature loss of on-orbit capability.

Solar radiation pressure can also be important for interplanetary missions. While its strength obviously drops off rapidly for outer planet missions, it may in the absence of internally generated disturbances be essentially the only torque acting upon the vehicle during interplanetary cruise. For missions to Venus and Mercury, solar torques will often define the spacecraft disturbance torque limits, and can even be used in an active control mode. This was first accomplished on an emergency basis following the shutdown of an unstable roll-control loop on the Mariner 10 mission to Venus and Mercury. In this case, differential tilt between opposing sets of solar panels was used to introduce a deliberate offset between the center of mass and center of pressure in such a way as to effect roll control.

Magnetic Torque

Earth and other planets such as Jupiter which have a substantial magnetic field exert yet another torque on spacecraft in low orbits about the primary. This is given by

$$T = M \times B \qquad (7.46)$$

M is the spacecraft magnetic dipole moment due to current loops and residual magnetization in the spacecraft. B is the Earth magnetic field vector *expressed in spacecraft coordinates*; its magnitude is proportional to $1/r^3$, where r is the radius vector to the spacecraft.

Few aerospace engineers are intimately involved with electromagnetic equipment, and so a brief discussion of measurement units for M and B is in order. Magnetic moment may be produced physically by passing a current through a coil of wire; the larger the coil, the greater the moment produced. Thus, in the SI system, M has units of ampere-turn-m^2 (Atm^2). B is measured in Tesla in the SI system. With M and B as specified, T of course has units of Nm.

An older but still popular system of units in electromagnetic theory is the centimeter-gram-second (CGS) system. In CGS units, M and B are measured in pole-cm and gauss, respectively, with the resultant torque in dyne-cm. Conversion factors between the two systems are:

$$1 \text{ Atm}^2 = 1000 \text{ pole cm} \qquad (7.47a)$$

$$1 \text{ Tesla} = 10^4 \text{ gauss} \qquad (7.47b)$$

Earth's magnetic field at an altitude of 200 km is approximately 0.3 gauss or 3×10^{-5} Tesla. A typical small spacecraft might possess a residual magnetic moment on the order of 0.1 Atm^2. The magnetic torque on such a spacecraft in low orbit would then be approximately 3×10^{-6} Nm.

Magnetic torque may well, as in this example, be a disturbance torque. However, it is common to reverse the viewpoint and take advantage of the planetary magnetic field as a control torque to counter the effects of other disturbances. We shall discuss this in more detail in a later section.

Miscellaneous Disturbance Torques

In addition to torques introduced by the spacecraft's external environment, a variety of other sources of attitude disturbance exist, many of them generated by the spacecraft during the course of its operation.

Effluent venting, whether accidental or deliberate, is a common source of spacecraft disturbance torque. When such venting must be allowed, as for example with propellant tank pressure relief valves, "T-vents" are typically used to minimize the resulting attitude perturbations. Jettisoned parts, such as doors or lens covers, will produce a transient reaction torque when released.

All of the effects we have discussed so far involve an actual momentum exchange between the spacecraft and the external environment, resulting from the application of an external torque. The momentum change in the spacecraft is the integral of this torque. Of major significance also in spacecraft attitude control are internal torques, resulting from momentum exchange between internal moving parts. This has no effect on the overall system angular momentum, but can and does influence the orientation of body-mounted sensors and hence the attitude control loops which may be operating. Typical internal torques are those due to antenna, solar array, or instrument scanner motion, or to other deployable booms and appendages. As these devices are articulated, the rest of the spacecraft will react so as to keep the total system angular momentum constant.

A major portion of the spacecraft ADCS designer's effort may be devoted to the task of coping with internally generated disturbances. If at sufficiently low bandwidth, they will be compensated by the ACS. Typically, however, internal torques are transient events with rather high-frequency content relative to the ADCS passband limits. When this is so, the ACS can remove only the low-frequency components, leaving the remainder to contribute to the overall system jitter. Such jitter can be a major problem in the design and operation of observatory or sensor spacecraft.

7.6 PASSIVE ATTITUDE CONTROL

The concept of passive attitude control follows readily from the discussion of the preceding sections. Passive stabilization techniques take advantage of basic physical principles and naturally occurring forces by designing the spacecraft so as to enhance the effect of one force while reducing others. In effect, we use the previously analyzed disturbance torques to control the spacecraft, choosing a design so as to emphasize one and mitigate the others.

An advantage of passive control is the ability to attain a very long satellite lifetime, not limited by onboard consumables or, possibly, even by wear and tear on moving parts. Typical disadvantages of passive control are relatively poor overall accuracy and somewhat inflexible response to changing conditions. Where these limitations are not of concern, passive techniques work very well. An excellent example is furnished by the Transit radio navigation satellite system,[7] for which the main operational requirement was a roughly nadir-pointed antenna. These satellites are gravity-gradient stabilized, with several having operational lifetimes of over 15 years.

A spacecraft design intended to provide passive control does not necessarily guarantee stability in any useful sense, and indeed we have seen that environmental and other effects can induce substantial unwanted attitude motion in a passively "stabilized" vehicle. For this reason, most such spacecraft include devices designed to augment their natural damping. Such "nutation dampers" can take a variety of forms, as we shall discuss, and include eddy current dampers, magnetic hysteresis rods, ball-in-tube-devices, and viscous fluid dampers.

Spin Stabilization

A basic passive technique is that of spin stabilization, wherein the intrinsic gyroscopic "stiffness" of a spinning body is used to maintain its orientation in inertial space. If no external disturbance torques are experienced, the angular momentum vector remains fixed in space, constant in both direction and magnitude. If a nutation angle exists, either from initial conditions or as the result of a disturbance torque, a properly designed energy damper will quickly (within seconds or minutes) remove this angle, so that the spin axis and the angular momentum vector are coincident.

An applied torque will, in general, have components both perpendicular and parallel to the momentum vector. The parallel component spins the spacecraft up or down; i.e., increases or decreases H. The perpendicular torque component causes a displacement of H in the direction of T. This is

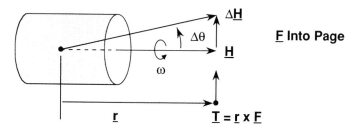

Fig. 7.13   Response of spin-stabilized spacecraft to external torque.

illustrated in Fig. 7.13, where the external force F causing the torque T is perpendicular to the plane containing H. Note then that ΔH, while parallel to T, is perpendicular to the actual disturbance force F, since $T = r \times F$. The magnitude of the angular momentum displacement is found from

$$\mathrm{d}H/\mathrm{d}t = T = rF \cong \Delta H/\Delta t \qquad (7.48)$$

where, from the geometry,

$$\Delta H = 2H \sin\Delta\theta/2 \cong H\Delta\theta = I\omega\Delta\theta \qquad (7.49)$$

hence

$$\Delta\theta \cong rF\Delta t/H = rF\Delta t/I\omega \qquad (7.50)$$

The gyroscopic stability to which we have alluded shows up in Eq. (7.50) with the appearance of the angular momentum in the denominator. The higher this value, the smaller the perturbation angle $\Delta\theta$ that a given disturbance torque will introduce.

Spin stabilization is useful in a number of special cases where reliability and simplicity are more important than operational flexibility. Satellites intended for geostationary orbit, for example, are usually spin stabilized for the two required transfer orbit burns (see Chapter 4). Some missions utilize spin stabilization as the best means of meeting scientific objectives. Notable examples in this regard are Pioneers 10 and 11, the first spacecraft to fly by Jupiter and Saturn. The primary scientific goal of these spacecraft was the investigation of interplanetary electromagnetic fields and particles; this was most easily done from a spinning platform.

Long-term stability of a spinning spacecraft requires, as we have said, a favorable inertia ratio. In visual terms, the vehicle must be a "wheel" rather than a "pencil." Also, most spin-stabilized satellites will require nutation dampers as mentioned earlier to control the effect of disturbance torques on the spin axis motion. Furthermore, if it is desired to be able to alter the inertial orientation of the spin axis during the mission, the designer must provide the capability for control torques to precess the spin axis. This is commonly done with magnetic coils or small thrusters.

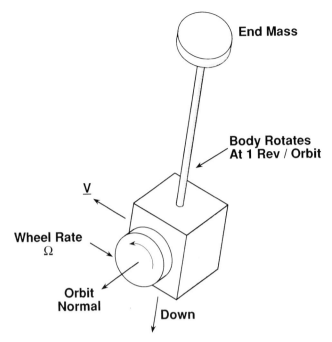

Fig. 7.14   Gravity gradient stabilization with momentum wheel.

Gravity-Gradient Stabilization

From our previous discussion, it is clear that a spacecraft in a reasonably low orbit will tend to stabilize with its minimum-inertia axis in a vertical orientation. This property can obviously be used to advantage by the designer when a nadir or zenith orientation is desired for particular instruments. The principal design feature of such a satellite again involves the inertia ratio; the vehicle must possess an axis such that $I_z \ll I_x, I_y$. As noted previously, even when the spacecraft is designed in this fashion the control torques are small, and additional damping is required to remove pendulum-like oscillations due to disturbances. These oscillations, or librations, are typically controlled through the use of magnetic hysteresis rods or eddy current dampers. Active "damping" (really active control) is also possible and, as might be expected, typically offers better performance.

The usual way of obtaining the required spacecraft inertia properties (i.e., long and thin) is to deploy a motor-driven boom with a relatively heavy (several kilograms or more) end mass. The "boom" will often be little more than a reel of prestressed metallic tape, similar to the familiar carpenter's measuring tape, which when unrolled springs into a more or less cylindrical form. Such an "open stem" boom will have substantial (for its mass) lateral stiffness, but little torsional rigidity. The possibility of coupling between easily excited, lightly damped torsional modes and the librational modes then arises, and often cannot be analytically dismissed. Again, careful selection of damping mechanisms is required.

Pure gravity-gradient attitude control provides no inherent yaw stability; the spacecraft is completely free to rotate about its vertical axis. When this is unacceptable, additional measures must be taken. One possibility is to add a momentum wheel with its axis perpendicular to the spacecraft vertical axis, as shown in Fig. 7.14. A stable condition then occurs with the wheel angular momentum aligned along the positive orbit normal.[8] Such a configuration has been flown on numerous satellites, though not with uniform success. Large amplitude librations are sometimes observed, often during particular orbital "seasons" (i.e., sun angles). Oscillations of sufficient magnitude to invert the spacecraft have occasionally occurred. These have been linked to long-period resonances in the spacecraft gravity-gradient boom which are excited by solar thermal input under the right conditions.[9]

Gravity-gradient stabilization is useful when long life on orbit is needed and attitude stabilization requirements are relatively broad. Libration amplitudes of 10 to 20 deg are not uncommon, although better performance can be obtained with careful design. An example is the GEOSAT spacecraft, a U.S. Navy radar altimetry satellite launched in 1984. Vertical stabilization to within 1 deg (1σ) was achieved through the use of a very stiff boom having an eddy current damper as its tip mass. In general, though, it will be found that gravity-gradient stabilization is too inflexible and imprecise for most applications.

Aerodynamic and Solar Pressure Stabilization

As with gravity gradient, the existence of aerodynamic and solar radiation pressure torques implies the possibility of their use in spacecraft control. This has in fact been accomplished, although the flight history is considerably reduced compared to the gravity-gradient case. The most prominent example of aerodynamic stabilization occurred with MAGSAT, a low altitude spacecraft intended to map the Earth's magnetic field.[10] This vehicle used an aerodynamic trim boom to assist in orienting the spacecraft.

The first use of solar radiation pressure to control a spacecraft occurred during the Mariner 10 mission, during which a sequence of Mercury and Venus flybys were executed. Nearly three months into its cruise phase between Earth and Venus, instability in the attitude control loop was encountered during a sequence of spacecraft roll and scan platform articulation maneuvers.[11] The resulting oscillations depleted 0.6 kg, about 16%, of the spacecraft's nitrogen control gas over the course of an hour, prior to shutdown of the roll loop by mission controllers. Subsequent analysis showed the problem to be due to an unforeseen flexible-body effect, driven by energy input from the scan platform and roll/yaw thrusters.

The roll/yaw thrusters were mounted on the tips of the solar panels in order to take advantage of the greater moment of force produced in this configuration. Preflight analysis had been done to alleviate concerns over potential excitation of the solar panels by the thrusters; however, the judgment was that only minimal interaction was possible due to the substantial difference between the ACS bandwidth and the primary solar array structural modes. Under certain flight conditions, however, it was found that higher-order modes could be excited by the thrusters and that energy in these modes could couple into lower-frequency modes that would alter the spacecraft body attitude. This would, of course, result in further use of the thrusters to correct the attitude error, followed by additional disturbances, etc., in a classic example of an unfavorable interaction between the structural and attitude control system designs.

In any event, various system-wide corrective measures were taken, and among them was a scheme to implement roll control by differentially tilting the separately articulated solar panels when necessary to implement a maneuver. The scheme worked well, albeit through intensive ground-controller interaction, and allowed sufficient fuel to be hoarded to carry the spacecraft through three encounters with Mercury.

7.7   ACTIVE CONTROL

Feedback Control Concepts

The basic concept of active attitude control is that the satellite attitude is measured and compared with a desired value. The *error signal* so developed is then used to determine a corrective torque maneuver, T_c, which is implemented by the onboard actuators. Since external disturbances will occur, and since both measurements and corrections will be imperfect, the cycle will continue indefinitely. Figure 7.15 illustrates the process conceptually for a very simple single-input, single-output (SISO) system.

This is not a text on feedback control; the subject is too detailed to be treated appropriately here. Excellent basic references include texts by Dorf,[12] Saucedo and Schiring,[13] and Kwakernaak and Sivan.[14] Kaplan[5] and Wertz[2] include brief reviews of basic feedback control concepts oriented toward the requirements of spacecraft attitude control. Nonetheless, a cursory overview of control system design concepts is appropriate before discussing the various types of hardware which might be used to implement them onboard a space vehicle.

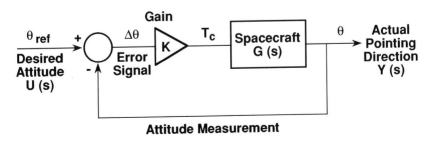

Fig. 7.15   Basic closed-loop control system block diagram.

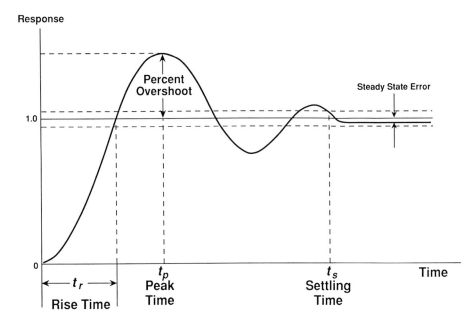

Fig. 7.16   Closed-loop control system response to step-function input.

The reader will recognize that most of the system-level blocks in Fig. 7.15 are fixed either by mission requirements (e.g., desired attitude at a given time) or by the vehicle hardware itself. The control system designer can expect to have a major role in the selection of attitude control actuators and attitude measurement devices, but once this is done, he must live with the result. The only flexibility remaining lies in the "gain" block. At the undergraduate level, a course in feedback control is nothing more than an introduction to various methods for determining the appropriate gain, K, and analyzing the resulting performance of the system.

The gain block, or *compensator*, is a *control law* which specifies the magnitude of the correction torque to be applied in response to a given error measurement. Conceptually, this could be a correction factor which is a constant multiple of the error magnitude; i.e., a 1-deg error requires 2 Nm of correction torque, while a 2-deg error calls for twice as much restoring torque. Reality is rarely this simple, and often the required compensator is somewhat more complex. Nonetheless, useful insight can be obtained even assuming constant gain, as we will see in a subsequent section.

There are several basic perfomance parameters commonly of interest to the designer. Figure 7.16 shows the temporal response of a typical closed-loop control system to a *unit step* input. This illustrates the system's behavior in response to a sudden disturbance at the input, such as an instantaneous shift in the desired attitude angle. The designer might have specifications for rise time, settling time, allowable overshoot, or allowable steady-state error. If the controller is expected to track a time-varying attitude profile, it would be of interest to examine the response to a *ramp* input. If it were necessary to follow an accelerating track, the system response to a parabola would be important. The requirement to follow more complex inputs requires control systems of correspondingly increased complexity.

The key to elementary control system analysis is that an electrical, hydraulic, or mechanical system (for us, the spacecraft) can usually be modeled over a useful operating range as a linear time invariant (LTI) system. Such systems can be represented mathematically by linear ordinary differential equations (ODE) having constant coefficients, an extremely useful property. Doebelin[15] provides an excellent treatment of the methods for and pitfalls of mathematically describing common physical systems. When there is only one variable to be controlled, such as the attitude of a single spacecraft axis, the system may be both LTI and SISO, and the design and analysis are relatively straightforward.

The advantage of describing a system with linear constant-coefficient ODEs lies in the utility of the Laplace transformation[16] in solving such equations. The transformed differential equation is a polynomial, allowing the solution to be obtained with algebraic rather than integro-differential manipulations. The subsequent analysis of the input/output relationship for the system is greatly simplified in the transform domain.

Referring to Fig. 7.15 and employing standard notation, we define the Laplace-transformed "output," the actual pointing direction, as $Y(s)$ and the desired pointing direction or "input" as $U(s)$. The spacecraft or "plant" dynamics are represented as polynomial $G(s)$. The input/output relationship is then

$$H(s) \equiv Y(s)/U(s) = G(s)K(s)/[1 + G(s)K(s)] \qquad (7.51)$$

where $H(s)$ is called the system *transfer function*. The time-domain signal $h(t)$, the inverse Laplace transform of $H(s)$, is the *impulse response* of the system. The denominator of the transfer function, $1 + G(s)K(s)$ as written here, is called the *characteristic equation*. Most performance characteristics of LTI SISO systems are determined by the locations of the system *poles* in the complex s-domain (i.e., $s = \sigma + i\omega$). These poles are the roots of the characteristic equation, leading to the use of *root locus* techniques in the design and analysis of control systems. The polynomial degree of the characteristic equation is referred to as the *order* of the system. For example, the damped simple harmonic oscillator used to model many basic physical systems, such as a simple pendulum with friction, a mass-spring-dashpot arrangement, or an RLC circuit, is the classic second-order system.

Reaction Wheels

Reaction wheels are a common choice for active spacecraft attitude control, particularly with unmanned spacecraft. In this mode of control an electric motor attached to the spacecraft spins a small, freely rotating wheel (much like a phonograph turntable), the rotational axis of which is aligned with a vehicle control axis. The spacecraft must carry one wheel per axis for full attitude control. Some redundancy is usually desired, requiring four or more wheels. The electric motor drives the wheel in response to a correction command computed as part of the spacecraft's feedback control loop. Reaction wheels give very fast response relative to other systems. Control system bandwidths can run to several tens of hertz.

Reaction wheels are fairly heavy, cumbersome, expensive, and are potentially complex, with moving parts. They are capable of generating internal torques only; the wheel and spacecraft together produce no net system torque.

With such a system, the spacecraft rotates one way and the wheel the opposite way in response to torques imposed externally on the spacecraft. From application of Euler's momentum equation, the integral of the net torque applied over a period of time will produce a particular value of total angular momentum stored onboard the spacecraft, resident in the rotating wheel or wheels, depending on how many axes are controlled. When it is spinning as fast as it can with the motor drive, the wheel becomes "saturated," and cannot compensate further external torques. If further torques are applied, the spacecraft will tumble. In practice it is desirable to avoid operation of a reaction wheel at speeds near saturation.

Because reaction wheels can only store, and not remove, the sum of environmental torques imposed on the spacecraft, it is necessary periodically to impose upon the spacecraft a counteracting external torque to compensate for the accumulated momentum onboard. Known as "momentum dumping," this can be done by magnetic torquers (useful in LEO) or by control jets (in high orbit or about planets not having a magnetic field). Magnetic torquing as a means of momentum dumping is greatly to be preferred, since when jets are used the complexities of a second system and the problems of a limited consumable resource are introduced. Indeed, in many cases reaction wheels will not be useful, and the designer must weigh their drawbacks against their many positive features, among which are precision and reliability, particularly in the newer versions that make use of magnetic rather than mechanical bearings.

A reaction wheel operating about a given spacecraft axis has a straightforward control logic. If an undesirable motion about a particular axis is sensed, the spacecraft commands the reaction wheel to rotate in a countervailing sense. The correction torque is computed as an appropriately weighted combination of position error and rate error. That is, the more the spacecraft is out of position, and the faster it is rotating out of position, the larger will be the computed correction torque.

As long as all of the axes having reaction wheels are mutually orthogonal, the control laws for each axis will be simple and straightforward. If full redundancy is desired, however, this approach has the disadvantage of requiring two wheels for each axis, bringing a penalty in power, weight, and expense to operate the system. A more common approach mounts four reaction wheels in the form of a tetrahedron, coupling all wheels into all spacecraft axes. Any three wheels can then be used to control the spacecraft, the fourth wheel being redundant, allowing failure of any single wheel while substantially increasing momentum storage when all wheels are working. Thus, the system can operate for a longer period before needing to dump momentum.

Although reaction wheels operate by varying wheel speed in response to the imposition of external torques, that does not mean that the average speed of the wheels must necessarily be zero. The wheels can also be operated around a nominal low speed (possibly a few rpm) in what is called a momentum-bias system. The momentum-bias configuration has several advantages. It avoids the problem of having the wheel go through zero speed from, say, a minus direction to a plus direction in response to torques on the spacecraft. This in turn avoids the problem of sticking friction (stiction) on the wheel when it is temporarily stopped.

Because of the nonlinearity of the stiction term, the response of wheels to a control torque will be nonlinear in the region around zero speed, imposing a jerking or otherwise irregular motion on the spacecraft as it goes through this region.[17] If this poses a problem in maintaining accurate, jitter-free control of the spacecraft, then the system designer may favor a momentum-bias system, which avoids the region around zero. As a disadvantage, the momentum-bias system lowers the total control authority available to the wheel before the saturation torque limit is reached, forcing momentum to be dumped from the spacecraft.

Momentum Wheels

When a reaction wheel is intended to operate at a relatively high speed (perhaps several tens of revolutions per minute), then a change of both terminology and control logic is employed. The spacecraft is said to possess a *momentum wheel*; a tachometer-based control loop maintains wheel speed at a nominally constant value with respect to the spacecraft body. This speed is adjusted slightly up or down in response to external torques. When the range of these adjustments exceeds what the control-system designer has set as the limit, momentum dumping allows the wheel speed to be brought back into the desired range. When magnetic coils are used to unload the wheel, this is done more or less continuously so that the tachometer circuit can operate around an essentially constant nominal value.

Use of a momentum wheel on a spacecraft offers the advantage of substantial gyroscopic stability. That is, a given level of disturbance torque will produce a much smaller change in desired nominal position of the spacecraft because of the relatively small percentage change it makes in the total spacecraft angular momentum vector. For this reason momentum-wheel systems are generally confined to use on spacecraft requiring a relatively consistent pointing direction. An example might be a low-orbit satellite where it is desired to have the vehicle angular momentum vector directed more or less continuously along the positive orbit normal, and to have the body of the spacecraft rotate slowly (i.e., 0.000175 Hz) to keep one side always facing the Earth. Use of a momentum wheel on the spacecraft aligned with its angular momentum vector along the orbit normal would be a common approach to such a requirement. The tachometer wheel control loop would function to keep the slowly rotating body facing correctly toward Earth.

The momentum-wheel system described here represents an attitude-control design referred to as a dual-spin configuration (Fig. 7.17). This configuration exists whenever a spacecraft contains two bodies rotating at different rates about a common axis. Then the spacecraft behaves in some ways like a spinner, but a part of it, such as an antenna or sensor on the outer shell, can be pointed more or less continuously in a desired direction.

As with simple spinners, dual-spin spacecraft require onboard devices such as jets or magnetic torquers for control of the overall momentum magnitude and direction. A reaction-wheel system offers the advantages of high-precision, independent control about all three spacecraft axes, whereas a simple spinner will have an extremely straightforward control-system design but minimal flexibility for pointing sensors or other devices on the spacecraft. Dual-spin design offers a combination of features from each—the dynamic advantages of simple spinners plus some of the precision-

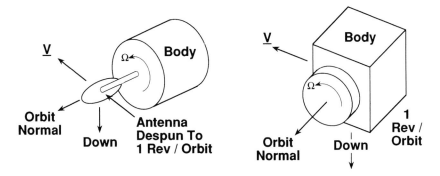

Fig. 7.17   Dual-spin spacecraft configuration concept.

pointing capability of a three-axis control system. Dual spin offers an even more important advantage under certain conditions: it allows relaxation of the major axis spin rule if the despun platform has more damping than the spinning portion of the spacecraft.[18,19]

As discussed in Section 7.4, no spacecraft can be characterized exactly as a rigid body. All objects will have some inherent flexibility, and under stress will dissipate energy. Objects with nonzero angular momentum thus tend to stabilize in a minimun-energy, flat-spin condition; i.e., rotating about the axis of maximum moment of inertia. But if the designer deliberately provides more energy dissipation on the despun platform than in the spinning portion of the spacecraft, the flat-spin condition may not be the only low-energy position of stable equilibrium. This can allow spinning the spacecraft about the minor axis of inertia with the assurance that, at least over some range, nutation angles resulting from disturbance torques will be damped rather than grow.

This important finding—arrived at independently by Landon and Iorillo in the early 1960s—has resulted in many practical applications, particularly to geostationary communication satellites, because it ameliorates the configuration limitations imposed by common launch vehicles. In terms of its control dynamics, a spacecraft is better when "pancake," as opposed to "pencil," shaped. In contrast, most launch vehicles (prior to the Space Shuttle) foster a pencil-shaped spacecraft configuration. With the realization that the major-axis spin stability rule could be relaxed with the dual-spin design, appropriate distribution of damping mechanisms allowed a better match between launch vehicle shroud and spacecraft configuration requirements.

Control Moment Gyros

Momentum wheels can be used in yet another configuration, as control moment gyros (CMG). The CMG is basically a gimballed momentum wheel, as shown in Fig. 7.18, with the gimbal fixed perpendicular to the spin axis of the wheel.[20] A torque applied at the gimbal produces a change in the angular momentum perpendicular to the existing angular-momentum vector H, and thus a reaction torque on the body. Control moment gyros are relatively heavy, but can provide control authority higher by a factor of 100 or more than can reaction wheels. Besides imposing a weight penalty, CMGs tend to be relatively noisy in an attitude control sense, with resonances at frequencies that are multiples of the spin-rate.

In many applications not requiring the ultimate in precision pointing, however, CMGs offer an excellent high-authority attitude control mechanism without the use of consumables such as reaction gas. The most notable use of control moment gyros in the U.S. space program has been the Skylab spacecraft launched in 1973 and occupied by three crews of Apollo astronauts during 1973 and 1974.

Magnetic Torquers

A spacecraft orbiting at relatively low altitude about a planet with an appreciable magnetic field can make effective use of magnetic torquers,

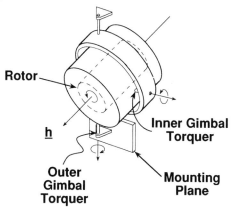

Fig. 7.18   Control moment gyro concept.

particularly for initial attitude acquisition maneuvers and for dumping excess angular momentum from reaction wheels. They prove particularly advantageous when the burden of carrying consumables, such as fuel for reaction jets, would be an impediment in spacecraft design or when exhaust gas flowing from such jets might contaminate or otherwise harm the spacecraft. A classic example in this regard, the HST, must have its primary mirror kept as clean as possible. As drawbacks, magnetic torquers have relatively low control authority and can interfere with certain components on the spacecraft, such as vidicon-tube star trackers.

Reaction Jets

Reaction-control jets are a common and effective means of providing spacecraft attitude control. They are standard equipment on manned spacecraft because they can quickly exert large control forces. They are also common on satellites intended to operate in relatively high orbit, where a magnetic field will not be available for angular-momentum dumping. Offsetting these advantages, reaction-control jets use consumables, such as a neutral gas (e.g., Freon or nitrogen) or hydrazine in either monopropellant or bipropellant systems. Normally on/off operated, they do not readily lend themselves to proportional control, although that is possible by using pulse lengths of varying duration or a mix of control jets, not all of which need to be used in every situation. It is usually not acceptable to have only one jet functioning for a given control axis, because its failure will leave the spacecraft disabled in that axis. Thus, jets usually require redundant systems, which leads to complex plumbing and control. Also, when attitude jets are used, there will likely be some coupling between attitude and translation control systems. Unless a pure couple is introduced by opposing jets about the spacecraft's center of mass, the intended attitude control maneuver will also produce a small component through the spacecraft's center of mass. This will result in an orbital perturbation.

Summary

Table 7.1 summarizes several different methods of spacecraft control. The column labeled "typical accuracy" should not be taken too literally; the intent is to provide a one-significant-digit comparison among the various methods, rather than a definitive statement of achievable accuracies.

We conclude this section with a simple example, control of the pitch angle θ on a spacecraft through the use of a reaction wheel. As shown in Fig. 7.19, the reaction wheel is assumed to be aligned with the pitch axis and spinning at a relatively slow angular velocity Ω and to have moment of inertia J about its rotational axis. The spacecraft has moment of inertia I about the pitch axis, and is assumed to be stabilized in roll and yaw; i.e., $\omega_\phi \cong \omega_\psi \cong 0$. Unknown external disturbance torques are presumed to act on the spacecraft. The rate of change of body angular momentum can be written as the sum of the wheel reaction torque and any external torques. Euler's equation under these circumstances is

$$\dot{H} = I\dot{\omega}_\theta = I\ddot{\theta} = T_{\text{wheel}} + T_{\text{ext}} \qquad (7.52)$$

Table 7.1   Attitude control techniques

Method	Accuracy, deg	Remarks
Spin stabilization	0.1	Passive, simple, low cost, inertially oriented
Gravity gradient	1–3	Passive, simple, low cost, central-body oriented
Reaction jets	0.1	Quick, high authority, costly, consumables
Magnetic torquers	1–2	Near-Earth usage, slow, lightweight, low cost
Reaction wheels	0.01	Quick, costly, high precision
Control moment gyros	0.1	High authority, quick, heavy, costly

The wheel torque profile is chosen by the attitude control system designer. Obviously, the designer seeks a stable, controlled response to external torques as well as satisfaction of certain performance criteria such as mentioned earlier. Based on intuition and experience, he might choose a feedback control law for the wheel that is the sum of position- and rate-error terms,

$$T_{\text{wheel}} = -K_p\theta - K_r\omega_\theta = -J(\ddot{\theta} + \dot{\Omega}) \tag{7.53}$$

The constants K_p and K_r represent appropriately chosen position- and rate-feedback gains. Combining Eqs. (7.52) and (7.53) yields

$$\ddot{\theta} + (K_r/I)\dot{\theta} + (K_p/I)\theta = T_{\text{ext}}/I \tag{7.54}$$

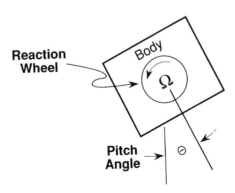

Fig. 7.19   Reaction wheel control of spacecraft pitch axis.

which has the form of the classical damped simple-harmonic oscillator, with a driving or forcing term on the right-hand side. Upon taking the Laplace transform of Eq. (7.54), the characteristic equation is found to be

$$s^2 + 2\zeta\omega_n s + \omega_n^2 = 0 \qquad (7.55)$$

where ζ and ω_n are the damping ratio and natural frequency, given by

$$\omega_n^2 = K_p/I \qquad (7.56)$$

$$\zeta = K_r/2I\omega_n \qquad (7.57)$$

The selection of a particular damping ratio ζ and natural frequency ω_n forces the choice of gains K_r and K_p. This choice can be made to a certain extent at the discretion of the control-system designer, depending on which performance criteria are most important in a given circumstance. However, for such simple systems there are long-standing criteria by which the feedback gains are chosen to obtain the most appropriate compromise among the various parameters such as overshoot, settling time, etc.[12]

To complete the example, suppose we desire the settling time to be less than 1 s following a transient disturbance. The settling time is[12]

$$T_s \equiv 4\tau = 4/\zeta\omega_n \leq 1 \text{ s} \qquad (7.58)$$

hence $\zeta\omega_n \geq 4$ rad/s. According to the ITAE performance index (integral over time of the absolute error of the system response), optimal behavior is obtained for a second order system when $2\zeta\omega_n = \sqrt{2}\omega_n$. Thus we find $\zeta = \sqrt{2}/2 \cong 0.707$, and $\omega_n \geq 4/\zeta = 4\sqrt{2}$ rad/s. The system gains K_p and K_r are found from Eqs. (7.56) and (7.57). Higher values of ω_n than the minimum would produce a shorter settling time, and might be desirable provided stability can be attained.

7.8　ATTITUDE DETERMINATION

Attitude Determination Concepts

We now consider spacecraft attitude determination, the process of deriving estimates of actual spacecraft attitude from measurements. Note that we use the term "estimates." Complete determination is not possible; there will always be some error, as discussed in the introductory sections of this chapter.

ADCS engineers treat two broad categories of attitude measurements. The first, single-axis attitude determination, seeks the orientation of a single spacecraft axis in space (often, but not always, the spin axis of either a simple spinner or a dual-spin spacecraft). The other, three-axis attitude determination, seeks the complete orientation of the body in inertial space. This may be thought of as single-axis attitude determination plus a rotational, or clock, angle about that axis.

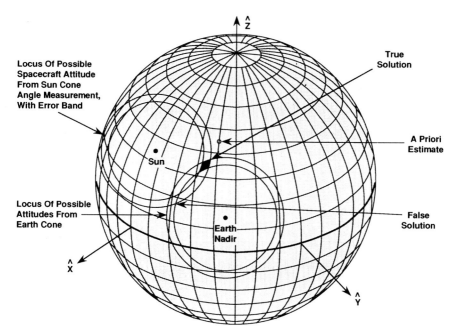

Fig. 7.20   Single-axis attitude determination.

Single-axis attitude determination results when sensors yield an arc-length measurement (see Fig. 7.7) between the sensor boresight and the known reference point. The reference point may be the sun, the Earth nadir position, the moon, or a star. The crucial point is that only an arc-length magnitude is known, rather than a magnitude and direction. Specification of the axis orientation with respect to inertial space then theoretically requires three independent measurements to obtain a sufficient number of parameters for the measurement. In practice, the engineer usually selects two independent measurements together with a scheme to choose between the true solution and a false (image) solution caused by the underspecification of parameters. The most common scheme entails using an a priori estimate of the true attitude and choosing the measurement that comes closest to the assumed value. Figure 7.20 illustrates the concept.

To effect the three-axis attitude determination requires two vectors that can be measured in the spacecraft body frame and have known values in the inertial reference frame. Examples of such potentially known vectors include, again, the sun, the stars, and the Earth nadir. The key lies in the type of sensor used to effect the measurement rather than in the nature of the reference point. The sensor must measure not merely a simple boresight error, as in single-axis attitude determination, but two angular components of the error vector. The third vector component is known since only unit vectors need be considered in spacecraft attitude control.

Consider that u and v are measured vectors to known stars in spacecraft body coordinates. We can define a unitary triplet of column vectors $\hat{i}, \hat{j},$

and $\hat{k}$, as

$$\hat{i} = u/|u| \qquad (7.59a)$$

$$\hat{j} = (u \times v)/|u \times v| \qquad (7.59b)$$

$$\hat{k} = i \times j \qquad (7.59c)$$

The unit vectors are measured in the spacecraft body frame but are also known, for example from star catalogs, in the inertial reference frame. The attitude matrix T, introduced in Eq. (7.4), rotates the inertial frame into the body frame, and can be obtained as

$$[\hat{i} \quad \hat{j} \quad \hat{k}]_b = T[\hat{i} \quad \hat{j} \quad \hat{k}]_i \qquad (7.60)$$

or

$$T = [\hat{i} \quad \hat{j} \quad \hat{k}]_b [\hat{i} \quad \hat{j} \quad \hat{k}]_i^{-1} \qquad (7.61)$$

Once the attitude matrix T is available, Eq. (7.5) can be used to obtain the individual pitch, roll, and yaw angles. Recall from Eq. (7.5) that T contains redundant information on spacecraft orientation, so the preceding system of equations is overdetermined. This allows a least-squares or other estimate[21] for spacecraft attitude, rather than a simple deterministic measurement, as the preceding equations would apply. Nonetheless, Eqs. (7.60) and (7.61) are useful to demonstrate the conceptual approach to three-axis attitude determination.

Attitude Determination Devices

The analytical approaches just discussed require measurements to be made as input data for the calculations. Attitude measurements are commonly made with a number of different devices, including sun sensors, star sensors, magnetometers, gyroscopes, and Earth-horizon scanners. These will be discussed briefly in the sections to follow.

The sun sensors most commonly used on spacecraft are digital sensors, an example of which is shown in Fig. 7.21. A given sensor measures the sun angle in a plane perpendicular to the slit entrance for the sunlight. These typically will be used in orthogonally mounted pairs to provide a vector sun angle in body coordinates, as discussed earlier. Sun sensors can be used on either spinning spacecraft or despun three-axis stable spacecraft. The sensor depicted in Fig. 7.21 has nine bits of resolution. A variety of choices are possible in trading off this resolution between total dynamic range of the sensor and the precision of the measurement associated with the least-significant bit. For example, in designing a coarse acquisition sensor it might be useful to specify a range of ± 64 deg and an accuracy of 0.125 deg in the least-significant bit. Such a sensor would not be useful for precision attitude determination measurements, but would provide a means of deter-

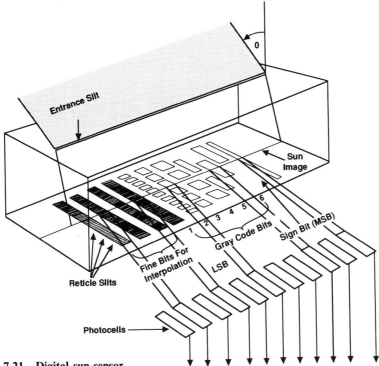

Fig. 7.21   Digital sun sensor.

mining the spacecraft's attitude from a wide variety of initially unknown configurations. Four such sensors mounted around the spacecraft can provide essentially hemispherical coverage in terms of the ability to find the sun from an initially unknown position. The importance of such a capability to mission designers and to the operations team is obvious.

At the other end of the scale, sun sensors can be designed to yield a precision of a few arcseconds in the least-significant bit, but at the price of a compromise in the overall dynamic range available with a given sensor.

For a satellite orbiting at an altitude below 1000 km, the Earth will subtend a cone angle greater than 120 deg as viewed from the spacecraft. The size of the target thus makes the Earth a tempting reference point for most LEO spacecraft attitude determination requirements. This is all the more true when, as is often the case, the fundamental purpose of placing the satellite in such an orbit is to observe a target on the Earth's surface or in its atmosphere. Earth-referenced attitude determination schemes therefore make substantial engineering sense in such cases.

The most common means of determining the Earth nadir vector employs horizon scanners. With the position of the horizon defined on each side of the spacecraft, and to its fore and aft positions, the subsatellite point or nadir will be readily defined. Although a variety of sensors have been used in this application, the most common today operate in the $14-16\,\mu m$ infrared band. This is the so-called CO_2 band, characteristic of the carbon dioxide layer in the Earth's upper atmosphere. This relatively well-defined atmospheric band is usable both day and night, irrespective of the cloud layer. For these reasons the CO_2 band makes an especially good attitude reference within the Earth's atmosphere.

The $15\,\mu m$ horizon varies by as much as 20 km from point to point on the Earth's surface, or between daylight and darkness, or at different times of the year. For low-orbiting spacecraft this alone produces an angle-accuracy limit of around 0.05 deg in either pitch or roll. When the various known and normally modeled effects, including the Earth's oblateness (discussed in Chapter 4) are taken into account, angular accuracies can be on the order of 0.02–0.03 deg using Earth horizon sensors.

The most common types of Earth horizon sensors feature small scanners attached to a wheel oriented so as to rotate around the spacecraft's pitch axis. The scanners angle outward somewhat from Earth's nadir direction, but not so far that they miss the outer edge of the Earth's horizon.

On each rotational scan in the pitch axis, therefore, each sensor will record a rising pulse and a falling pulse as the Earth's horizon is encountered going from cold space, across the Earth, and then back into cold space. Since the wheel speed is known and controlled by a tachometer circuit, the timing of these pulses, when compared against the reference time at which the pulses should occur, can be used to measure spacecraft pitch angle.

Additionally, if the scanners are mounted symmetrically on the spacecraft and if the roll angle is zero, then each scanner will have exactly the same duration between pulses. Any difference between the periods of pulse separation on either side of the spacecraft can be used to deduce the spacecraft roll angle. Figure 7.22 illustrates the geometry.

Spacecraft yaw angle, the rotational position around the radius vector from the Earth, cannot be determined using the measurements described earlier, because the Earth appears circular from the spacecraft no matter what the yaw angle might be. In inertial space, however, the yaw angle at a given moment is the same angle that will be observed as a roll angle one-fourth of an orbit later. Thus, spacecraft roll information can be used to estimate yaw, albeit on a very low bandwidth basis.

We have discussed horizon sensors for use in pairs, but a single scanner can provide excellent pitch information and adequate roll information. This allows graceful degradation in the event of a failure in one of the pair. However, extraction of roll information from one scanner requires knowledge of the orbital altitude by the onboard control logic. This is required since the reference Earth width as seen by the scanner depends upon this altitude.

The use of sun sensors and Earth horizon scanners together can provide a very powerful attitude determination and control system for a LEO spacecraft. A system can be configured using these sensors in a scheme that uses a succession of single-axis attitude determination measurements. If the spacecraft carries a slightly more sophisticated computer, with sufficient capacity to store spacecraft orbital ephemeris data periodically uplinked

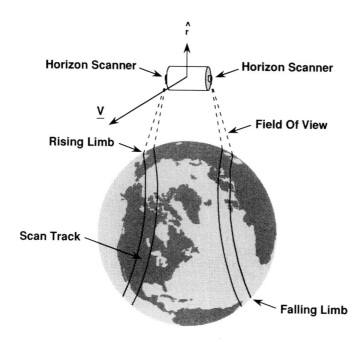

Fig. 7.22   Earth horizon scanner attitude determination concept.

from the ground, then more sophisticated processing is possible. The Earth nadir vector will be known in inertial space, as will the position of the sun at any time during the year. From the sensors, the two unit vectors necessary for three-axis attitude determination will be available in the spacecraft body frame. The spacecraft attitude can then be determined in the inertial frame and, since Earth's position is known, relative to the Earth as well. Of course, most low-orbiting satellites will have the sun available as a reference point during only part of the orbit around the Earth. For this reason, any onboard logic designed to use the sun vector as a reference must also be designed to cope with periods when the sun is not available.

A useful approach to compensating for the lack of a sun vector is to use an onboard magnetometer. Measurements are made of the three mutually orthogonal components of the ambient magnetic field. These components are then compared with the known reference components for that point in the orbit, as determined by standard magnetic field models. The difference between the measured components in spacecraft body coordinates and the known components yields the spacecraft's rotational attitude. Various models of Earth's magnetic field describe minute variations from point to point over Earth's surface. These high-order models are mainly of scientific interest because the variability of the field from one time to another and because of small perturbing magnetic effects onboard the spacecraft render their use in attitude determination somewhat problematic.

The most common magnetic field model for use in spacecraft attitude determination is the so-called tilted-centered dipole model (Fig. 7.23). This can be expressed as

$$\begin{bmatrix} B_{\text{north}} \\ B_{\text{east}} \\ B_{\text{down}} \end{bmatrix} = -(6378 \text{ km/r})^3 \begin{bmatrix} -C\phi & S\phi C\lambda & S\phi S\lambda \\ 0 & S\lambda & -C\lambda \\ -2S\phi & -2C\phi C\lambda & -2C\phi S\lambda \end{bmatrix} \begin{bmatrix} 29900 \\ 1900 \ nT \\ -5530 \end{bmatrix}$$

(7.62)

where r is the magnitude of the radius vector and (ϕ,λ) are the subsatellite latitude and longitude.

Equation (7.62) provides a vector known in inertial space as a function of the satellite's position in orbit and against which a measurement made in spacecraft body coordinates may be compared. The typical precision of a magnetic field based attitude determination measurement is on the order of $1-2$ deg.

The most accurate source for a reference vector to use in spacecraft attitude determination is a fixed star of known catalog position. Star trackers offer the potential of absolute attitude determination accuracy down to the order of approximately an arcsecond, roughly the precision of most star-catalog data. However, this is obtained at relatively high cost, not only in dollars but also in power, weight, and onboard processing required to use the information returned by the tracker. Star trackers impose additional operational penalties in that they are obviously sensitive to light from the sun and reflected light from the Earth, the moon, and stray objects that may appear in the field of view.

Many different types of star trackers have been built. Gimballed trackers point at a star and maintain the star in a centered position. The star angles in body coordinates are then read from the gimbals. Such trackers offer great precision and a very wide effective field of regard, but use many moving parts and are quite cumbersome. More commonly used today, so-called fixed-head star trackers scan the star field either electronically or by means of the spacecraft's motion. These trackers use no moving parts but have a relatively narrow field of view, on the order of $5-10$ deg.

The operational problem of star identification and verification is not trivial. Modern trackers employ processing logic internal to the tracker and can catalog stars according to both brightness and spectral type. The ADCS designer typically prefers to use only the brighter stars, both to minimize tracking error due to noise and to minimize potential confusion between similar stars in the catalog. But to use stars sufficiently bright (third or fourth visual magnitude) to minimize confusion, the time gaps between appropriate star observations may be as much as 15 min to half an hour, depending on the spacecraft orbit and the number and orientation of star trackers. If dimmer stars—down to seventh or eighth magnitude—are used, then there are nearly always appropriate stars in the field of view, but by the same token their identification and verification presents a problem.

Moreover, even if the stars can be unambiguously identified, a much larger catalog will be required for onboard storage, at a substantially greater investment of time in loading and debugging the attitude determination software.

Long gaps between suitable star observations pose problems for an attitude control system not unlike those associated with sun outage in attitude determination using sun sensors. As a practical matter, if the attitude control designer must handle gaps of several minutes or more between suitable stars, then gyroscopes will be needed on the spacecraft to provide an attitude reference during periods when no star is in the field of view of one of the trackers. Indeed, in such a case the preferred control algorithm will normally feature the use of rate gyros to measure the rotational rates around the various spacecraft axes. These rates will then be integrated over a period of time to establish the spacecraft's rotational position starting from a known baseline. When a suitable star is in the field of view of one of the trackers, or if Earth horizon scanners or sun sensors are also used, angular position updates are available to recalibrate the gyros. The system operates in this fashion more or less continuously, using gyros for high-bandwidth vehicle control and other attitude sensors to update the angular position reference. With this procedure, one is actually flying the spacecraft from a gyro reference platform and updating the navigational accuracy of the platform as targets of opportunity for attitude reference come into the field of view of one or more spacecraft sensors.

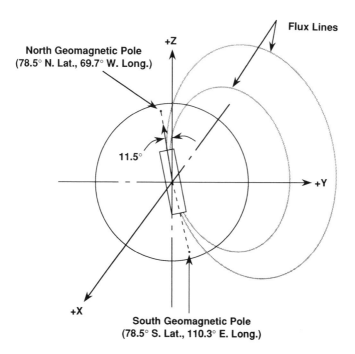

Fig. 7.23   Tilted-centered dipole model for Earth's magnetic field.

A classical gyroscope makes use of the fact that the angular momentum vector of a spinning body is constant unless an external torque is imposed on it. The gyroscope may be suspended in a nested set of gimbals and the whole assembly inserted into the spacecraft. The rotational motion of the spacecraft with respect to the gyroscope, which is fixed in space, is noted by measuring the change in gimbal angles with respect to the spacecraft. As with gimballed star trackers, this approach provides the highest accuracy available.

It is more common, however, and nearly as accurate to use a "strapdown" platform. In such a platform the gyroscope is kept in more or less the same position relative to its assembly and to the spacecraft body by means of a control loop that supplies the torque necessary to keep the rotor in the correct position. The applied torque needed to maintain equilibrium is known by the electronic logic and used as a measure of the torques which have actually been imposed on the spacecraft body. Gyros may be configured to measure either rate or integrated rate (position), and may provide either one or two degrees of freedom for making a measurement.

Gyroscopes can provide extremely high bandwidth and extremely sensitive attitude or attitude-rate information. In many ways they make ideal components in an attitude-determination and control system. As noted earlier, they do require periodic recalibration from star trackers, sun sensors, or Earth horizon scanners, to remove the effects of drift and other systematic errors resulting from the fact that the gyro rotor cannot be maintained in truly torque-free motion. The models of highest accuracy are heavy, and all models have traditionally been rather expensive.

Gyro systems have been designed to last many years on orbit. However, it is fair to say that individual unit failures are not uncommon, and that system level robustness is generally obtained through the use of individually redundant units.

The limitations of conventional electromechanical gyroscopes have been well understood for many decades. For this reason, since the mid-1960s, there has been interest in replacing conventional gyros with inertial rate-measuring devices that use different underlying principles. These include ring-laser, laser fiber-optic, and hemispherical-resonator gyroscopes (the name "gyroscope" applied to such devices is a misnomer, but has wide currency).

Fiber-optic and ring-laser gyros provide a closed path around which laser light can be sent in opposite directions from a laser source. When the closed path (loop) is rotated relative to inertial space, the counter-rotating beams of light experience slightly different path lengths depending on which direction they have traveled. When the two beams are brought together at the end of the path, one will be slightly out of phase with respect to the other. The amount of the phase difference depends on the rotational rate applied to the closed loop.

The advantages of laser gyros have sparked development of them for various airborne and spacecraft applications. They have few moving parts, very high levels of reliability, and are inherently robust, capable of withstanding much rougher treatment than most mechanical systems. Fiber-optic gyros in particular can be made quite compact. For these and other

Table 7.2   Attitude determination techniques

Sensor	Accuracy, deg	Remarks
Sun sensor	0.01–0.1	Simple, reliable, cheap, intermittent use
Horizon scanner	0.02–0.03	Expensive, orbit-dependent, poor in yaw
Magnetometer	1	Cheap, low altitude only, continuous coverage
Star tracker	0.001	Expensive, heavy, complex, high accuracy
Gyroscope	0.01/h	Best short-term reference, costly

reasons, aircraft and spacecraft designers have for some time been moving away from electromechanical gyros and toward such newer concepts. This is exemplified by the standard use of ring-laser gyros in aircraft such as the Boeing 757 and 767 series, as well as the inclusion of laser gyros as the baseline inertial navigation platform for the Orbital Sciences Corporation Transfer Orbit Stage discussed in Chapter 5. Nevertheless, laser and other innovative gyro types have yet to see full acceptance in the aerospace industry, and electromechanical gyros can be expected to remain in heavy application for a long time.

Table 7.2 summarizes the range of attitude determination approaches discussed in this chapter and provides a rough quantitative level of accuracy obtainable for each.

7.9   SYSTEM DESIGN CONSIDERATIONS

Attitude determination and control system design does not end, nor does the designer's responsibility end, with the selection of measurement and control methods and components to implement them; that may indeed be the easiest part of the task. Many system-level considerations confront the ADCS designer. These must be addressed before the spacecraft can be ready to fly, or even before the ADCS can be properly related to the remaining subsystems aboard the spacecraft.

Throughout this text we have mentioned the difficulty of coping with errors inevitable in translating any idealized approach into practice, or which arise as a consequence of simplifying the real world with a model for use in design and analyses. Nowhere is this more true than in treating attitude determination and control systems. For example, throughout this chapter it has been assumed that the spacecraft coordinate system will be anchored in the spacecraft's center of mass. This may be true in principle, because the coordinate system can always be redefined if necessary to fit the actual center-of-mass location; but the attitude control system actuators and the attitude determination system sensors will have been located with respect to the *intended* spacecraft-body-axis frame *supposedly* located in the center of mass.

Since offsets between the center of mass and the geometric center will always exist, it follows that there will inevitably be a coupling between

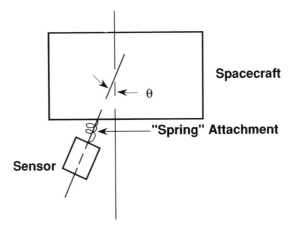

Fig. 7.24   Conceptual model for a spacecraft control-structure interaction.

attitude control and translational maneuvers. That is, thrusters intended to effect translational motion (to impart ΔV) will inevitably alter the attitude. This must be compensated by the attitude control system during ΔV maneuvers; and indeed it may well be that the attitude control system's overall control authority is determined on the basis of expected center-of-mass and center-of-thrust offsets for the ΔV system. Similarly, the use of attitude thrusters to control the rotational position of a spacecraft will usually impart a residual ΔV to the spacecraft, probably causing an undesired orbital perturbation.

Another assumption throughout this chapter (and usually in the real world) is that of an essentially rigid spacecraft body to which are attached the various parts and components that make up the complete system. In practice no spacecraft structure or component attached to it will ever be perfectly rigid. This has obvious and well-understood effects with regard to the major-axis spin rule, as discussed earlier. Additional and more subtle effects are present, the most important being the potential for interaction between structural flexibility and the attitude determination and control system.

This situation has been depicted conceptually in Fig. 7.24, showing a sensor mounted through a spring-like attachment to the spacecraft body. As the spacecraft moves in response to attitude control maneuvers there will be relative motion between the attitude determination sensor and the spacecraft body. The sensor will not register the correct angle between the spacecraft's body axis and the desired reference source, and its error will be fed back into the spacecraft control system and used to cause an additional maneuver, which will again deflect the sensor relative to the spacecraft. The process continues *ad infinitum*. In short, the feedback signal has become corrupted by the relative deflection between the spacecraft body and the attitude determination sensor. The phasing of this error can cause the entire loop to become unstable. In practice, the spacecraft designer wants, as much as possible, to remove the effects of structural flexibility. This is done by using the most rigid mounting possible between various spacecraft components and also by using attachments that will provide as much damping as possible.

However, as spacecraft primary structures become larger and larger, the fundamental modal frequency inevitably becomes lower and lower, to a point where not uncommonly the fundamental structural modes fall near or within the desired attitude control system bandwidth. This is not an acceptable situation. The spacecraft fundamental structural modes and the attitude control system bandwidth must be separated to the maximum extent practical; this usually means a factor of five or more separation between the ADCS passband and the lowest-order structural mode. Even greater separation is desirable when it can be achieved. If this cannot be done either by stiffening the spacecraft or lowering the control bandwidth, then the spacecraft structural modes themselves must be modeled as part of the overall attitude control loop. Development of control systems for flexible spacecraft has been an important research topic for two decades, and as spacecraft have become much larger it has acquired great practical moment. The literature is replete with examples of both theoretical and applied developments relating spacecraft attitude control and structural interaction.

So far, this chapter has mainly concerned design of systems to measure spacecraft attitude and to control it about some nominal point, which has implicitly been assumed to be constant. This in fact may be the easiest part of the mission operator's problem. In contrast, how and in what way to maneuver the spacecraft between one desired attitude and another can be a major issue. Provision in the spacecraft design for such attitude maneuvers impacts spacecraft control through development of both onboard and ground operations software. At the very least, provisions must be made for initial acquisition maneuvers following deployment of the spacecraft from its launch vehicle.

Furthermore, if the spacecraft has a very flexible operating profile, with more than one intended target of observation (such as would be the case, for example, with an astronomical telescope), then the designer's goal must include the ability to execute an equally flexible set of attitude maneuvers. The attitude-maneuver design must consider other spacecraft subsystems as well as the attitude system. A solar power system may need to have the arrays pointed toward the sun; and the arrays may need articulation individually to compensate for spacecraft maneuvers. Thermal radiators may need to be oriented toward dark space, or at least not toward the sun, while antennas may be required to point continuously toward Earth. As an excellent example of Murphy's Law, ground controllers want uninterrupted communication with the spacecraft at the exact time, during an attitude maneuver, when it is most difficult to achieve.

Similarly, the vehicle's scientific sensors or other systems often cannot look at the Earth or the sun without at least interrupting operation or, in the worst case, being irreparably damaged. Furthermore, attitude maneuvers usually must be conducted with reasonable fuel efficiency. In other cases it may be required to minimize the time to execute a maneuver. Thus, design of optimal attitude maneuvers subject to known constraints on the spacecraft has been the subject of much theoretical and applied interest.[22,23]

We close this chapter with a few comments on testing. Few tasks are more challenging to the ADCS designer than testing his system. To be as realistic as possible, the system must be tested in essentially its operational configuration. However, the spacecraft is intended to operate in zero gravity (g) and vacuum, not at 1 g in the atmosphere. Air bearing tables can be useful in simulating zero-g behavior, but cannot replicate the extremely low damping of the vacuum environment. It is also difficult (though not impossible for small systems) to devise an air bearing arrangement to allow simultaneous testing of all three axes. Attitude sensor inputs are also difficult to replicate on the ground, even in space simulation chambers. Given the uncertainties which are present, it is unsurprising that getting the attitude determination and control system to behave properly is often the major task following initial deployment of the spacecraft in orbit.

References

[1]Hughes, P. C., *Spacecraft Attitude Dynamics*, Wiley, New York, 1986.

[2]Wertz, J. R. (ed.), *Spacecraft Attitude Determination and Control*, D. Reidel, Boston, MA, 1978.

[3]Noether, G. E., *Introduction to Statistics* (2nd ed.), Houghton Mifflin, Boston, MA, 1976.

[4]Taff, L. G., *Computational Spherical Astronomy*, Wiley, New York, 1981.

[5]Kaplan, M., *Modern Spacecraft Dynamics and Control*, Wiley, New York, 1976.

[6]Bracewell, R. N., and Garriott, O. K., "Rotation of Artificial Earth Satellites," *Nature*, Vol. 82, No. 4638, Sept. 1958, pp. 760–762.

[7]Mobley, F. F., and Fischell, R. E., "Orbital Results from Gravity-Gradient Stabilized Satellites," NASA SP-107, 1966.

[8]Thomson, W. T., "Spin Stabilization of Attitude Against Gravity Gradient Torque," *Journal of the Astronautical Sciences*, Vol. 9, 1962, pp. 31–33.

[9]Goldman, R. L., "Influence of Thermal Distortion on Gravity Gradient Stabilization," *Journal of Spacecraft and Rockets*, Vol. 12, No. 7, 1975, pp. 406–413.

[10]Tossman, B. E., Mobley, F. F., Fountain, G. H., Heffernan, K. J., Ray, J. C., and Williams, C. E., "MAGSAT Attitude Control System Design and Performance," AIAA Paper 80-1730, Aug. 1980.

[11]Dunne, J. A., and Burgess, E., *The Voyage of Mariner 10: Mission to Venus and Mercury*, NASA SP-424, Washington, DC, 1978.

[12]Dorf, R. C., Modern Control Systems (3rd ed.), Addison-Wesley, Reading, MA, 19

[13]Saucedo, R., and Schiring, E. E., *Introduction to Continuous and Digital Control Systems*, MacMillan, New York, 1968.

[14]Kwakernaak, H., and Sivan, R., *Linear Optimal Control Systems*, Wiley-Interscience, New York, 1972.

[15]Doebelin, E. O., *System Modeling and Response*, Wiley, New York, 1980.

[16]Wylie, C. R., and Barrett, L. C., *Advanced Engineering Mathematics* (5th ed.), McGraw-Hill, New York, 1982.

[17]Dahl, P. R., "A Solid Friction Model," Aerospace Corp., TOR-0158 (3107-18)-1, El Segundo, CA, May 1968.

[18]Landon, V., and Stewart, B., "Nutational Stability of an Axisymmetric Body Containing a Rotor," *Journal of Spacecraft and Rockets*, Vol. 1. Nov.-Dec. 1964, pp. 682–684.

[19]Likins, P. W., "Attitude Stability for Dual-Spin Spacecraft," *Journal of Spacecraft and Rockets*, Vol. 4, Dec. 1967, pp. 1638–1643.

[20]O'Connor, B. J., and Morine, L. A., "A Description of the CMG and Its Application to Space Vehicle Control," *Journal of Spacecraft and Rockets*, Vol. 6, March 1969, pp. 225–231.

[21]Gelb, A., *Applied Optimal Estimation*, MIT Press, Boston, MA, 1974.

[22]Li, F., and Bainum, P. M., "Numerical Approach for Solving Rigid Spacecraft Minimum Time Attitude Maneuvers," *Journal of Guidance, Control, and Dynamics*, Vol. 13, Jan.-Feb. 1990, pp. 38–45.

[23]Byers, R. M., Vadali, S. R., and Junkins, J. L., "Near-Minimum Time, Closed-Loop Slewing of Flexible Spacecraft," *Journal of Guidance, Control, and Dynamics*, Vol. 13, Jan.-Feb. 1990, pp. 57–65.

8
CONFIGURATION AND
STRUCTURAL DESIGN

8.1 CONFIGURATION DESIGN

Of all the subsystem areas discussed in this book, configuration design may most closely approximate systems engineering as a whole. The configuration designer must be involved in detail with every other subsystem in the spacecraft. The configuration must accommodate all the disparate requirements and desires of the various subsystems and, where those are in conflict, reach a suitable compromise. For a complex spacecraft, the wide variety of requirements, desires, and constraints and the conflicts that inevitably arise among them provide a substantial challenge. A variety of innovative solutions have evolved in various projects. These will be discussed in some detail, not as the final answers, but simply as examples of working solutions. Before we discuss solutions, we need to understand the factors that drive the design.

Design Drivers

It can be stated as an axiom that configuration design is *always* a compromise. A variety of requirements, which invariably involve some conflict, drive the design of every spacecraft. As with any complex system, there usually exist a variety of solutions, each of which can result in a more or less satisfactory design. As a result, this section will not present solutions but will instead discuss the design drivers and some of the considerations involved in developing solutions.

Mission goals. A variety of typical missions can be listed to illustrate the types of missions that can be carried out by any spacecraft. Table 8.1 lists the common generic classes of spacecraft missions. It may be possible to think of missions that do not precisely fit the list, but the general characteristics that are encompassed in the list will cover most missions.

(1) *Communications satellites.* Communications satellites have mostly been located in geostationary orbit because of the wide area coverage available and the simplicity of communicating with an object that remains stationary in the sky. Since ground stations require no tracking capability, construction and operating costs are substantially reduced. The need to simply point accurately in one direction and perform only a relatively

simple relay function also simplifies the spacecraft. On the other hand, the very large number of channels handled by a modern communications satellite effectively complicates the design while the large investment involved, and the importance of the function dictates very high reliability and long life.

Recently, interest has been growing in a network of low-altitude satellites that replace the geosynchronous type. The low-altitude satellites will involve lower unit costs but also a very large number of satellites and possibly some increase in operational complexity. The major advantage of this approach is in its robustness. Loss of a substantial percentage of the satellites will result in a functional albeit degraded system, whereas loss of a single big satellite will shut down the entire system.

Because much of their territory is at high latitudes poorly served from geostationary orbit, the Soviets have evolved the Molniya communications spacecraft. These spacecraft operate in highly elliptic synchronous orbits oriented so that the apoapsis is located over the regions of interest. Thus, the spacecraft spends most of its time above the horizon of the Soviet Union. By careful selection of inclination and periapsis latitude and altitude, the plane of the orbit is "locked" against precession so as to maintain the desired orientation relative to the Soviet Union throughout the year.

(2) *Earth observation satellites.* With the possible exception of communications satellites, Earth observation satellites are probably the most common general type of spacecraft currently in existence. Both military and civilian versions exist. Operating orbits range from low circular through elliptic to geostationary and beyond. Even though they all carry out the same generic function (i.e., they observe the Earth and its near environment), the variety is huge, with many types of sensors operating in

Table 8.1   Example of spacecraft mission goals

Communications relay
Earth observation
Civilian
Military
High-altitude
Low-altitude
Solar observation
Astronomical
Fields and particles
Planetary observation
Flyby
Orbiter
Lander

a variety of wavelengths. Most are passive; i.e., they conduct their observations using naturally emitted radiation. A few conduct active observations using radar.

The type of Earth observation spacecraft most familiar to the casual observer is the weather satellite. Both military and civilian agencies operate networks of these spacecraft. The military and civilian spacecraft are generally similar, although specific requirements may result in some differences in sensors or operations. Two generic types of weather satellites exist: those located in geostationary orbit to provide wide area coverage (almost 40% of the Earth from one satellite) and those in low circular polar orbit that provide high-resolution data. The latter are usually in sun-synchronous orbit (see Chapter 4) so that a given locality is viewed at the same local time each day. The low-orbit spacecraft are nadir pointing (see Chapter 9). The geosynchronous types may be nadir pointers or spinners with despun or spin scan instrumentation.

Military reconnaissance spacecraft constitute a large percentage of Earth observation spacecraft. Some of these are at high altitude for wide area surveillance, whereas others operate at low altitude for detailed study. Among the latter are some of the highest-resolution spacecraft imagers yet flown. The performance is classified, but the open literature describes identifying specific individual aircraft by tail number. In some cases, the spacecraft descend to relatively low altitudes in order to improve resolution. Circular orbits at such altitudes would not be stable; therefore, the spacecraft operate in elliptic orbits with very low periapsis altitudes so that the excess momentum can overcome the drag that would render a circular orbit impractical.

Earth resources satellites such as the U.S. Landsat and the French Satellite Probatoire d'Observation de la Terre (SPOT) are invaluable in the study of the surface composition of the Earth. Both scientific and commercial interests are served by the data from these spacecraft, which generally employ sensors operating in a variety of spectral bands. Again, near-polar sun-synchronous orbits are most commonly used.

(3) *Solar observation.* Solar observation is among the oldest disciplines in space science, going back to the sounding rocket observations that began just after World War II. The advantages of solar observation without atmospheric filtering are obvious. For some observations it is desirable to get away from the Earth altogether; thus, many solar observation spacecraft have been in solar rather than Earth orbit. The sun emits huge amounts of energy in virtually every known wavelength from infrared to x ray plus particulate radiation. Thus, the sensors for solar observation are by no means restricted to optical wavelengths. Such phenomena as the decay of solar-emitted neutrons make it necessary to approach as close to the sun as possible if those particles are to be detected. To date, no spacecraft has come much closer than the orbit of Mercury, but several mission concepts have been studied for grazing or impact missions. An interesting possibility, applied to the International Sun-Earth Explorer (ISEE) mission, is to place the spacecraft in a "halo" orbit about the L-1

libration point (see Chapter 4), thus locating the spacecraft in line between the Earth and the sun.

(4) *Astronomical.* Astronomical spacecraft to date have been operated in low Earth orbit. Observations in the infrared, visible, and ultraviolet are of interest. Some instruments used for broad area surveys will have relatively generous pointing constraints, whereas others designed for detailed observation will have extremely tight constraints. The Hubble Space Telescope is a case in point, requiring the most difficult pointing accuracy yet flown. The reason for discussing this seeming attitude control topic here is that pointing accuracy constraints translate into alignment accuracy and control requirements on the spacecraft structure, which will be strong drivers on configurations, structural design, and material choice.

(5) *Fields and particles.* Spacecraft devoted to the observation of magnetic fields and particulate radiation are generally less concerned with accurate pointing than are other types of spacecraft. In many cases, a rotating spacecraft is desired to allow widespread coverage of the sky. Spacecraft designed to conduct this type of investigation as well as those requiring high-accuracy pointing often have some difficulty meeting all the requirements and desires. During interplanetary cruise the three-axis stabilized Voyager spacecraft are occasionally commanded into a roll-and-tumble sequence in order to provide the fields and particles payload with a survey of the celestial sphere.

(6) *Planetary observation.* Spacecraft designed for planetary observation from orbit differ little from their counterparts at Earth except for requirements edicted by differing environments. Some planetary spacecraft are on flyby rather than orbital missions. In such a case, a scan platform for narrow field of view instruments is highly desirable if not mandatory. This will allow multiple scans and photomosaic generation. This would be very difficult to accomplish by maneuvering the spacecraft during the few minutes available in a typical encounter. Planetary landers, of course, require aerodynamic deceleration and/or rocket propulsion for descent and landing.

Payload and instrument requirements. Table 8.2 presents a typical list of requirements that may be levied on the spacecraft by the payload. This list primarily addresses a payload of observational instruments, but many of the requirements are typical of essentially any payload.

Payload items may demand a specific location on the spacecraft in order to meet the other requirements listed. This can often be a problem when more than one instrument wants the same piece of spacecraft "real estate," or when the requirement conflicts with those of other subsystems.

Pointing accuracy requirements can drive configuration and structural design far more substantially than might appear to the casual observer. For example, stringent requirements may dictate extreme rigidity and temperature stability in order to minimize distortion in alignment between the

Table   8.2   Payload/instrument requirements

Location
Pointing accuracy
Temperature
Magnetic field
Radiation
Field of view

instrument mount and the attitude control reference. This can in turn dictate structural design, material choice, and configuration design.

Many payload elements have relatively tight temperature constraints because of delicate components. This will require attention but is usually not a major driver. However, when a particular sensor requires very low temperature, as is often the case with infrared sensors, the need to provide a clear view of space while eliminating the sun, planetary surface, or illuminated or hot spacecraft parts from the radiator field of view can be a major design driver.

In some cases a magnetically sensitive component can simply be shielded from spacecraft-generated magnetic fields and thus will not offer any particular configuration problem. However, if the component is a sensor for detecting and measuring planetary magnetic fields, it must be isolated from the spacecraft fields without compromising its function. Generally, the answer is distance—often a fairly large distance. This in turn usually dictates some sort of deployable structures.

Many components are sensitive to radiation dosage. Although shielding is possible, it costs weight. Clever configuration design may be called upon to minimize exposure to radiation sources such as radioisotope-based power generators and heaters.

Field of view requirements on configuration are obvious since the payload has to be able to see its target without interference from other parts of the spacecraft. This requirement is more easily stated than satisfied and will often tax the designer's ingenuity to achieve an acceptable compromise.

Environment.   Environmental drivers (Table 8.3) on configuration and structure design are fairly obvious. The variable intensity of solar energy with distance is primarily of concern for thermal control and solar to electric conversion. In a discussion of spacecraft, one might assume that atmosphere would only be of concern for planetary landers and entry systems. Recall, however, that all spacecraft have to survive in the Earth's atmosphere first, and concerns of chemical attack (oxygen and humidity), pressure fluctuations, wind, etc., must be considered. Of particular concern is the rapid pressure drop during ascent and passage through the pressure regime conducive to corona discharge.

Table 8.3   Environment

Solar distance
Atmosphere
Radiation
Thermal
Vibration
Acoustic

Environmental radiation is usually not a major concern to the designer of configuration and structure, except that, on occasion, it may be necessary to accommodate shielding of sensitive components. In severe environments, where one might shield the entire spacecraft, the configuration may be driven toward a very compact design to maximize self-shielding and minimize the external area that must be shielded.

The impact of local thermal environment can range from minimal to substantial. For a spacecraft operating in deep space, the sun is essentially the entire thermal environment, and, unless it is very close, it is relatively easy to deal with. On the other hand, a spacecraft in low orbit about Mercury not only experiences solar intensity on the order of 10 times that of Earth but is also exposed to radiation from the hot surface of the planet. The temperature of the hot side of Mercury (up to 700°C) is such that the reradiation of the absorbed solar energy takes place in the infrared. Since spacecraft are usually designed to radiate in the infrared to dispose of absorbed and internally generated heat, they are also fairly good infrared absorbers. Thus, the surface of Mercury, radiating infrared at a rate nearly comparable to the sun itself, is a major source of thermal input. Very clever configuration and mission design is required to maintain a spacecraft within acceptable limits.

The design requirements for vibration and acoustics are sufficiently obvious to require little comment. As a caveat, however, the engineer should keep in mind that *which* environment is the driver may be less clear. Launch or atmospheric entry may be the most severe; however, they are brief compared to, for example, a 4-h cross-country flight or a 4-day truck ride. Designing for the mission without considering how the hardware is handled on the ground frequently causes major problems.

Power source.   Table 8.4 lists various types of power sources that may impact configuration and structure. Notably absent are batteries and fuel cells, which, except for accommodating bulk and mass, pose few constraints as a rule. The same cannot be said for those on the list. Solar photovoltaic systems require large areas with an essentially unobstructed view of the sun and, at least in the case of large flat arrays, the ability to maintain the array surface normal to the sun. Drum-shaped, spin-stabilized craft require even larger areas because only a part of the area can be exposed to the sun. All the preceding factors would not cause great

Table 8.4   Power source

Solar photovoltaic
Radioisotope thermoelectric generators
New technology
Reactor-based
Solar dynamic
Radioisotope dynamic

problems except that it is also necessary to mount and accurately point antennas, science instruments, attitude control sensors, etc. These requirements are often in conflict regarding which item of hardware occupies a particular area on the vehicle and because of possible shadowing, field of view interference, etc.

Radioisotope thermoelectric generators (RTGs) generally relieve some of the location problem and the demand for large area; however, they bring their own set of problems. Because of the need to reject heat from the outer surface of the RTG and because of the radiation from the decaying isotopes, it is usually not practical to mount them inside the spacecraft or even extremely close to it. In most applications, RTGs are boom mounted at some distance from the spacecraft to reduce the effect of both nuclear and thermal radiation. Launch volume constraints dictate that such mounting structure be deployable. Examples of this sort of installation will be seen later.

Newer technology systems that have not flown (except for one experimental reactor) on U.S. spacecraft are listed. The radiation output from a nuclear reactor is far more energetic and damaging than that from an RTG. Also, since reactors emit far more power, the waste heat to be disposed of is greater. This latter requirement leads to very large radiator areas with all the predictable problems in launch stowage, thermal input, view of space, etc. The radiation requires great distances and/or massive shielding. In proposed designs using reactors a compromise is usually reached, placing the reactor as far from the spacecraft as practical and then shielding to reduce the radiation flux at the spacecraft distance to an acceptable level. Because of the thickness and great weight of the shield, "shadow shielding" is employed. That is, the shield is placed between the reactor and spacecraft rather than shielding the full 4π sr as is done for Earth-bound installations. The very long boom with large masses at either end and possibly in the middle (as in ion propulsion units) introduces some major challenges in structure and mechanism as well as in dynamics and control.

Solar dynamic systems require the same sun pointing and access as photovoltaic arrays, possibly with somewhat tighter accuracy constraints depending on the type of collector used. Most dynamic conversion systems have very large waste heat radiators because the low-temperature end of the thermodynamic cycle is relatively cool in order to achieve good efficiency. The radiators require a good view of space and a minimum view of the sun, nearby planets, etc.

An additional problem involving any unit using dynamic conversion is the introduction into the structure of a "hum" at the frequency of the rotating machinery. It may be necessary to design the structure and select materials to damp out the vibration as much as possible to reduce the impact on attitude control.

Radioisotope dynamic systems are, as one would expect, a hybrid of the problems of the solar dynamic system and the RTG. The radiation problem persists combined with the need for large radiator area and the potential vibration problems inherent in dynamic conversion. The much greater efficiency of these dynamic units compared to RTGs or solar photovoltaic is the incentive to use them. However, they do introduce some challenges to the configuration and structure designer.

Launch vehicles. The first and most obvious constraint forced on the designer is that of launch mass capability (Table 8.5). Next is payload dimension, not only the length and diameter of the payload volume available but also dimensions of the attachment interfaces. The mass and volume available dictate the size of the basic structure, drive the selection and design of deployable structures, and, in many cases, strongly influence the choice of materials.

The acoustic and vibration environment imposed by the launch vehicle is generally the most intense that the spacecraft will encounter, although, as mentioned earlier, because of the relative brevity of the powered flight, the cumulative effect of prelaunch environments may be severe. The measured or calculated launch environments are used to define qualification test criteria. These criteria are defined by drawing curves that envelop the actual environment. Factors are then applied to these curves to define flight qualification or flight acceptance (FA) criteria and higher factors to define type acceptance (TA) criteria. A factor of 1.5 applied to the actual might be used to define FA levels and 2.0 to define TA. Actual flight articles would be tested to FA levels to demonstrate workmanship and margin over expected actuals. Flight articles would be expected to withstand FA without damage or unaccceptable response. TA levels are used to demonstrate design values and may push the structure to near failure. Yielding or other responses may occur that would render the article unaceptable for flight use. TA levels are only applied to nonflight prototypes. Chapter 3 provides environmental data for several launch vehicles.

Table 8.5 **Launch vehicle constraints**

Mass
Dimension
Vibration
Acoustic
Safety

Launch safety constraints are generally not a major driver for structure and configuration design when expendable launch vehicles are used. The primary constraint is that the spacecraft not fail, a criterion to which everyone involved will subscribe. Occasionally, however, there will be issues revolving around a launch abort. In the event of an errant launch vehicle being destroyed by range safety, there may be the desire that the spacecraft break up in certain ways or not break up at all upon re-entry. This situation might obtain in the case of a spacecraft bearing RTGs or a nuclear reactor, for example. For the most part, the constraints will be on pad operations involving personnel. The most common example is imposition of minimum safety factors and possibly fracture mechanics criteria on pressure vessels that must be pressurized in the presence of personnel.

Spacecraft destined for launch on the Space Shuttle come under much more severe scrutiny. Because the Shuttle is manned and because there are few of them, strict safety constraints are levied to ensure that no problem with the payload will cause a hazard to the Shuttle or crew. With the exception of higher safety factors and more emphasis on fracture mechanics, there is not a great difference in the engineering side of designing for the Shuttle vs an expendable launcher. The real difference is in the extensive review and certification process designed to prove compliance.

Communication. The primary impact of the spacecraft communication system on configuration (Table 8.6) lies in antenna size and required location relative to the major attitude control references. Relatively small lower-gain antennas usually do not present a major problem, whereas a large high-gain dish, especially one that is required to move during the course of a mission, can require considerable attention. In the latter case, the antenna will generally be latched down during launch and deployed after placement on orbit, placing additional requirements on the structure design and on related mechanical deployment devices.

Not only is pointing direction a concern to the configuration designer, but the required pointing accuracy may be as well. Pointing accuracy requirements may drive structural design and the choice of structural materials in an effort to minimize the effect of distortion from thermal effects or structural loads. Such distortions, by introducing errors between attitude control references and the antenna mount, can cause problems in accurate pointing of very tight beams.

Finally, the radiated power of the communications system may impose certain requirements. In the case of very-high-power systems, it may be necessary to avoid placing components where they can be illuminated by sidelobes of the antenna. Also, if a very large amount of power is being

Table 8.6   Communications

Antenna size
Pointing accuracy
Radiated power

radiated from the antenna, then even more is being dissipated within the spacecraft. The configuration design must be able to accommodate the conduction of this internally generated heat to the appropriate radiating surface of the spacecraft for rejection to space.

Spacecraft Design Concepts

This section presents several spacecraft design concepts as an illustration of how various design teams have dealt with the design drivers discussed previously. Both overall configuration and internal packaging concepts are presented. Some of the pros and cons of each concept are discussed. Concepts for deployable booms and scan platforms are also discussed.

Spacecraft configuration.

(1) *Voyager.* Figure 8.1 shows the Voyager spacecraft. Two of these vehicles, built and tested by the Jet Propulsion Laboratory (JPL) for NASA, were launched in 1977 to explore the outer planets of the solar system. The baseline mission was to be a 4-yr trip involving a flyby of Jupiter and Saturn by each spacecraft. In the event, by use of planetary swingby techniques, Voyager 2 was redirected to extend its mission to encounters of Uranus and Neptune. The latter encounter, in Aug. 1989, was some 12 yr after launch. Voyager 1, its trajectory bent out of the plane of the ecliptic by its encounter with Saturn, will encounter no more planets but continues to send back data concerning regions of space not previously visited. Both spacecraft have considerably exceeded solar system escape velocity and will continue indefinitely into interstellar space.

The great distances from the sun at which the Voyager spacecraft are designed to operate dictate two of the most prominent features of the configuration. Since solar power is impractical, power is provided by RTGs. Since great distance from the sun also connotes great distance from Earth, a large antenna is required to support the data rates desired.

However, the sun does provide a useful attitude reference. A sun sensor peers through an opening in the antenna dish. Since the antenna must point at the Earth, there is a slight offset between the antenna boresight and that of the sun sensor. But this bias was designed for the baseline Jupiter/Saturn mission and is too large at the huge distance from the sun at which both spacecraft now operate. A star tracker is used to provide the reference for the third axis.

The RTGs are mounted on a rigid hinged boom that was latched to the final launch stage and then swung into final position after stage burnout. The in-line arrangement is used to minimize radiation to the bus, since the inboard RTG, although itself a source of radiation, is also a shield for the radiation from the other two.

With one exception, the science instruments are mounted on the boom radially opposite to the RTG boom. This has the desirable feature of placing them as far as possible from the RTGs. The fields and particles instruments, concerned mostly with the sun, solar wind, and planetary trapped radiation, are mounted on the boom outside the shadow of the

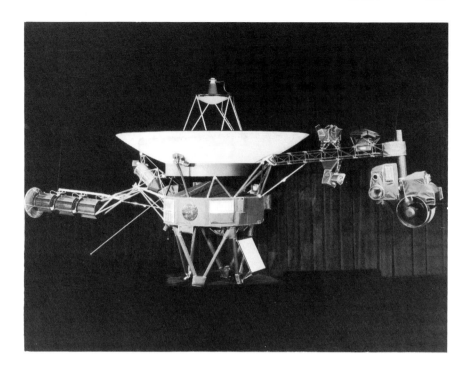

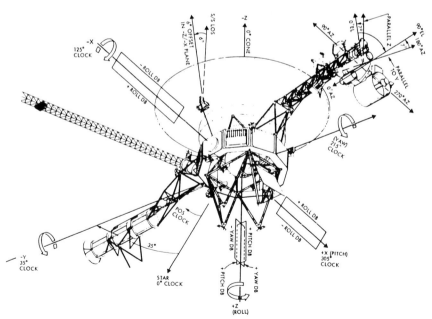

Fig. 8.1 Voyager spacecraft. (Courtesy of Jet Propulsion Laboratory.)

antenna. The visual imaging, infrared, and ultraviolet instruments are located on a two-axis scan platform located at the end of the boom. These instruments are primarily concerned with the planets and satellites and require accurate pointing and quick retargeting capability.

The one science instrument not located on the science boom is the magnetometer. This unit needs to be as far as possible from the spacecraft magnetic fields. It is mounted on a deployable Astromast boom, which at full deployment is 15 m (50 ft) in length but which retracts for launch into a can less than 1 m in length. Two magnetometers are mounted: one at the outer end of the boom and the other halfway out. The outer one is used for obtaining science data, whereas the other is primarily intended to help evaluate spacecraft fields in support of data analysis.

With both the RTG boom and the science boom folded aft for launch, these items are clearly much closer together than when deployed. Both are latched to the solid-propellant final stage, and the case and propellant provide considerable shielding. Nevertheless, more radiation dose was accumulated in the few weeks between final assembly and launch than in months of spaceflight.

The 10-sided bus contains the majority of engineering subsystems. The large spherical tank contains the hydrazine propellant for the combined attitude control and propulsion subsystem. This tank is located in the center of the bus not only to maintain the hydrazine at a satisfactory temperature but also to aid in shielding the instruments from the RTGs.

The four legs descending from the bottom of the spacecraft, often taken for landing gear, are actually the truss that supported the solid-propellant rocket motor, which was the final launch stage. These structures form the basis of a story that is repeated here not with any intent to embarrass but to provide an object lesson concerning how a seemingly innocuous decision in one subsystem can have an unlooked for detrimental effect in another area.

Originally, the separation plane was to be at the bottom of the bus and the truss was to be jettisoned along with the rocket motor. However, some concern was expressed regarding the effect of shock generated by the explosive release nuts on nearby electronics. An easy solution was adopted: separate at the motor and leave the truss legs attached to the bus. The small additional weight of the truss would have no significant effect on the mission, and this approach would save much effort in shock isolation, etc.

All was well until, a few days after launch, the aft-facing array of thrusters was fired to correct the trajectory. The thrusters fired for the proper amount of time and appeared to have operated properly. Unfortunately, the change in velocity was low by a substantial percentage. The technical detective work will not be detailed here, but the conclusion was that the exhaust jets from the thrusters, expanding rapidly in the vacuum, impinged upon the nearby truss members. The effect was generation of drag on the truss by the supersonic flow, thus reducing the effective thrust.

Because the propellant supply was calculated with a substantial margin, the effect of this mistake was not a catastrophe but rather the relatively minor annoyance of a reduction in propellant margin for the mission.

However, one can easily see how a similar situation could be catastrophic to another mission with less margin. Again, the lesson to be learned is that no subsystem is an island. Any decision made in one area must be assessed for its impact upon others. This is the essence of systems engineering.

The primary structure of the Voyager is made of aluminum. Composites are used in various locations. The largest composite structure is the antenna dish. The Astromast boom makes use of very thin fiberglass members with parallel orientation of the fibers to obtain the strength and elasticity to allow it to coil into its canister and still maintain reasonable rigidity when deployed.

(2) *Galileo.* The JPL Galileo spacecraft, shown in Fig. 8.2, has a history of frustration almost unparalleled in the history of the space program. Originally scheduled for launch in 1982, the program has been buffeted by Shuttle delays, changes in upper stages, the Challenger accident, and a variety of other problems. It was finally launched in 1989 on an upper stage of much lower capability than the originally intended Centaur. This will greatly extend the planned flight time to Jupiter.

The purpose of the mission is to place the spacecraft into a highly elliptic orbit about Jupiter from which it will intensively study Jupiter, most of its satellites, and the surrounding environment. Prior to orbit insertion a probe will be dropped into the atmosphere of Jupiter. The Orbiter will receive and record the probe data for playback to Earth.

Galileo is unique among planetary spacecraft to date in that it is a dual spinner. That is, one portion of the spacecraft spins while the other spins at a different rate or not at all. This concept is discussed in some detail in Chapter 7. The idea of using it in the Galileo was to incorporate the best aspects of spin-stabilized and three-axis-stabilized spacecraft. The spinning portion would provide the global coverage desired by the fields and particles instruments, and the fixed portion would provide a stable base for the high-resolution imaging which a spinning spacecraft cannot provide. The spinning portion of the spacecraft would provide attitude stability with minimal expenditure of attitude control propellant. The concept has been used successfully in a variety of Earth-orbiting spacecraft, most notably the Hughes communication satellites. Dual spin, however, is less well adapted to this application. Large amounts of power and data must pass across the spin-bearing interface, greatly complicating the design. The spinning portion of the spacecraft includes the multisided bus, similar to that of Voyager, the communications antenna, the booms supporting the RTGs, and the magnetometer. The bus contains the engineering subsystems of the spacecraft. The antenna, of metal mesh, is 15 ft in diameter when deployed. For launch it is folded, using its rigid ribs, around the center feed support.

The lower portion, as depicted in the illustration, is the fixed portion. It mounts the scan platform for science, the small dish antenna for receiving data from the Jupiter probe, and the probe itself. The probe is the cone-shaped object on the centerline of the vehicle. When the probe is separated from the spacecraft some 150 days prior to encountering Jupiter, it must be spinning in order to provide stability. This requires that the

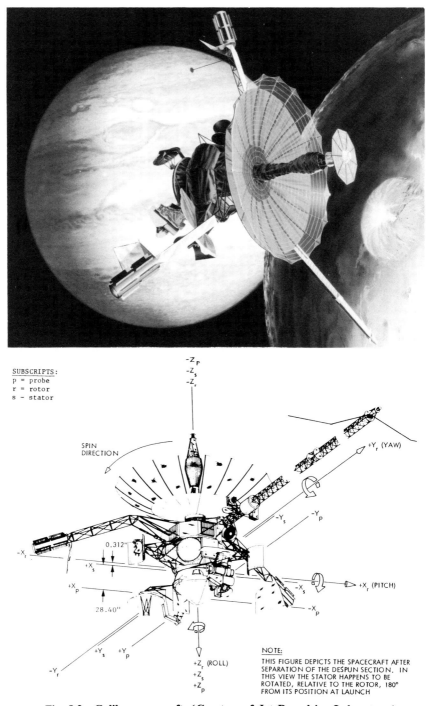

SUBSCRIPTS:
p = probe
r = rotor
s = stator

-Z_p
-Z_s
-Z_r

SPIN
DIRECTION

+Y_r (YAW)

-Y_s   -Y_p

-X_r

0.312"

+X_s

+X_p

-X_s

+X_r (PITCH)

-X_p

28.40"

+Y_s +Y_p

-Y_r

+Z_r (ROLL)
+Z_s
+Z_p

NOTE:
THIS FIGURE DEPICTS THE SPACECRAFT AFTER
SEPARATION OF THE DESPUN SECTION. IN
THIS VIEW THE STATOR HAPPENS TO BE
ROTATED, RELATIVE TO THE ROTOR, 180°
FROM ITS POSITION AT LAUNCH

Fig. 8.2   Galileo spacecraft. (Courtesy of Jet Propulsion Laboratory.)

nonspinning portion of the spacecraft be spun for probe separation and then despun again.

The main orbit insertion rocket motor is located on the centerline of the vehicle behind the probe. If the probe fails to separate, orbit insertion will not be possible. Also mounted on this section, on outriggers, are attitude control thrusters.

The original idea for this approach to the Galileo configuration was to simplify and to reduce cost while, as noted earlier, obtaining the advantages of two different types of spacecraft. From the preceding discussion it is not at all clear that things worked out as intended.

Some changes, e.g., sunshades were required for the Venus swingby that results from the low energy stage discussed earlier. These are not shown in Fig. 8.2b, but do appear in Fig. 8.2a.

In general, the structure of the spacecraft is quite similar to that of the Voyager.

(3) *Fltsatcom.* Turning from planetary spacecraft to geostationary communications satellites, Fig. 8.3 depicts the Fltsatcom. This is a satellite made by TRW for the U.S. Navy for communications purposes. Of fairly conventional aluminum construction, this configuration is interesting in that it consists of two identical hexagonal buses mounted one above the other. The lower bus contains all of the engineering subsystems (note that the solar panels are mounted on that portion) that perform all of the functions required by the spacecraft (e.g., power, attitude control, etc.). The second bus contains all the payload related hardware (e.g., transponders, etc.). The antennas related to its communications relay function are mounted on this bus.

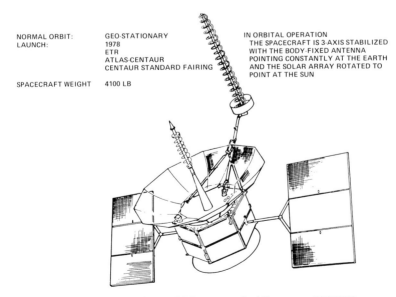

NORMAL ORBIT:
LAUNCH:

GEO-STATIONARY
1978
ETR
ATLAS-CENTAUR
CENTAUR STANDARD FAIRING

SPACECRAFT WEIGHT   4100 LB

IN ORBITAL OPERATION
THE SPACECRAFT IS 3-AXIS STABILIZED
WITH THE BODY-FIXED ANTENNA
POINTING CONSTANTLY AT THE EARTH
AND THE SOLAR ARRAY ROTATED TO
POINT AT THE SUN

Fig. 8.3   FLTSATCOM spacecraft. (Courtesy of TRW.)

Since the antennas point at the Earth, the solar array arms are oriented normal to the orbit plane and the arrays rotate to maintain sun pointing at all times. For launch, the arrays fold so that their long axes are parallel to the antenna boresight and fold between segments to form a hexagonal cylinder around the spacecraft. This configuration is maintained until after orbit insertion so that the deployed antennas do not have to withstand the insertion g loads. Only a portion of the arrays is illuminated in this configuration, but power requirements are so low in cruise mode compared to the operational relay mode that ample power is available.

The advantage of the two-bus configuration is adaptability. If an entirely different payload is desired, the communications relay bus can be replaced by a different package while still retaining the tested and reliable engineering spacecraft essentially unchanged. For example, this same spacecraft was proposed as a low-altitude Mars polar Orbiter with minimal changes in engineering subsystems. One change was a reduction in size of the solar arrays, since, even at the greatly reduced illumination at Mars, the small power usage of a science payload vs the massive communications relay made the full-size arrays unnecessary.

(4) *HS 376.* Figure 8.4 shows a different approach to geosynchronous communications satellites. The HS 376 series by Hughes Aircraft is typical of the drum-shaped dual-spin satellites that have been the mainstay of that company for many years. The dual-spin terminology refers to the fact that the main bus spins at a moderate rate while the antenna assembly spins once per orbit, in other words, maintaining orientation toward the Earth at all times. The electronics is mounted on shelves within the rotating bus, as is the attitude control propellant and other equipment. Only radio frequency energy crosses the spin bearing between the bus and the antennas. (Note the difference between this and the Galileo, which must send both power and data signals through slip rings or equivalent devices at the spin bearing.)

The orbit insertion or apogee kick solid-propellant rocket motor is mounted in the bus firing out the antiantenna end. Waste heat rejection is also primarily out this end of the bus.

Earlier spacecraft of similar configuration were of fixed geometry. However, these spacecraft quickly encountered one of the major weaknesses of the rotating, drum-shaped configuration. Since only 30–40% of the exterior of the spacecraft can be effectively illuminated at any time, high-powered spacecraft of fixed geometry begin to experience power limitations. There is simply not enough fixed surface area on the exterior of the spacecraft to provide sufficient solar array area to generate the required power. Launch volume constraints preclude simply making the spacecraft larger in diameter or longer. There are several possible solutions to this problem involving deployable vanes, paddles, etc. The solution chosen in this instance maintains the general configuration of the vehicle. Additional solar panel area is incorporated in a cylindrical shell that surrounds the main bus. This provides power during cruise to orbit. Once the apogee motor has fired, the shell is deployed down, exposing the solar cells on the

Fig. 8.4   HS-376 spacecraft. (Courtesy of Hughes Aircraft Co.)

main bus structure. This approximately doubles the length of the spacecraft and the available solar array area.

(5) *Defense Meteorological Satellite Program.* The Defense Meteorological Satellite Program (DMSP) spacecraft built by RCA (now a part of General Electric) is a weather satellite designed to operate in low circular, sun-synchronous orbit. The civilian Television and Infrared Observation Satellite (Tiros), although not identical, is sufficiently similar that most comments made here apply with equal force to both. The spacecraft is depicted in Fig. 8.5. The main electronics bus is the large, boxlike structure covered with circular temperature-control devices. These are thin, polished sheets with pie-shaped cutouts that rotate under control of a bimetallic element to expose insulated or uninsulated skin areas depending on the amount of heat to be rejected. The spacecraft is nadir pointing in operation. The flat face not visible in the illustration faces the planet, and

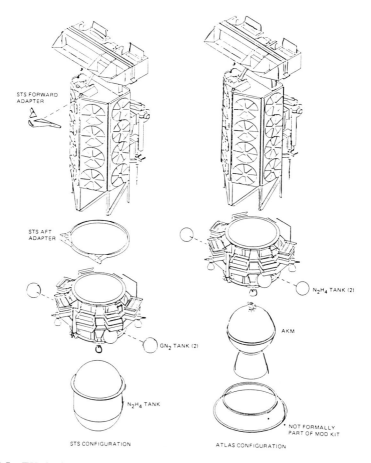

Fig. 8.5 TIROS/DMSP spacecraft. (Courtesy of General Electric Astro. Division.)

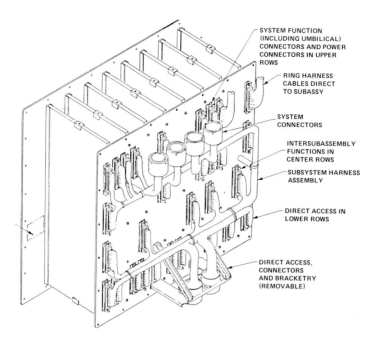

SYSTEM FUNCTION
(INCLUDING UMBILICAL)
CONNECTORS AND POWER
CONNECTORS IN UPPER
ROWS

RING HARNESS
CABLES DIRECT
TO SUBASSY

SYSTEM
CONNECTORS

INTERSUBASSEMBLY
FUNCTIONS IN
CENTER ROWS

SUBSYSTEM HARNESS
ASSEMBLY

DIRECT ACCESS IN
LOWER ROWS

DIRECT ACCESS,
CONNECTORS
AND BRACKETRY
(REMOVABLE)

Fig. 8.6   Dual shear plate packaging.

instrumentation is mounted in that area. Instruments are also located in the large, transversely mounted structure on the end of the main body.

Two versions of the spacecraft are shown, one with a large, solid-propellant motor and auxiliary monopropellant system, and the other with a monopropellant only system, a large hydrazine tank replacing the solid rocket motor. The solid-propellant version was launched using a refurbished Atlas missile as the lower stage, whereas the other version was intended to be launched in the Shuttle. The Tiros/DMSP is a very capable spacecraft, providing guidance and control functions for its own launch vehicle in the expendable launch case.

One virtue of this concept is ease of internal access. The large, Earth-facing side opens like a door, exposing equipment mounted on that side as well as that mounted on the other walls. With the spacecraft in a vertical position as shown, the size and configuration make internal access quite easy compared to some.

This spacecraft was also evaluated for adaptation to planetary missions and showed excellent promise.

Design and packaging concepts. A variety of internal structural design and electronic packaging concepts have evolved in conjunction with the configuration designs discussed previously. Table 8.7 describes three basic types along with some of the good and bad features of each. These points are discussed further in the following paragraphs.

It should be noted that various organizations will use their own variations of these approaches, and the names applied to the various concepts will not necessarily agree between companies or with the terminology of this book.

(1) *Dual shear plate.* This approach, most prominently used by JPL in the Mariner, Viking, and Voyager family of spacecraft, mounts the electronics on flat honeycomb plates or, in the case of such things as gyroscope packages, in especially tailored boxes compatible with the overall packaging scheme. The plates are then bolted to inner and outer shear plates as shown in Fig. 8.6. These shear plates are inserted into the bus frame from the outside, and both shear plates are bolted to the bus. The shear plates provide closure to the bus frame, resulting in a final structure that is very rigid and sturdy. Because the electronics is customized for the application, the packaging density can be very efficient and quite high, probably the best among the concepts discussed.

Table 8.7   Structural/packaging design concepts

Dual shear plate
Bus frame
Shear plates close frame inside and out
Custom electronic modules or mounting plates tie to shear plates
Pros/cons
Strong, rigid structure
Good thermal contact
Requires custom electronics packaging and cabling
Efficient volumetric packaging
Examples: Mariner, Viking, Voyager
Shelf
Shelf structure inside spacecraft skin
Electronic packages mount on shelf
Pros/cons
Can use standard "black boxes"
Less efficient volumetric packaging
More difficult heat-transfer-path-
Example: HS 376
Skin panel/frame
Bus frame
Large skin panels (often hinged) close frame
Electronics mounted on skin
Pros/cons
Can use standard "black boxes"
Good heat-transfer contact
Easy access
Examples: Fltsatcom, Tiros/DMSP

Because the electronics is distributed over the aluminum honeycomb sheets that are then tightly mounted to the shear plates, which are also heat rejection surfaces, thermal transfer capability is generally good, although special provisions may be required for high-power dissipation items.

The negative aspects of this concept are the relatively large piece part count: shear plates, honeycomb sheets, very large fastener count, etc. This all equates to high manufacturing cost and labor-intensive operations. Furthermore, if one wishes to use an already existing electronics subsystem, it must be repackaged to be compatible with this approach. Cabling also may be more complex.

In summary, the negative aspects equate to an expensive approach compared to other schemes. However, the virtues of strong rigid structure and high-density custom packaging may be worth the cost in some applications.

(2) *Shelf.* The arrangement referred to as shelf-type packaging could equally well be referred to by other descriptions. It refers to an arrangement wherein shelves or bulkheads mounted orthogonal to the axis of a cylindrical spacecraft provide support for the electronics and other internal systems. This arrangement is typical, for example, of the interior structure of the dual-spin spacecraft described earlier.

This approach generally is less volumetrically efficient in terms of the amount of electronics per unit volume of spacecraft than some others. On the other hand, this is often not a major disadvantage, since the volume of the spacecraft is driven by the required solar array area and more internal volume is available than is required. The use of a basic flat mounting structure is more adaptable to the use of standard electronics in existing "black box" configuration without requiring customizing.

Rejection of large amounts of internally generated heat can be a problem, since components mounted near the centerline are far from the walls of the cylinder. If heat is rejected from these walls, the conduction path may be long. In the case of the HS 376, the end opposite the antennas is essentially open for heat rejection. This works well for items with a clear view of the open end but will be less satisfactory if a stack of shelves or bulkheads is used.

(3) *Skin panel/frame.* This concept, of which examples are shown in Figs. 8.7 and 8.8, uses a basic structural frame or bus. The faces of this structure are closed with plates or panels that may in some cases form part of the load-bearing structure as do the shear plates in the other configuration discussed earlier. In the examples shown, Fltsatcom and Tiros/DMSP, the panels are hinged along one side to swing open for easy access. The panels provide mounting structure for electronic equipment, cabling, and other hardware.

The ability to use standard, uncustomized electronics assemblies is an advantage of this configuration. Emplacement of the boxes directly on the plates that can directly reject heat to space is an advantage for thermal control. This approach is in general somewhat less rigid and structurally efficient than the dual shear plate concept.

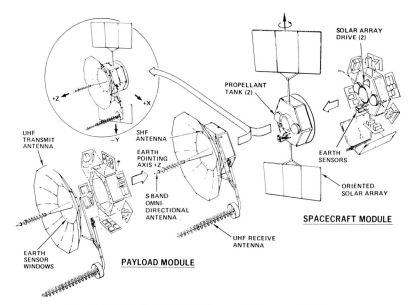

Fig. 8.7   FLTSATCOM skin panel frame packaging.

(4) *Factors in structural concept selection.*   All of the structural concepts discussed earlier have significant virtues and some undesirable features. Which one is chosen will depend on a variety of factors, including overall configuration mission, payload, and, occasionally, organizational prejudice. However, there are a number of factors that should be considered in any basic choice and in the subsequent implementation of that choice.

Almost as an axiom, it is desirable to minimize parts count. The larger the number of individual parts, particularly small ones like fasteners, the higher the cost is likely to be. This is true not only of manufacturing but also of test and operations, which are likely to be much more labor-intensive and time-consuming. The desire to minimize fasteners in particular will be viewed with alarm by structural analysts who tend to prefer many small fasteners to a few large ones in order to improve structural load transmission and distribution. As always, the design will be a compromise.

With rare exceptions, there will be pressure to minimize structural mass. This is usually a tradeoff against cost of materials and qualification testing. A very sophisticated structure designed with tight margins to achieve minimum mass will often require expensive materials and be expensive to design and fabricate because of the more detailed and sophisticated analysis required. Even less obvious and therefore often a cost trap for the unwary is the fact that such structures are often more expensive to test for qualification and workmanship demonstration.

In most cases, the spacecraft structure design will be driven by a

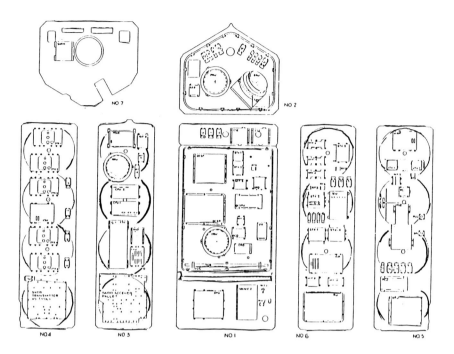

Fig. 8.8   TIROS/DMSP skin panel frame packaging.

requirement for structural stiffness rather than strength. This is because excessive deflection under load, even though there is no permanent yielding, usually cannot be tolerated.

The degree of understanding and the maturity of the concept are important in minimizing development cost as is minimization of complexity. A well-understood, reasonably simple structure with characteristics that can be firmly predicted is highly desirable in the minimization of both risk and cost.

Operational aspects of the design are another important area that is sometimes overlooked. Whatever the mission, it is first necessary that the spacecraft be integrated and tested on the ground. Ease of subsystem integration, work access, means of handling, and ease of repair in the event of damage should all be considered. Finally, with the growing importance of on-orbit repair and servicing, the special needs of extravehicular activity (EVA) will be important for some missions.

Deployable structures. The requirement for deployable structures arises from the limitations on dimensions and geometry of the launch volume vs the need for large antennas and solar arrays, the requirements for instrument field of view and isolation, and the requirement to isolate radiation producing objects such as RTGs and reactors. A variety of concepts have been developed, of which a representative few will be discussed here.

Fig. 8.9   Mariner IV spacecraft. (Courtesy of Jet Propulsion Laboratory.)

(1) *Solar arrays.*   Deployable solar arrays have for the most part been flat rigid panels. The simplest have been the single-hinged type, which are launched pinned to the spacecraft structure and/or to one another to form a compact rigid assembly. The JPL Ranger and Mariner series of spacecraft typify this approach, which is illustrated in Fig. 8.9.

More complex flat panel folding schemes such as those for Fltsatcom discussed earlier have also been applied. An approach used on a variety of spacecraft and favored by the Soviets for their various manned craft is an extendable linkage concept.

Arrays made up of a series of flat plates that fold into a long narrow box or arrays flexible enough to roll up like an old-fashioned window shade have been designed and demonstrated and have seen some operational use. These designs are deployed by an extendable boom of the Astromast type discussed later and can be retracted for high-*g* maneuvers or entry.

This by no means exhausts the variety of concepts. These are limited only by the ingenuity of the designer. From the viewpoint of the spacecraft systems engineer, the concerns include realizing the required area, obtaining reliable deployment (and retraction if required), and ensuring that the lightweight flexible structure does not detrimentally interact with the vehicle attitude control subsystem.

(2) *Deployable booms.*   A number of concepts exist and have been successfully flown. The first, simplest, cheapest, and (usually) heaviest is the hinged rigid boom. This is simply a long rigid boom, usually of tubular construction, with one or more hinged joints. The boom is folded and latched in place for launch. Upon release, springs deploy the boom until it latches into the proper configuration. The virtues of low cost and simplicity

make this boom attractive where there is room to stow it (stowed length is typically one-half to one-third deployed length) and the weight is acceptable. Many booms of this type have been flown. Use of stiff composite materials to keep the natural frequency as high as possible to minimize attitude control interactions is attractive. Close attention to joint design, particularly with regard to rigidity, is required for precise location and natural frequency control.

The Astromast boom is an extremely sophisticated deployable structure. The illustration of the Voyager spacecraft in Fig. 8.1 depicts an Astromast boom. The full deployed length is 15 m, but it is contained for launch in a canister about 1 m long. The boom consists of three fiberglass longerons stiffened at intervals by fiberglass intercostals and beryllium-copper cables. Stowed, the longerons are coiled into the canister, taking on the appearance of a coil spring. The intercostals and cables stack in the interior of the canister. The boom provides its own deployment force using the extensive amount of strain energy stored in the coiled longerons. The problem is to restrain the deployment to prevent damage from too rapid movement. This is done by a cable running up the center. The cable is attached to a motor through an extremely high ratio, anti-backdrive gear system that pays out the cable at the desired deployment rate. This motor can also be used to retract the boom as desired. This boom has an excellent deployed-to-stowed-length ratio and is very light in weight for its length. Considering the length and light weight, it is also fairly rigid, but typically cannot be deployed horizontally in a *1-g* gravity field without support. As one would expect for such a sophisticated device, these booms are fairly expensive compared to the hinged rigid types. Deployment and retraction cycles may be limited because of the large amounts of strain experienced by the longerons when retracted.

Stem-type booms come in several variants. In general they consist of two metal or composite strips formed to a particular cross section. These may be welded along the edges or may mechanically join via a series of teeth along the edges. In any case, the two strips are stowed by rolling up on a reel where they are deformed to a flat shape and thus stow rather like tapes. As they are reeled out, the two strips return to their originally formed shape to provide a cross section for stiffness. The edges, if they are not already welded, interlock during this process. These booms are capable of many cycles. Beryllium-copper is a favored material to allow high cycle life and precise repeatability. Such a boom was used as the manipulator arm on the Viking Lander. Length limitations on booms of this type depend on the loads. They are fairly rigid even at fairly high length-to-cross-section ratios. Some experiments have been done by JPL and the World Space Foundation with lower cost variants using stainless steel shim stock. These are less precise and have lower cycle life, but may be satisfactory for many applications.

(3) *Articulating platforms.* For many missions involving high-resolution imaging and/or where rapid retargeting of instruments is required, a scan platform is a desirable approach. Platforms of this type are usually free to

move in two orthogonal axes, although simple single-axis platforms have been flown. Platforms of this type on the various Mariner, Viking Orbiter, and Voyager spacecraft have been highly effective in obtaining maximum coverage during brief flybys and providing detailed mosaics in areas of interest.

The usual practice has been to move the platform using precision stepper motors. Attitude reference is that of the spacecraft, and the platform motion is controlled relative to that reference frame. Occasionally this required special effort in control of spacecraft attitude and platform pointing in order to achieve high-precision results.

Predictably, the techniques just described have become inadequate for some applications. A more recent high-precision approach involves providing the scan platform with its own attitude references including a gyroscope package and celestial sensors. This eliminates any error that might be introduced between the reference frame and the platform, e.g., in the joints, angular position sensors, etc., in the older system.

In essence the platform points itself relative to its own reference frame by reacting against the greater mass and inertia of the spacecraft. The spacecraft then stabilizes itself using its own attitude references.

For many instruments, the lower precision of the spacecraft referenced platform is quite satisfactory. The Mariner MkII spacecraft concept, depicted in Fig. 8.10, has two scan platforms on diametrically opposed booms. One is a high-precision platform with inertial references and a star tracker, and the other is a conventional low-precision type.

Fig. 8.10   Mariner Mark II spacecraft. (Courtesy of Jet Propulsion Laboratory.)

8.2 MASS PROPERTIES

Spacecraft mass properties that are usually of interest to the spacecraft designer are vehicle mass, center of mass, moment of inertia, and moment-of-inertia ratio. Depending on the spacecraft type and mission, some may be of less interest than others, but the first two are always important.

Vehicle Mass

Spacecraft mass is always of substantial importance. Even if weight is not particularly critical in the absolute sense, because of ample launch vehicle performance margins, it is still important to know vehicle weight accurately. Generally, however, mass is critical, and control must be exercised to ensure that acceptable values are not exceeded. This requires maintenance of a detailed list at least to the major component level. Table 8.8 is a typical example of such a list.

The list will change constantly all the way up to launch. Early in the program, the list will be composed mostly of estimates and may contain some factors that are to be determined (TBD) or tentative allocations. As the design matures, the estimates will improve and the TBDs disappear. Eventually prototype hardware will become available and actual values will appear in the list. As the hardware is refined toward actual flight units, the values will continue to change somewhat, but the uncertainties are much smaller. Occasionally, testing results will mandate a hardware modification that will cause a substantial change, but such occurrences should be rare, and the mass list should remain fairly stable in the later phases of the program.

Some of the reasons that an accurate knowledge of mass is required include launch vehicle performance, propellant loading requirements for maneuvers, and determination of the other mass properties.

The significance of spacecraft mass relative to launch vehicle performance is obvious and is the one that usually springs to mind in discussing spacecraft mass, but other factors are important as well.

In order to ensure that adequate propellant is loaded for propulsion maneuvers, knowledge of the spacecraft mass is essential. This is especially true when a solid-propellant rocket is used, since the total impulse is fixed once the propellant is cast and trimmed. The only control over velocity change is total mass, which is achieved by ballasting. This requires accurate knowledge of the basic mass.

Finally, mass of the total vehicle and its major subassemblies must be known in order to compute the other mass properties, which will subsequently be discussed.

The final check on mass is a very accurate weighing of the entire spacecraft. This is usually done once or twice during assembly and test to verify the mass list as it then stands. The final weighing will be done shortly before launch with the spacecraft as complete as possible. An accurate knowledge of what components are on the spacecraft and a list of deviations (i.e., missing parts, attached ground support equipment) is mandatory. Weighing is usually done with highly accurate load cells.

Table 8.8 Galileo subsystem mass allocations

	Subsystem	Allocated mass (kg)			
		Orbit module	Upper S/C adapter	Lower S/C adapter	Airborne support equipment
2001 STRU	Structure[a]	237.4	38.3	61.8	0
2002 RFS	Radio frequency	45.9	0	0	0
2003 MDS	Modulation/demodulation	9.4	0	0	0
2004 PPS	Power/pyro	154.2	2.3	4.2	0
2006 CDS	Command and data	34.4	0	0	0
2007 AACS	Attitude & articulation control[b]	113.9	0	0	0
2009 CABL	Cabling	60.4	4.4	1.4	0
2010 RPM	Propulsion (RPM burnout)	215.68	0	0	0
2011 TEMP	Temperature control	37.6	2.0	4.3	0
2012 DEV	Mechanical devices	38.5	5.0	1.6	0
2016 DMS	Data memory	8.9	0	0	0
2017 SXA	S/X Band antenna	6.1	0	0	0
2042 XSDC	X/S Downconverter	2.5	0	0	0
2070 BAL	Ballast	20.0	0	0	0
2071 OPE	Orbiter purge equipment	0.9	1.6	0.1	0
2080 SAH	System assembly hardware	3.9	1.5	1.1	0
2023 PWS	Plasma wave[b]	7.62	0	0	0
2025 EPD	Energetic particles	9.37	0	0	0
2027 PPR	Photopolarimeter	4.91	0	0	0

2029	DDS	Dust detector	4.15	0	0
2032	PLS	Plasma	11.99	0	0
2034	UVS	Ultraviolet spectrometer[b]	5.16	0	0
2035	MAG	Magnetometer[b]	5.86	0	0
2036	SSI	Solid state imaging	27.71	0	0
2037	NIMS	Near infrared mapping spectrometer	18.23	0	0
2040	SCAS	Science calibration	3.43	0	0
2002	USO	Ultra stable oscillator	2.05	0	0
2052	RRH	Relay radio hardware[c]	23.3	0	0
2060		Probe adapter[c]	7.0	0	0
2072	PPE	Purge purification equipment	0	0	38.0

[a]Includes HGA structural elements and RHUs
[b]Includes RHUs
[c]Includes System Mass Contingency

NOTE: In addition to the Subsystem Mass Allocations given above, the following System Mass Contingency breakdown exists:

Orbiter Engineering	11.6 kg
Orbiter Science	1.62 kg
Upper S/C Adapter	4.9 kg
Lower S/C Adapter	10.5 kg
Airborne Supt Equip	2.0 kg

Vehicle Center of Mass

For any space vehicle, accurate knowledge of the location of the center of mass is vital. It is essential for attitude control purposes, since, in space, all attitude maneuvers take place around the center of mass. Placement of thrusters, size of thrusters, and the lever arms upon which they act are all designed relative to the center of mass. When thrusters are used for translation, it is important that the effective thrust vector pass as nearly as possible through the center of mass in order to minimize unwanted rotational inputs and the propellant wasted in correcting such inputs.

Launch vehicles frequently impose relatively tight constraints on the location of payload center of mass in order to limit the moment that may be imposed on the payload adapter by the various launch loads.

From the preceding discussion, it is clear that the payload center of mass must be both well controlled and accurately known. From the beginning, the configuration designer works with the design to place the center of mass within an acceptable envelope and locates thrusters, etc., accordingly. It is often necessary to juggle the location of major components or entire subsystems in order to achieve an acceptable location. This will sometimes conflict with other requirements such as thermal control, field of view, etc., resulting in some relatively complex maneuvering in order to achieve a mutually acceptable arrangement.

As noted earlier, the center of mass is computed from the beginning of the design process using the best weights and dimensions available. As with the mass, the information is updated as the design matures and actual hardware becomes available. Actual measurement is used to verify the center-of-mass location of the complete assembly. This usually takes place in conjunction with the weighing process with all of the same constraints and caveats regarding accurate configuration knowledge as discussed earlier. Often the center-of-mass location is measured in all three spacecraft axes. Sometimes, however, it will be acceptable to determine it only in the plane normal to the launch vehicle thrust axis (parallel to the interface plane in an expendable launch vehicle) and compute it in the third axis if the tolerance on accuracy is acceptable.

Vehicle Moment of Inertia

An accurate knowledge of vehicle moment of inertia is vital for design of attitude control effectors (e.g., thrusters, magnetic torquers, momentum wheels) in order to achieve the desired maneuver rates about the spacecraft axes. This, together with mission duration, expected disturbance torques, etc., is used to size the tank capacity in a thruster-based system.

Moment of inertia is computed based on knowledge of component mass and location. Reasonable approximations usually provide satisfactory accuracy. Examples of this include using point masses for compact items and rings, shells, or plates in place of more complex structures.

In most cases, moment of inertia is not directly measured, particularly in large, complex spacecraft, since experience has shown that careful calculations based on measured mass and location data provide satisfactory

```
REPORT NUMBER CASD/LVP 75-078A
DATED 15 SEPTEMBER 1976

TC-6/CENTAUR D-1T/MJS 1977 MISSION, REVISION A

    WEIGHTS TAKEN FROM CASD/LVP 73-022-11 TITAN IIIE/CENTAUR
    D-1T CONFIGURATION, PERFORMANCE, AND WEIGHT STATUS
    REPORT DATED JUNE 1976, EXCEPT THE PAYLOAD WEIGHT AND
    ACTUAL WEIGHT OF TC-6 CENTAUR VEHICLE ARE THE
    LATEST AVAILABLE
    VERSUS TIME DATA BASED ON CASD/LVP 73-022-11
    APPENDIX A - BENCHMARK TRAJECTORY DATA TC-6
    MJS 1977 MISSION.


MJS 77 MM+PM (WITHOUT SPACECRAFT ADAPTER)
WEIGHT      Z BAR       Y BAR       X BAR       IZZ         IYY         IXX
PRIZY       PRIZX       PRIXY

4454.80     -.01        -.47        2562.91     1422.1      1184.5      570.9
-3.89       -.78        8.72


SPACECRAFT ADAPTER
WEIGHT      Z BAR       Y BAR       X BAR       IZZ         IYY         IXX
PRIZY       PRIZX       PRIXY

102.70      -.45        .08         2502.50     11.8        12.4        21.1
-.10        .20         -.00


MJS 77 MM+PM+S/C ADAPTER
WEIGHT      Z BAR       Y BAR       X BAR       IZZ         IYY         IXX
PRIZY       PRIZX       PRIXY

4557.50     -.02        -.46        2561.55     1513.0      1276.0      592.0
-4.00       .00         8.00


TC-6 DRY CENTAUR
WEIGHT      Z BAR       Y BAR       X BAR       IZZ         IYY         IXX
PRIZY       PRIZX       PRIXY

3824.90     .28         3.05        2320.24     14245.5     14220.7     1759.7
20.55       -139.93     -249.72
```

Fig. 8.11   Typical mass properties listing.

accuracy. Direct measurement of moment of inertia has occasionally been done on programs for which it was considered necessary.

Moment-of-Inertia Ratio

For a spinning spacecraft, the moment-of-inertia ratio between the three major axes is usually more important than the actual values of moment of inertia. (However, moment of inertia about the spin axis is certainly necessary in computing spinup requirements.) The reason is that, for a spinning body in free space, the spin is most stable about the axis of maximum moment of inertia. A spacecraft set spinning about one of the other axes will eventually shift its spin axis until it is spinning about the maximum moment-of-inertia axis. If there are no energy-dissipating mechanisms (e.g., flexible structures such as whip antennas or liquids) in the spacecraft, then spin about the lesser moment axis may be maintained for an extended period. However, the presence of energy-dissipating mechanisms will cause the shift. The classical example is the Explorer I satellite, a long, thin spinner with four wire whip antennas. After a relatively short time on orbit, spin shifted from a "bullet-like" spin about the long axis to a flat or "propellor-like" spin. This was merely an annoyance in the Explorer case, but such a flat spin or the coning motion that occurs in the transition from one axis to another can prove fatal to the mission in some cases. Active nutation control can prevent the shift or delay its onset, but of course this costs mass. Knowledge and control of moment-of-inertia ratio is therefore a major factor in design of spinning spacecraft.

Mass Properties Bookkeeping

Figure 8.11 presents the first page of a typical mass properties listing such as that which would be maintained for a three-axis-stabilized spacecraft launch configuration. Note that the characteristics of each component are listed and then the cumulative total as the configuration is built up so that mass properties at any given stage in the flight profile may be known.

8.3 STRUCTURAL LOADS

Sources of Structural Loads

Table 8.9 lists the primary sources of structural loads that may be imposed on a spacecraft. Although most are concerned with launch and ground handling, some affect the vehicle throughout its operating lifetime.

Linear acceleration is usually a maximum at staging, often of the first stage, which often has a higher thrust-to-weight ratio than the upper stages. The exception to this would be a vehicle such as the three-stage Delta, where the solid third stage as it approaches the end of burn probably causes the highest acceleration.

Even though it is the factor most associated with space launch in the eyes of the layman, linear acceleration is often not the most significant design driver. This is especially true for an all-liquid-propellant launch vehicle where acoustic and vibration loads may well overshadow linear accelera-

Table 8.9   Sources of structural loads

Linear acceleration
Structurally transmitted vibration
Shock
Acoustic loads
Aerodynamic loads
Internal pressure
Thermal stress

tion as design factors. In the case of a vehicle that re-enters the atmosphere in a purely ballistic mode, the loads imposed during entry may well exceed those of launch. Lifting entry substantially reduces the loads.

Structurally transmitted vibration is one of the major design drivers. Main propulsion is usually the primary source in flight during the launch phase, although aerodynamic and other forces may also contribute. The Space Shuttle, which was designed to minimize longitudinal loads, is especially bad in terms of structurally transmitted vibration because the payload is mounted immediately above the engines without the isolation afforded by a long, flexible tank assembly in between.

In addition to flight loads, however, the more prosaic ground handling and transportation loads may be significant drivers as well. Although typically less intense, these inputs will be of longer duration. Several hours on a truck vs 8–10 min of launch typifies the nature of the problem. Thus, although the launch inputs are more intense, the cumulative cyclic loads from ground activities may be almost as bad in terms of detrimental effect.

Shock loads in flight are usually associated with such functions as firing of pyrotechnic devices, release of other types of latches, or engagement of latches. Ground handling again can be a contributor, since such activities as setting the spacecraft on a hard rigid surface even at relatively low speeds can cause a significant shock load. Ground-handling problems can be minimized by proper procedures and equipment design. In-flight shocks may require isolation or relocation of devices farther from sensitive components.

Acoustic loads are most severe at liftoff because of reflection of rocket engine noise from the ground. They may also be fairly high in the vicinity of maximum dynamic pressure because of aerodynamically generated noise. This is especially true of the Space Shuttle with its large, flexible payload bay doors. Acoustic loads are especially damaging to structures fabricated of large areas of thin-gage material such as solar panels.

Aerodynamic load inputs to the payload come about as a result of their effect on the launch vehicle, since the payload is enclosed during passage through the atmosphere. Passage through wind shear layers or aerodynamic loads due to vehicle angle of attack caused by maneuvering can cause abrupt changes in acceleration. They may also cause deflection of the airframe of the vehicle. Since, in general, payloads of expendable launchers are cantilevered off the forward end of the vehicle, airframe deflection has little impact on the payload except in extreme cases. In the case of the Space Shuttle, long payloads are attached at points along the length of the cargo bay. Deflection of the airframe can therefore induce loads into the payload structure. Some load alleviation provision is built into the attach points, and in many cases it is possible to design a statically determinant attachment that at least makes the problem reasonably easy to analyze. Very large and/or complex payloads may require attachment at a number of points that can lead to a complex analytical problem. In some cases airborne support equipment (ASE) is designed to interface with the Shuttle and take the loads from the airframe and protect the payload. This can be costly in payload capability, since all such ASE is charged against Shuttle cargo capacity.

Internal pressure is a major source of structural loads, particularly in tanks, plumbing, and rocket engines. It may also be a source of loads during ascent in inadequately vented areas. Early honeycomb structures, especially nose fairings, sometimes encountered damage or failure because pressure was retained inside the honeycomb cavities while the external pressure decreased with altitude. Adhesive weakening caused by aerodynamic heating allowed internal pressure to separate the face sheets. Careful attention to venting of enclosed volumes is important in preventing problems of this type.

Internal pressure or the lack of it can also be a problem in handling and transportation. Some operations may result in reduced pressure in various volumes that must then resist the externally applied atmospheric pressure. A common but by no means unique example is in air transport of launch vehicle stages, especially in unpressurized cargo aircraft. If the internal tank pressure is reduced during high-altitude flight, either deliberately or because of a support equipment malfunction, then during descent the pressure differential across the tank walls can be negative, resulting in the collapse of the tank. Prevention of this simply requires attention and care, but the concern cannot be ignored.

Thermal stress usually results from differential expansion or contraction of structures subjected to heating or cooling. It may also arise as a result of differential heating or cooling. The former effect can be mitigated to some degree by selection of materials with compatible coefficients of thermal expansion.

Once the vehicle is in space, the primary sources of heat are the sun and any internally generated heat. The latter is usually the smaller effect, but cannot be ignored, especially in design of electronic components, circuit boards, etc. Differential heating caused by the sun on one side and the heat sink of dark space on the other can result in substantial structural loads. These are most easily dealt with by thermal insulation or by simply designing the structure to withstand the stress. Note that in a rotating spacecraft the inputs are cyclic, possibly at a fairly high rate. In massive structures the thermal inertia of the system tends to stabilize the temperature. However, if the material being dealt with is thin, substantial cyclic stress can be generated, possibly leading to eventual failure.

Entry into eclipse results in rapid cooling of external surfaces and low thermal mass extremities. These can quickly become quite cool without solar input. Upon re-emergence into the sunlight, the temperature rapidly increases. This can cause not only substantial structural load but sufficient deformation that accurate pointing of sensors may be difficult.

Thermal inputs to long booms of various types can couple with the structural design and possibly with shadow patterns to cause cyclic motion of the boom, which can cause instability in spacecraft pointing or at least increase the requirements on the attitude control system.

The presence of cryogenic materials onboard the spacecraft for propulsion or sensor cooling is a major source of thermally induced stress. The problem is complicated by the need to thermally isolate the cryogenic system from the spacecraft structure to minimize heat leakage.

Structural Loads Analysis

Detailed analysis of structural loads usually involves complex computer software such as NASTRAN. For preliminary purposes, however, loads inputs can usually be approximated using factors and formulas empirically derived from previous launches. Structural elements may then be sized in preliminary manner using standard statics techniques. These preliminary sizings and material choices may then be refined by more sophisticated techniques.

It should be borne in mind that, although the sources of structural loads were discussed separately, they generally act in combination and must be used that way for design purposes. As an example, a cryogenic tank, pressurized during launch, will be subjected to thermally induced loads, internal pressure loads, and the vibration, linear acceleration, and acoustic loads of launch. Similarly, a deployable structure may encounter release and latching shocks while still under differential thermal stress resulting from exiting the Earth's shadow.

Load Alleviation

Various means are used to somewhat alleviate structural loads. The Shuttle throttles back the main engines during passage through maximum dynamic pressure. (Although this is done out of concern for the structural integrity of the Orbiter, it can be beneficial to the payloads as well.) Most expendable vehicles lack this capability, although solid motor thrust profiles and angle-of-attack control may be practiced to moderate the loads during this critical period.

Acoustic inputs can probably best be dealt with by design of the launch facility to minimize reflection of engine exhaust noise back to the vehicle. The payload must be designed to withstand whatever acoustic inputs the launch vehicle and launch facility impose. Use of stiffeners and/or dampening material on large, lightweight areas can help to minimize the structural response to these inputs.

The Shuttle payload attachment system is designed to minimize input of airframe structural loads into the payload. Figure 8.12 presents the basic attachment concept. By providing one or more degrees of freedom at each attach point, a statically determinant attachment is created. However, for some payloads, which may be very long and flexible or otherwise not able to accept the loads, it will be necessary to design a structural support that interfaces to the Orbiter attach points and isolates the payload itself from the Orbiter airframe deflection.

Fracture Mechanics

Fracture mechanics is a highly specialized field and will not be dealt with in any detail here. It is important, however, that the spacecraft designer be aware of the existence and purpose of the discipline.

Although fracture mechanics can be applied to any highly stressed part, its greatest application is to pressure vessels. The most desirable characteristic of a pressure vessel, especially for man-rated applications, is leak

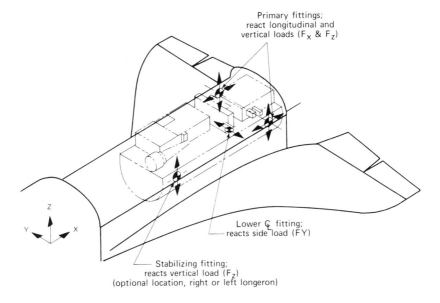

Primary fittings;
react longitudinal and
vertical loads (F_X & F_Z)

Lower ₵ fitting;
reacts side load (FY)

Stabilizing fitting;
reacts vertical load (F_Z)
(optional location, right or left longeron)

Fig. 8.12   Shuttle payload attachment.

before burst. In other words, if a crack forms, it is desirable that it propagate *through* the tank wall before it reaches the critical length, which will result in the crack propagating *around* the tank. The leak thus provides warning and possibly pressure relief before catastrophic failure occurs.

Fracture mechanics computes the probability of leak before burst based on a number of factors that include the following: material, vessel size, wall thickness, pressure, contained fluid, environment, vessel history (particularly pressure cycles and exposure to various substances), and extensive empirical data on crack propagation under similar circumstances. All of this information allows computation, to some level of confidence, of probability of failure and of leak before failure. Use of this technique is especially important for Shuttle payloads.

Stress Levels and Safety Factors

In a great many cases, material choice and thickness of spacecraft structures will be driven by factors other than strength. The primary factors typically will be stiffness, i.e., minimizing deflection under load and/or minimum gage of material that is available or that will allow it to be handled safely. In some cases, however, pressure vessels and some major structure being classical examples, the actual strength of the material to resist yielding or breakage is important. At this point safety factors come into play.

A typical safety factor for general use is about 1.5 for yield. That is, the structure is designed to yield only when subjected to loads 1.5 times the maximum expected to be encountered in service. "Yield" is defined in this case as undergoing a deformation in shape from which the structure does not recover when the load is removed. For all except very brittle materials, actual failure, i.e., structural breakage, takes place at stresses somewhat higher than yield. The ratio of yield stress to failure stress varies from one material to another, but typically if the factor of safety on yield is 1.5, the factor of safety to failure will be about 2.0.

For some applications in manned spacecraft or man-rated systems, the factors of safety may be higher than those discussed earlier, especially for items critical to flight safety.

For noncritical components the safety factors may be lower than those discussed earlier. In the case of a component that is not safety related or critical to mission success, lower factors may be acceptable. In some cases a factor of 1.0 on yield might even be accepted, meaning that a small, permanent deformation is acceptable as long as the part does not break.

An important factor to be considered is the nature of the load, particularly whether it is steady or cyclic and how long it lasts. The factors previously discussed assume steady loads or very few cycles. If the load is cyclic, then the fatigue characteristic of the material is the major consideration. If many cycles are expected, then it is important to keep the stress in the material at a level that allows an acceptable fatigue life. Typically this will result in a structure substantially overdesigned compared with a static load of the same magnitude.

If the load is steady but will be applied for extremely long periods, the creep characteristics may become important. An example might be a bolted joint, which is expected to maintain the same tension for years. However, the bolts, if very highly loaded, might lose tension over long periods because of creep if subjected to stress that allows inadequate creep life, even though there is no immediate danger of failure due to overload.

In considering safety factors, a frequently overlooked point is that all of the data, both the loads and the material characteristics, have some associated uncertainty. This may be of secondary importance when designing ground equipment with safety factors of 5 or 10. It can be very critical, as we shall see, when designing for the small safety factors typical of aerospace hardware. This is most easily demonstrated by an example.

In this example, let us assume that we have a structural material with a quoted yield strength of 35,000 psi. This might be typical of an aluminum alloy, for example. If we simply apply a safety factor of 1.5, the allowable stress in the structure in question would then be 23,333 psi. However, it is known that there exists a significant spread in the strength data available, the standard deviation being 1850 psi. Thus, to minimize the probability of failure, the 3σ low strength should be used. This amounts to $35,000 - 3(1850) = 29,450$ psi. (Some may be inclined to consider that "aluminum is aluminum" and use the handbook value; however, there can be lot-to-lot variation or within-lot variations due to handling, processing, or environmental history that can be significant. In the case of many composite materials, the effect of environment and process is even more pronounced, and considerable attention must be paid to possible variations in characteristics.)

Similarly, we find that, due to a variety of factors, there is a standard deviation about the average load. If we use a 3σ high load of just less than the 3σ low strength, i.e., 29,250 psi, then the average load is 21,000 psi. Use of this value for design of the component will result in 3σ confidence that no combination of worst case strength and loads will yield.

Based on the average numbers, the factor of safety is 35,000/21,000 = 1.67. This is significantly higher than the originally targeted 1.5. However, if we used average strength and the 1.5 factor, there would have been a statistically significant probability of a worst case combination causing yielding. For components of low criticality, that risk might very well be acceptable, but clearly it is not for critical components. Figure 8.13 illustrates the example. Note that for a 4 or 5σ case the curves actually cross, and yielding would occur. The probability of this is sufficiently low that it is unlikely that it would be a matter of concern.

It can be seen from the preceding discussion that a relatively small increase in safety factor can have a substantial impact on probability of failure. Less apparent but equally true is the fact that increased safety factor can allow substantial cost saving. With a larger safety factor it may be possible to reduce the amount of testing and detailed analysis required with a resulting reduction in costs. Thus, availability of substantial mass margins can translate into a much lower cost program if program management is clever enough to take advantage of the opportunity thus offered. This requires management to avoid the pitfall of proceeding with a sophisticated test and analysis effort simply because "that's the way we have always done it." The JPL/Ball Aerospace Solar Mesosphere Explorer (SME) is a textbook example of a program that took advantage of ample mass margin in order to keep spacecraft costs low.

Figure 8.14 presents a handy means of estimating failure rate based on the average loads and strengths and the standard deviation about each. The vertical axis plots number of combined standard deviations, i.e., standard

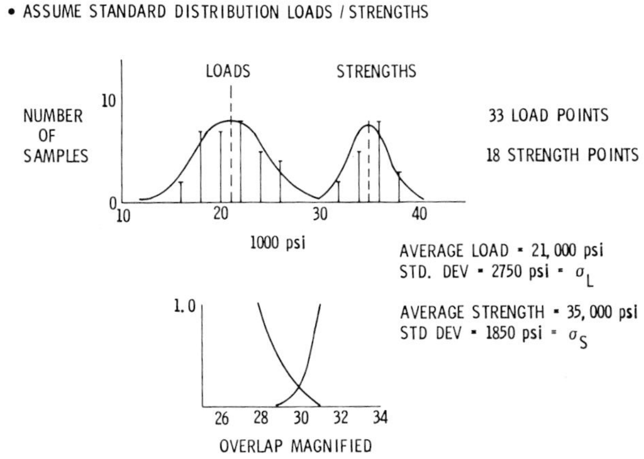

Fig. 8.13   Uncertainty distributions of loads and strength.

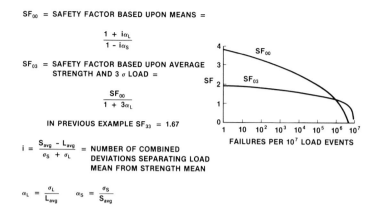

SF_{00} = SAFETY FACTOR BASED UPON MEANS =

$$\frac{1 + i\alpha_L}{1 - i\alpha_S}$$

SF_{03} = SAFETY FACTOR BASED UPON AVERAGE STRENGTH AND 3 σ LOAD =

$$\frac{SF_{00}}{1 + 3\alpha_L}$$

IN PREVIOUS EXAMPLE SF_{33} = 1.67

$i = \dfrac{S_{avg} - L_{avg}}{\sigma_S + \sigma_L}$ = NUMBER OF COMBINED DEVIATIONS SEPARATING LOAD MEAN FROM STRENGTH MEAN

$\alpha_L = \dfrac{\sigma_L}{L_{avg}}$ $\alpha_S = \dfrac{\sigma_S}{S_{avg}}$

Fig. 8.14   Safety factor vs failure rate.

deviation in load plus standard deviation in strength times the number on the vertical axis. The horizontal axis plots number of failures per 10^7 load events. For example, a failure rate of one per million load events requires 3.5 combined standard deviations separating the average values. Also plotted are the related safety factors where the upper curve, SF_{avg}, is based on the average values of both load and strength, whereas the lower curve is based on average strength divided by average load plus three standard deviations. Note that the horizontal axis is load events. This may be one per mission in some cases, and in others it could be hundreds, thousands, or even millions per mission if the member is loaded in a cyclic or vibratory fashion.

This has been a very light touch on a complex subject that is generally not well understood. The point is that a simple statement of a value as "safety factor" is meaningless without understanding the basis from which it is derived. Furthermore, safety factor and failure rate are a tradeoff with mass and cost and should be considered in that light.

8.4 LARGE STRUCTURES

As space activities increase in variety and complexity, it is to be expected that there will be increased interest in very large structures. In fact, proposals have already been made for solar arrays and microwave antennas on a scale of kilometers to beam power from geostationary orbit to Earth. The popular view, encouraged by those of entrepreneurial bent, is that, in a weightless environment, structures can be arbitrarily large and light in structure. Although there is an element of truth in this, there are major practical limitations with which the designer of such structures must deal.

Most large structures such as solar arrays, antennas, telescopes, etc., must maintain shape to a fairly tight tolerance if they are to function effectively. This is often difficult on small structures that are currently in use and becomes far more so on a scale of tens of meters or kilometers. Thermal distortion for a given temperature differential is a direct function of the dimensions of the structure. Bigger structures distort more in an absolute sense. In most applications, attitude control maneuvers are required. A very lightweight flexible structure will distort during maneuvers and in response to attitude hold control inputs because of the inertia of the structure. As the force is removed, the structure springs back, but the low-mass, low restoring force and (probably) near-zero damping tend to give rise to low-frequency oscillations that die out very slowly and, in fact, may excite control system instability.

Obviously, the control force distortion concern can be partially alleviated by using relatively small forces and maneuvering very slowly. However, operational needs will dictate some minimum maneuver rate and settling time, and the system must be able to deal with anticipated disturbances. These requirements will set a lower limit to the control forces required.

Even the assumption of weightlessness is not entirely valid for very large structures. In such large structures, the forces caused by the radial gradient in the gravity field can cause distortion, at least in lower-altitude orbits.

Unfortunately, the accuracy requirements do not change as the size of the structure increases. A microwave antenna still requires surface accuracy on the order of a wavelength, whether it is 1 m or 1 km in diameter. Although solar arrays need not maintain the same degree of surface control as an antenna, it is still important to maintain shape with reasonable accuracy. In any case, excessive structural distortion makes accurate pointing almost impossible.

For very large space structures, it is simply not practical to maintain shape via designing strength and stiffness into the structure. The mass of such a structure would be enormous, increasing transportation costs to an intolerable level. The greater complexity of assembling a more massive structure is also a matter of concern. In any case, it is not at all clear that a brute force approach could solve the problem. New materials, particularly composites that offer the possibility of tailoring characteristics such as stiffness and thermal response, can contribute greatly but do not offer a total solution.

The concerns just listed clearly indicate that large space structures are by no means as simple as their proponents, enamored by the tremendous promise offered by such structures, have indicated. Worthwhile large structures must be relatively easy to deploy and assemble and must have predictable, repeatable, controllable characteristics. In order to properly design control systems, it is necessary to be able to model the response with satisfactory fidelity. This capability is now becoming available through the use of modern, high-capacity computers.

One way to deal with the problem is active shape control. This concept is now being explored for large, Earth-based optical telescopes. Sensing the shape of the surface using laser range finders or by measuring the energy distribution in the beam leaving an antenna is used as input to an active control system that mechanically or thermally distorts the surface to compensate for structural irregularities. In the case of a phased array antenna, the phasing can be altered to accomplish the same end. Note that this is a potential solution to the surface control problem. The operation of the structure as a spacecraft, i.e., attitude control and gross pointing, remains and demands adequate modeling and solution to the control input, response, settling time, and flexibility problems.

The control of large, flexible structures is a complex issue involving optimization among material characteristic choices, structural design approach, and control system design. Adding to the complexity is the probable requirement to launch in many separate pieces that are themselves folded into a compact shape. Each piece must be deployed, checked out, and joined to its mating pieces in as straightforward and automatic a fashion as possible to create the structure that the system was designed to control.

8.5 MATERIALS

Structural Materials

Most materials used in space applications to date have been the conventional aerospace structural materials. These will continue to dominate for the foreseeable future, although steady growth in the use of newer materials is to be expected.

Among the conventional structural materials, aluminum is by far the most common. A large variety of alloys exist providing a broad range of such characteristics as strength and weldability. Thus, for applications at moderate temperature in which moderate strength and good strength-to-weight ratio are desirable, aluminum is still most often the material of choice. This popularity is enhanced by ready availability and ease of fabrication. A number of surface-coating processes exist to allow tailoring of surface characteristics for hardness, emissivity, absorbtivity, etc.

Magnesium is often used for applications in which higher stiffness is desired than can be provided by aluminum. It is somewhat more difficult to fabricate and, being more chemically active than aluminum, requires a surface coating for any extensive exposure to the atmosphere. Several coatings exist. Environmental constraints in recent years have limited the availability of certain desirable alloys containing zirconium.

Steel, in particular stainless steel, is often used in applications requiring higher strength and/or higher temperature resistance. A variety of steels may be used, but stainless steel is often preferred because its use eliminates concern about rust and corrosion during the fabrication and test phase. Additionally, if the part may be exposed to low temperature, the low ductile-to-brittle transition (DBT) temperature of stainless steel and similar alloys is an important factor.

Titanium is a lightweight, high-strength structural material with excellent high-temperature capability. It also exhibits good stiffness. Some alloys are fairly brittle, which tends to limit their application, but a number of alloys with reasonable ductility exist. Use of titanium is limited mostly by higher cost, lower availability, and fabrication complexity to applications that particularly benefit from its special capabilities. Pressure vessels of various types and external skin of high-speed vehicles are typical applications.

Beryllium offers the highest stiffness of any naturally occurring material along with low density, high strength, and high temperature tolerance. Thermal conductivity is also good. Beryllium has been used in limited applications where its desirable characteristics have been required. The main limitation on more extensive use of this apparently excellent material is toxicity. In bulk form, beryllium metal is quite benign and can be handled freely. The dust of beryllium or its oxide, however, has very detrimental effects on the human respiratory tract. This means that machining or grinding operations are subject to extensive safety measures to capture and contain dust and chips. This renders normal fabrication methods unusable without resorting to these intensive (i.e., expensive) measures.

Glass fiber-reinforced plastic, generically referred to as fiberglass, was the first composite material used for space structure and is probably still the most common. The matrix material may be epoxy, polyester, or other material, and the glass can range from a relatively low-quality fiber all the way to highly processed quartz fiber. Fiberglass is desirable because of the relative ease with which complex shapes can be fabricated. It also exhibits good strength and offers the ability to tailor strength and stiffness both in absolute value and direction in the material by choice of fiber density and orientation.

The inconel family of alloys and other similar alloys based on nickel, cobalt, etc., are used for high-temperature applications. Typical application is as a heat shield in the vicinity of a rocket nozzle to protect the lower-temperature components from thermal radiation or hot gas recirculation. These alloys are of relatively high density, equal to steel or greater so that weight can be a problem. However, inconel in particular lends itself to processing into quite thin foils, which allows its use as a shield, often in multiple layers, with minimum mass penalty.

New materials coming into use are mostly composites of various types, although some new alloys have also appeared. Among the alloys, aluminum-lithium is of considerable interest, since the addition of the lithium results in alloys generally similar in strength to the familiar aluminum alloys but several percent lighter. This material is already seeing extensive use in commercial aviation.

High-temperature refractory metals have been available for many years but have seen limited use because of high density, lack of ductility, cost, and other factors. Tungsten, tantalum, and molybdenum fall into this category. These materials are actually somewhat less available than they were some years ago. A great many suppliers have dropped out of the field. This may in part be related to the collapse of the commercial nuclear power industry in the U.S. One exception is niobium (formerly called columbium). This material is useful to temperatures as high as 1300 K but has a density only slightly higher than steel. It is available in commercial quantities. Like all the refractory metals, it oxidizes rapidly if heated in air, but a silicide coating offers substantial protection in this environment.

Metal matrix composites involve use of a metal matrix, e.g., aluminum, stiffened and strengthened by fibers of another metal or nonmetallic material. In aluminum, for example, fibers of boron, silicon carbide, and graphite have been used. Some difficulties have been encountered, such as the tendency of the molten aluminum to react with the graphite during manufacture of the composite. Work on protective coatings continues. Boron-stiffened aluminum is well developed and is used in the tubular truss structure that makes up much of the center section of the Shuttle Orbiter. This entire area is one of enormous promise. As yet, we have hardly scratched the surface of the potential of this type of composite.

Carbon-carbon composite consists of graphite fibers in a carbon matrix. It has the ability to hold shape and resist ablation and even oxidation at quite a high temperature. At the present level of development, however, carbon-carbon is not suitable for a load-bearing structure. For example, it is used in the nose cap and wing leading edges of the Shuttle Orbiter where it must resist intense re-entry heating but it does not form a part of the load-bearing structure. Progress is being made in the development of structural carbon-carbon, and it is expected to have a bright future as a hot structure for high-speed atmospheric and entry vehicles.

Graphite-epoxy is in such general use that it is hardly appropriate to group it with new materials. The use of high-strength and stiffness graphite fiber within an epoxy matrix offers an excellent high-strength structural material. Proper selection of the cloth and/or unidirectional fibers offers the ability to tailor strength and stiffness directionally and to the desired levels to optimize it for the purpose. The low density of graphite offers a weight advantage as well. High-temperature characteristics are improved by use of graphite instead of glass, although the matrix is the final limiting factor.

Films and Fabrics

By far the most commonly used plastic film material in space applications has been Mylar. This is a strong, transparent polymer that lends itself well to fabrication into sheets or films as thin as 0.00025 in. Coated with a few angstroms of aluminum to provide reflectivity, Mylar is well suited to the fabrication of the multilayer insulation extensively used on spacecraft.

A new polymeric film material with higher strength and the ability to withstand higher temperature than Mylar is the polyimide Kapton. These characteristics have made Kapton a desirable choice for outer layers of thermal blankets. A problem has arisen with the discovery that, in low Earth orbits, polymer surfaces undergo attack and erosion by atomic oxygen, which is prevalent at these altitudes. Kapton seems to be more susceptible to this sort of attack than Mylar. In any case, for long-life use in low orbit, metallization or coating with a more resistive polymer such as Teflon will probably be required. This erosion rate is sufficiently low that, for shorter missions, the problem may not be serious.

Teflon and polyethylene have been used extensively as bearings, rub strips, and in various protective functions because of their smoothness, inertness, and particularly for Teflon, lubricative ability.

Fiberglass cloth, which is strong and flexible, has seen use as an insulator and as protective armor against micrometeoroids. A commercially available cloth of fiberglass coated with Teflon called Betacloth has been used as the external surface of spacecraft thermal blankets for this purpose.

A series of new woven clothlike materials superficially similar to fiber-glass but of much higher temperature capability is becoming available. These cloths are woven from fibers of high temperature ceramic materials. The most well-known application of such materials is as the flexible reusable surface insulation (FRSI) used on the upper surfaces of the later-model Shuttle Orbiters. They can also be useful as insulators of high-temperature devices such as rocket engines.

Future Trends

As has been the case in the past, future trends in materials will be characterized by a desire for increased specific strength and specific stiffness. The latter will tend to dominate because, as observed earlier, most space structure designs are driven by stiffness more than strength. Higher thermal conductivity with lower coefficient of thermal expansion is also highly desirable for obvious reasons. The data in Fig. 8.15, adapted from

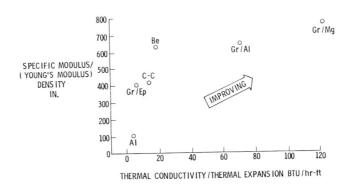

Fig. 8.15 Desired structural and thermal characteristics.

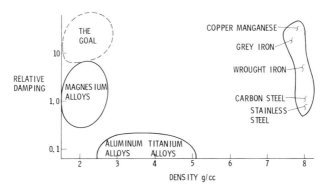

Fig. 8.16 Damping capability.

a curve by J. Garibotti of Textron, indicate desirable trends in stiffness and thermal characteristics. The currently available materials are grouped to the left with beryllium still showing an edge even over the composites. Graphite-aluminum offers substantial improvement once its problems are solved and graphite-magnesium shows even better sometime in the future. It is quite probable that other candidates will emerge as research continues.

Damping capability is also important as a means of reducing sensitivity to vibration and shock. Figure 8.16 rates damping ability vs density. The common aerospace alloys are generally poor, magnesium being the best. Excellent dampers are available as indicated toward the upper right hand; however, they tend to be heavy, dirty, and relatively weak and have a high DBT temperature. All of these characteristics make them unusable for space applications. The developing field of composites may offer the best hope of achieving the goal, although the present trend to use high-stiffness fibers may make this difficult.

Refractory metals stiffened with high-temperature fibers, structural carbon-carbon, and other new material developments should open new avenues for entry thermal protection. This will allow replacement of the existing fragile Shuttle tiles with hardier versions and offer improved capability in future entry systems.

9.1 INTRODUCTION

The thermal subsystems engineer's task is to maintain the temperature of all spacecraft components within appropriate limits over the mission lifetime, subject to a given range of environmental conditions and operating modes. Thermal control as a space vehicle design discipline is unusual in that, given clever technique and reasonable circumstances, the thermal "system" may involve little or no actual spacecraft hardware, whereas more demanding missions may require onboard equipment. In all cases, however, the required analysis will involve the thermal control engineer in the design of nearly all other onboard subsystems.

As with attitude control, thermal control techniques may be broadly grouped within two classes, passive and active, with the former preferred when possible because of simplicity, reliability and cost. Passive control includes the use of sunshades and cooling fins, special paint or coatings, insulating blankets, and tailoring of the geometric design to achieve both an acceptable global energy balance and local thermal properties. When the requirements are too severe for passive techniques, active control of spacecraft temperatures on a local or global basis will be employed. This may involve the use of heating or cooling devices, heat pipes, louvers or shutters, or alteration of the spacecraft attitude to attain suitable conditions.

The basic heat-transfer mechanisms are conduction, convection, and radiation. Broadly generalizing, it may be said that the overall energy balance between a spacecraft and its environment is dominated by radiative heat transfer, that conduction primarily controls the flow of energy between different portions of the vehicle, and that convection is unimportant in space vehicle design. As with all generalizations this is an oversimplification, useful to a point but allowing numerous exceptions. This will be seen in the following sections.

Passive Thermal Control

The techniques applied for passive thermal control include use of geometry, coatings, insulation blankets, sun shields, radiating fins, and heat pipes. "Geometry" refers to configuring the spacecraft to provide the required thermal radiating area, placing low-temperature objects in shadow, and exposing high-temperature objects to the sun or burying them deeply within the structure, or similar manipulation of spacecraft configuration to optimize thermal control.

Insulation blankets are, for the most part, multilayer insulation consisting of layers of aluminized Mylar or other plastic spaced with nylon or Dacron mesh. External coverings of fiberglass, Dacron, or other materials may be used to protect against atomic oxygen attack, micrometeoroids, etc.

Sun shields may be as simple as polished aluminum sheet or perhaps gold-plated aluminum. More sophisticated is silvered Teflon, which essentially acts as a second-surface mirror with the silver on the back providing reflectivity and the Teflon providing high infrared emissivity. Along the same line are actual glass second-surface mirrors, which are generally more efficient but at the cost of greater weight and possible problems with the brittle glass.

Fins are often used where it is necessary to reflect large amounts of heat, or smaller amounts at low temperature, thus requiring large surface area. Large numbers of fins in circular configurations tend to have problems obtaining an adequate view factor to space. Very long fins may be limited in effectiveness by the ability to conduct heat through the fins.

Heat pipes are tubular devices containing a wick running the length of the pipe, which is partially filled with a fluid such as ammonia. The pipe is connected between a portion of the spacecraft from which heat is to be removed and a portion to which it is to be dumped. The fluid evaporates from the hot end, with the vapor being driven to condense (releasing its heat of vaporization) at the cold end. Condensed fluid is drawn by capillary action back to the hot end.

One may take issue with placing heat pipes in the passive category since there is an actual circulation of fluid within the heat pipe driven by the heat flow. We consider heat pipes passive from the viewpoint of the spacecraft designer in that there is no direct control function required nor is there a requirement for the spacecraft to expend energy. The heat pipe simply conducts energy when there is a temperature differential and ceases to do so if the differential disappears. Control of heat pipes is possible by means of loaded gas reservoirs or valves. This of course reduces the advantage of simplicity and reliability.

Caution is required in using heat pipes to make sure that the hot end is not so hot as to dry the wick completely, thus rendering capillary action ineffective in transporting new fluid into that end. Similarly, the cold end must not be so cold as to freeze the liquid. Also, heat pipes work quite differently in $0\,g$ because of the absence of free convection, making interpretation of ground test results a problem unless the heat pipe is operating horizontally.

Active Thermal Control

Active thermal control of spacecraft may require devices such as heaters and coolers, shutters or louvers, or cryogenic materials. Thermal transport may be actively implemented by pumped circulation loops.

Heaters usually are wire-wound resistance heaters, or possibly deposited resistance strip heaters. Control may be thermostatic, by command, or both. For very small heaters where on/off control is not required, radioisotope heaters are sometimes used. The usual size is 1 W thermal output. It might be argued that such devices are passive, since they cannot be commanded and do not draw spacecraft power.

Various cooling devices have been applied or are under consideration. Refrigeration cycles such as those that are used on Earth are difficult to operate in $0\,g$ and have seen little or no use. Thermoelectric or Peltier cooling has been used with some success for cooling small, well-insulated objects. The primary application is for cooling detector elements in infrared observational instruments that are operated for long periods. The Villaumier refrigerator is of considerable interest for similar applications, and development of such devices has been in progress for some years.

A straightforward cooling device that has seen considerable use is the cryostat. This device depends on expansion of a high-pressure gas through an orifice to achieve cooling. To achieve very low temperature, two-stage cryostats using nitrogen in the first stage and hydrogen in the second have been used. The nitrogen precools the hydrogen so that its initial temperature is in the range where it will approach the liquid temperature upon expansion, affording quite a low temperature. The obvious disadvantage of such systems is the use of a consumable gas, which limits the number of operating cycles to a few, perhaps only one.

For long-term cooling to low temperature, an effective approach is to use cryogenic fluid. The principal use has been onboard spacecraft designed for infrared measurements. An example is the infrared astronomy satellite (IRAS), launched in January 1983. Cooling by expansion of liquid helium through a porous plug reduced the telescope optical temperature to $4\,K$, allowing observations at very long infrared (IR) wavelengths without interference to the telescope from its own heat. (As we shall see, a telescope mirror at $4\,K$ emits blackbody radiation at a peak wavelength of about $750\,mm$, thus preventing observations at or near this wavelength.) IRAS performed the first all-sky IR survey, expiring after nearly 11 mo of operation, when all of its helium was depleted. The fact that the observing life of IRAS ended upon exhaustion of the helium even though all other systems were still functioning provides a strong argument for both cryogenic refrigerators and on-orbit servicing.

Shutters or louvers are among the most common active thermal control devices. Common implementations are the louver, which essentially resembles a venetian blind, or the flat plate with cutouts. The former may be seen in the Voyager spacecraft illustration in Chapter 8. The blades are individually controlled by bimetallic springs so arranged that, when the surface warms up, the louvers open, allowing more heat to be rejected to space. Conversely, when the surface cools, the louvers close, occluding some of the radiating surface and reducing the radiated heat. The flat-plate variety is shown on the Television and Infrared Observation Satellite/Defense Meteorological Satellite Program (Tiros/DMSP) spacecraft illustration in Chapter 8. The flat plate is rotated by the bimetallic element. The plate has cutout sectors that are placed over insulated areas to decrease heat flow and rotate over uninsulated areas to increase heat flow and cool the spacecraft. The flat-plate variety is much simpler and less costly but allows less efficient use of surface area and fine tuning of areas on a given surface. Although the automatic control described is most common and usually satisfactory, it is obviously possible to provide commanded operation as well, either instead of the thermostatic approach or as an override.

Heat-Transfer Mechanisms

Heat-transfer mechanisms affecting spacecraft are of course the same as those with which we are familiar on Earth: conduction, convection, and radiation. The primary difference is that convection, which is often the overriding mechanism on Earth, is usually nonexistent in space. Still, convection will be encountered on the surface of planets with atmospheres, during atmospheric flight, and inside sealed pressurized spacecraft and pumped cooling loops. All three mechanisms will be discussed in this chapter.

9.2 CONDUCTIVE HEAT TRANSFER

Conduction occurs in solids, liquids, and gases. It is usually the primary mechanism for heat transfer within a spacecraft (although radiation may be important in internal cavities). Since all electronic devices generate at least some heat as part of their operation, there is a risk of overheating if care is not taken to provide adequate paths for conducting heat from the component to the appropriate heat rejection surface. Of course, the same concern exists with ground-based equipment. However, thermal design of such equipment is usually much less of a problem because of the efficiency of free convection in providing heat relief. It is also largely self-regulating. In special cases, such as cooling the final amplifier stage of a radio transmitter, the ground-based designer can provide a small fan to ensure forced convection over a particular area. Free convection is unavailable in space, even in pressurized spacecraft, because of the lack of gravity, whereas fan cooling is generally found only in manned spacecraft. Deliberate provision of adequate conduction paths is therefore a key requirement for the spacecraft thermal engineer.

Design practice in providing thermal conduction involves more than selecting a material with suitable conductivity. For example, unwelded joints, especially in vacuum, are very poor thermal conductors. Worse yet, they may exhibit a factor of two or more variability in conduction between supposedly identical joints. This situation can be substantially improved by use of conduction pads, thermal grease, or metal-loaded epoxy in joints that are mechanically fastened. Obviously this is only done where high or repeatable conductivity is essential to the design.

Referring to material selection, it should be noted that high thermal conductivity and high electrical conductivity normally are closely related. Therefore, a situation in which high thermal conductivity is required while electrical isolation is maintained is often difficult. One substance that is helpful here is beryllium oxide (BeO), which has high thermal conductivity but is an excellent electrical insulator. Care must be taken in the use of BeO, which has a powder form that is highly toxic if breathed.

The basic mathematical description of heat conduction is Fourier's law, written one-dimensionally as

$$Q = -\kappa A(dT/dx) \qquad (9.1)$$

and shown schematically in Fig. 9.1. Q is the power (energy per unit time), expressed in watts, British thermal units per second, or the equivalent. A is the area through which the heat flow occurs, and κ is the thermal conductivity in units such as watts per meter per Kelvin or British thermal units per hour per foot per degrees Fahrenheit. T is the temperature in absolute units such as Kelvins or degrees Rankine, and x is the linear distance over the conduction path. Qualitatively, Eq. (9.1) expresses the commonly observed fact that heat flows from hot to cold and that the more pronounced the temperature difference, the more energy transfer is observed.

It is often more useful to consider the power per unit area, or energy flux, which we denote as

$$q \equiv Q/A = -\kappa(dT/dx) \qquad (9.2)$$

with units of watts per square meter. Vectorially, Eq. (9.2) may be extended for isotropic materials to

$$q = -\kappa \nabla T \qquad (9.3)$$

Equation (9.3) may be applied to the energy flux through an arbitrary control volume; invoking Gauss' law and the law of conservation of energy yields the conduction equation,

$$\rho C \partial T/\partial t = \kappa \nabla^2 T + g(r,t) \qquad (9.4)$$

which allows the temperature in a substance to be calculated as a function of the position vector r and time. Internal heat generation (power per unit volume) is accounted for by the source term $g(r,t)$. C is the heat capacity of the substance, with units such as joules per kilogram per Kelvin, and ρ is

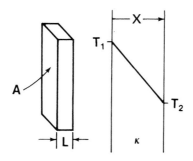

Fig. 9.1 Conduction in one dimension.

its density. The ∇^2 is the Laplacian operator, which in Cartesian coordinates is

$$\nabla^2 = \partial/\partial x^2 + \partial/\partial y^2 + \partial/\partial z^2 \qquad (9.5)$$

and is given for other coordinate systems of interest in standard references.[1]

The conduction equation is interesting mathematically for the range of solutions that are exhibited in response to differing initial and boundary conditions. Except in simple cases, which are outlined in standard texts,[2] a numerical solution is usually required to obtain practical results. As always, a discussion of numerical techniques is outside the scope of this text.

Generally, one wishes to solve the conduction equation to obtain the temperature distribution in some region. This region will be defined by the coordinates of its boundary, along which certain conditions must be specified to allow a solution to be obtained. In the example of Fig. 9.1, the infinite slab is defined as a region by faces at $x = 0$ and $x = L$, with no specification on its extent in the y and z directions. (Equivalently, the slab may be considered to be well insulated at its edges in the y and z directions, so that no heat flow is possible.) One might wish to know the temperature at all points within $(0,L)$ given knowledge of the slab's properties and the conditions on either face.

Boundary conditions for the conduction equation may be of two general types. Either the temperature or its derivative, the heat flux (through Fourier's law) may be specified on a given boundary. For a transient problem, the initial temperature distribution throughout the region must also be known. Let us consider the simple case of Fig. 9.1 and assume the faces at $x = (0,L)$ to have fixed temperatures T_0 and T_L. Then Eq. (9.4) reduces to

$$\mathrm{d}^2 T/\mathrm{d}x^2 = 0 \qquad (9.6)$$

which has the general solution

$$T(x) = ax + b \qquad (9.7)$$

Upon solving for the integration constants, we obtain

$$T(x) = (T_L - T_0)x/L + T_0 \qquad (9.8)$$

and from Fourier's law, the heat flux through the slab is found to be

$$q = -\kappa(\mathrm{d}T/\mathrm{d}x) = \kappa(T_L - T_0)/L \qquad (9.9)$$

Note that, instead of specifying both face temperatures, we could equally well have specified the heat flux at one face (which in this constant-area steady-state problem must be the same as at the other face) and a single boundary temperature. Assuming that T_L and the heat flux q_w are known, we obtain after twice integrating Eq. (9.6),

$$T(x) = -(q_w/\kappa)x + b \qquad (9.10)$$

and upon solving for the constant of integration,

$$T(x) = (L - x)q_w/\kappa + T_L \qquad (9.11)$$

It is seen that T_0 is now obtained as a solved quantity instead of a known boundary condition. Clearly, either approach can be used, but it is impossible to specify both of the face temperatures and also the heat flux. Moreover, two boundary conditions are always required; specification of one face temperature, or the heat flux alone, is insufficient.

This is a simple but useful example to which we shall return. In the transient case, or when internal sources of energy are present, the solutions rapidly become more complicated and are beyond the intended analytical scope of this book. The interested reader is referred to any introductory heat-transfer text[2] for treatment of a variety of useful basic cases.

9.3 CONVECTIVE HEAT TRANSFER

Of all the heat-transfer mechanisms, convection is the most difficult one to analyze, predict, or control. This is because it is essentially a fluid dynamic phenomenon, with behavior dependent on many factors not easily measurable or predictable. The convective heat-transfer coefficients h_c that we use are, for the most part, empirical and highly variable. Part of the problem arises because convection is in truth not a heat-transfer mechanism at all. The energy is still transferred by conduction or radiation, but the conditions defining the transfer are highly modified by mass transport in the fluid. This is illustrated schematically in Fig. 9.2.

So-called free convection is driven entirely by density differences and thus occurs only in a gravitational field. It does not occur in space except when the spacecraft is accelerating. However, it does occur most unavoidably on Earth and thoroughly skews any heat-transfer or thermal control data that might be obtained from testing the spacecraft in the atmosphere. This fact is a primary (but not the sole) reason for conducting spacecraft thermal vacuum tests prior to launch. It is literally the only opportunity available to the thermal control analyst to verify his results in something approximating a space environment.

If convective heat transfer is required in $0\,g$, it must be forced convection, driven by a pump, fan, or other circulation mechanism. The interior of a manned spacecraft cabin is one example. Another might be a propellant or pressurization tank where good thermal coupling to the walls is required. Forced convection is not commonly used as a signficant means of unmanned spacecraft thermal control in U.S. or European spacecraft. However, the Soviets have made extensive use of sealed, pressurized unmanned spacecraft with fans for circulation as a means of achieving uniform temperature and presumably to avoid the concerns of operating

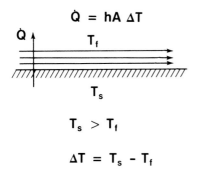

$$\dot{Q} = hA \; \Delta T$$

$T_s > T_f$

$\Delta T = T_s - T_f$

Fig. 9.2   Thermal convection.

some components in a vacuum. There is an obvious tradeoff here; design is much easier, but overall reliability may be lower since the integrity of the pressure hull is crucial to spacecraft survival.

Convection is important in various types of pumped cooling loops such as cold plates for electronics, regeneratively cooled rocket engines, and waste heat radiators. This of course is forced convection involving the special case of pipe or channel flow.

Convective heating is the critical mechanism controlling entry heating. It completely overpowers the radiative component until the entry velocity begins to approach Earth escape velocity. Even then, convection is still the more significant contributor. Similarly, it is the major mechanism in ascent aerodynamic heating. We have discussed this special case rather thoroughly in Chapter 6. In Table 9.1 we include a summary of several common entry vehicle thermal protection materials.

Convective heat flux may be written according to Newton's law of cooling as

$$Q = h_c A \Delta T = h_c A (T_2 - T_1) \qquad (9.12)$$

where Q is the power, h the convection or film coefficient, A the area, and ΔT the driving temperature differential.

The crucial element in this equation is the coefficient h_c, and, as implied earlier, our knowledge of its value in most situations is largely empirical. Engineering handbooks publish charts or tables giving ranges of values for the coefficient under varying sets of conditions, but the variance is usually significant, and tests under the specific conditions being considered may be required if high accuracy is to be obtained. Since convective heat transfer is a mass transport phenomenon as well as a thermal one, the coefficient depends strongly on whether the flow is laminar or turbulent. Generally the turbulent heat-transfer coefficient will be much higher. Thus, a laminar-to-turbulent transition along the surface of an entry body may result in a substantial increase in heating downstream of the transition.

Newton's law of cooling is of course an approximation. The problem of heat transfer from a moving fluid to a wall is a fluid dynamic problem, usually one that may be analyzed through the use of boundary-layer theory. In effect, the analytical solution of such a problem allows us to

Table 9.1 Thermal protection materials

TPS[a]	Type	Advantages	Disadvantages
AVCOAT 5025	Low-density charring ablator	Low density ($\rho = 34\,\text{lb/ft}^3$) Thoroughly tested Man rated Low thermal conductivity	Manual layup in honeycomb matrix Erosion capability estimated only
HTP-12-22 fibrous refractory composite insulation (FCRI)	Surface reradiation	Low density ($\rho = 12\,\text{lb/ft}^2$) Does not burn Good thermal shock tolerance Can maintain shape and support mechanical loads Low thermal conductivity	May melt under certain flight conditions ($T_{\text{melt}} = 3100^\circ\text{F}$) Uncertain erosion capability
ESM 1030	Low-density charring ablator	Low density ($\rho = 16\,\text{lb/ft}^3$)	Erosion capability unknown
Carbon-carbon over insulator	Surface reradiation and heat sink	Erosion capability known	High conductivity Possible thermal expansion problems Oxidation resistance unknown
Silica phenolic	High-density charring ablator	Erosion capability known Low thermal conductivity	High density ($\rho = 105\,\text{lb/ft}^3$)
Carbon phenolic	High-density charring ablator	Erosion capability known	High density ($\rho = 90\,\text{lb/ft}^3$) Oxidation resistance uncertain

[a]Thermal protection system.

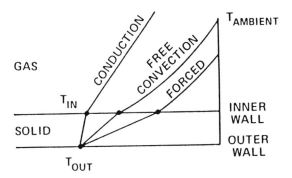

Fig. 9.3   Comparison of heat transfer mechanisms.

compute the value of h_c in the convection law. However, all of our comments elsewhere in this text concerning the intractability of fluid dynamics problems apply here as well; thus, direct solution for the film coefficient is restricted to a few special cases.

In all cases, convective heat transfer will maintain a higher flux than conduction. Forced convection is more effective than free convection, which is driven entirely by difference in density caused by the heat transfer. This relationship is illustrated qualitatively in Fig. 9.3. The film coefficient for free convection depends strongly on the orientation of the surface relative to the local vertical and, as noted earlier, does not occur in $0\,g$.

9.4 RADIATIVE HEAT TRANSFER

Radiation is the only typically practical means of heat transfer between a vehicle in space and its external environment. Mass expulsion is obviously used as a spacecraft coolant when open-cycle cryogenic cooling, as with IRAS, is performed, but this should be regarded as a special case. As noted previously, radiation becomes important as a heat-transfer mode during atmospheric entry at speeds above about 10 km/s. Even at entry speeds of 11.2 km/s (Earth escape velocity), however, it is still on the order of only 25% of the total entry heat flux. At very high entry speeds such as those that will be encountered by the Galileo atmospheric probe at Jupiter, radiative heat transfer will dominate.

Radiative energy transfer can strongly influence the design of certain entry vehicles, particularly those where gliding entry is employed. Assuming convective heating to be the major source of energy input, the entry vehicle surface will continue to heat until the point where energy dissipation due to thermal radiation exactly balances the convective input. This illustrates the reason for and importance of a good insulator (such as the Shuttle tiles) for surface coating of such a vehicle. It is essential to confine the energy to the surface, not allowing it to soak back into the primary structure. Tauber and Yang[3] provide an excellent survey of design tradeoffs for maneuvering entry vehicles.

Radiative heat transfer is a function of the temperature of the emitting and receiving bodies, the surface materials of the bodies, the intervening medium, and the relative geometry. The intensity, energy per unit area, is proportional to $1/r^2$ for a point source. If distance is sufficient, almost any object may be considered a point source. An example is the sun, which subtends a significant arc in the sky as viewed from Earth but may be considered a point source for many analytical purposes.

The ability to tailor the aborptivity and emissivity of spacecraft internal and external surfaces by means of coatings, surface treatment, etc., offers a simple and flexible means of passive spacecraft thermal control. Devices such as the louvers and movable flat-plate shades discussed previously may be looked upon as active means of varying the effective total emissivity of the spacecraft.

It will be seen that the heat flux from a surface varies as the fourth power of its temperature. Thus, for heat rejection at low temperature a relatively large area will be required. This may constitute a problem in terms of geometry, where one must simultaneously provide an adequate view factor to space, launch stowage, and minimal weight.

Stefan-Boltzmann Law

Radiative heat transfer may be defined as the transport of energy by electromagnetic waves emitted by all bodies at a temperature greater than 0 K. For purposes of thermal control, our primary interest lies in wavelengths between approximately 200 nm and 200 μm, the region between the middle ultraviolet and the far infrared. The Stefan-Boltzmann law states that the power emitted by such a body is

$$Q = \varepsilon \sigma A T^4 \qquad (9.13)$$

where T is the surface temperature, A the surface area, and ε the emissivity (unity for a blackbody, as we will discuss later). The Stefan-Boltzmann constant σ is 5.67×10^{-8} W/m^2 K^4.

Notation conventions in radiometry are notoriously confusing and are often inconsistent with those used in other areas of thermal control. To the extent that a standard exists, it is probably best exemplified in the text by Siegel and Howell,[4] and we shall adopt it here. Using this convention, we define the hemispherical total emissive power e as

$$e \equiv q = Q/A = \varepsilon \sigma T^4 \qquad (9.14)$$

The name derives from the fact that each area element of a surface can "see" a hemisphere above itself. The quantity e is the energy emitted, including all wavelengths, into this hemisphere per unit time and per unit area.

The Blackbody

The blackbody, as the term is used in radiative heat transfer, is an idealization. By definition, the blackbody neither reflects nor transmits incident energy. It is a perfect absorber at all wavelengths and all angles of incidence. As a result, provable by elementary energy-balance arguments, it also emits the maximum possible energy at all wavelengths and angles for a given temperature. The total radiant energy emitted is a function of temperature only.

Although true blackbodies do not exist, their characteristics are closely approached by certain finely divided powders such as carbon black, gold black, platinum black, and Carborundum. It is also possible to create structures that approximate blackbody behavior. For example, an array of parallel grooves (such as a stack of razor blades) or a honeycomb arrangement of cavities can be made to resemble a blackbody. Such structures may be used in radiometers. The actual emissivity ε and absorptivity α that characterize how real bodies emit and absorb electromagnetic radiations often differ in value and are dissimilar functions of temperature, incidence angle, wavelength, surface roughness, and chemical composition. These differences can be used by the spacecraft designer to control its temperature. As an example, a surface might be chosen to be highly reflective in the visible light band to reduce absorption of sunlight and highly emissive in the infrared to enhance heat rejection. Silver-plated Teflon was mentioned earlier as one material having such properties. Figure 9.4 shows α/ε values for a variety of common thermal control materials.

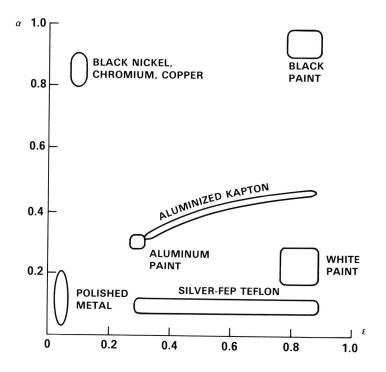

Fig. 9.4 Typical solar absorptivity and emissivity.

For analytical convenience, real bodies are sometimes represented as blackbodies at a specific temperature. The sun, for example, is well represented for thermal control purposes by a blackbody at 5780 K and the Earth by a blackbody at 290 K.

The equation describing blackbody radiation is known as Planck's law, after the German physicist Max Planck, who derived it in 1900. Because this development required the deliberate introduction by Planck of the concept of energy quanta, or discrete units of energy, it is said to mark the initiation of modern, as opposed to classical, physics. Planck's law is

$$e_{\lambda b} = e_{\lambda b}(\lambda, T) = 2\pi hc^2/[\lambda^5(e^{hc/\lambda kT} - 1)] \qquad (9.15)$$

where

$h = 6.626 \times 10^{-34}$ Js = Planck's constant
$k = 1.381 \times 10^{-23}$ J/K = Boltzmann's constant
$c = 2.9979 \times 10^8$ m/s = speed of light

The subscript b implies blackbody conditions, and e_λ denotes the hemispherical spectral emissive power, i.e., the power per unit emitting surface area into a hemispherical solid angle, per unit wavelength interval. Care with units is required in dealing with Eq. (9.15) and its variations. Dimensionally, $e_{\lambda b}$ has units of power per area and per wavelength; however, one should take care that wavelengths are expressed in appropriate units, such as micrometers or nanometers, whereas area is in units of m^2 or cm^2. If care is not taken, results in error by several orders of magnitude are easily produced.

Planck's law is for emission into a medium with unity index of refraction, i.e., a vacuum. It must be modified in other cases.[4]

Planck's law as given finds little direct use in spacecraft thermal control. However, it is integral to the development of a large number of other results. Included among these is Wien's displacement law, readily derivable from the Planck equation, which defines the wavelength at which the energy emitted from a body is at peak intensity. This may be considered the principal "color" of the radiation from the body, found from

$$(\lambda T)_{max} = (1/4.965114)hc/k = 2897.8 \ \mu m \ K \qquad (9.16)$$

The Earth's radiation spectrum peaks at $\lambda = 10 \ \mu m$. Applying this fact and Eq. (9.16) yields the result given earlier that Earth is approximately a blackbody at a temperature of 290 K.

The important fourth-power relationship empirically formulated by Stefan and confirmed by Boltzmann's development of statistical thermodynamics may be derived by integrating Planck's law over all wavelengths. When this is done, one obtains

$$e_b = \int_0^\infty e_{\lambda b}(\lambda, T) \ d\lambda = \sigma T^4 \qquad (9.17)$$

Table 9.2 Blackbody emissive fraction in range (0,λT)

λT, μm K	$e_{0,\lambda T}/e_b$
1448	0.01
2191	0.10
2898	0.25
4108	0.50
6149	0.75
9389	0.90
23,220	0.99

It is usually of greater practical interest to evaluate the integral of Eq. (9.17) between limits λ_1 and λ_2. This is most readily done by noting from Planck's law that an auxiliary function $e_{\lambda b}/T^5$ can be defined that depends only on the new variable (λT). Tables of the integral of $e_{\lambda b}/T^5$ may be compiled and used to evaluate the blackbody energy content between any two points $\lambda_1 T$ and $\lambda_2 T$. A few handy values for the integral over $(0,\lambda T)$ are found in Table 9.2.

9.5 RADIANT HEAT TRANSFER BETWEEN SURFACES

The primary interest in radiative heat transfer for spacecraft thermal control is to allow the energy flux between the spacecraft, or a part of the spacecraft, and its surroundings to be computed. This requires the ability to compute the energy transfer between arbitrarily positioned pairs of "surfaces"; the term is in quotes because often one surface will be composed totally or partially of deep space. The key point is that any surface of interest, say A_i, radiates to and receives radiation from all other surfaces A_j within its hemispherical field of view. All of these surfaces together enclose A_i and render a local solution impossible in the general case; the coupling between surfaces requires a global treatment. The problem is relatively tractable, though messy, when the various surfaces are black. When they are not, computer solution is required in all but the simplest cases. Fortunately, a few of these simple cases are of great utility for basic spacecraft design calculations.

Black Surfaces

Figure 9.5 shows two surfaces A_1 and A_2 with temperatures T_1 and T_2 at an arbitrary orientation of one with respect to another. If both surfaces are black, the net radiant interchange from A_1 to A_2 is

$$Q_{12} = \sigma(T_1^4 - T_2^4)A_1 F_{12} = \sigma(T_1^4 - T_2^4)A_2 F_{21} \qquad (9.18)$$

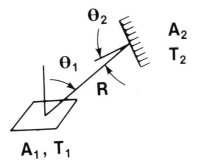

Fig. 9.5 Radiant heat transfer between black surfaces.

where F_{ij} is the view factor of the jth surface by the ith surface. Specifically, F_{12} is defined as the fraction of radiant energy leaving A_1 which is intercepted by A_2. Note the reciprocity in area-view factor products that is implicit in Eq. (9.18). View factors, also called configuration or angle factors, are essentially geometric and may be easily calculated for simple situations. In more complex cases, computer analysis is required. Extensive tables of view factors are available in standard texts.[4]

When the surfaces of an enclosure are not all black, energy incident on a nonblack surface will be partially reflected back into the enclosure; this continues in an infinite series of diminishing strength. The total energy incident on a given surface is then more difficult to account for and includes contributions from portions of the enclosure not allowed by the view factors F_{ij} for a black enclosure. Moreover, nonblack surfaces can and generally will exhibit variations in absorptivity, reflectivity, and emissivity as a function of the azimuth and elevation angle of the incident beam relative to the surface. Variations in all these characteristics with color will also exist. These complications render an analytical solution essentially impossible in most cases of interest. Excellent computational methods exist for handling these cases, mostly based on or equivalent to Hottel's net radiation method.[5]

Diffuse Surfaces

The simplest nonblack surface is the so-called diffuse gray surface. The term "gray" implies an absence of wavelength dependence. A "diffuse" surface offers no specular reflection to an incident beam; energy is reflected from the surface with an intensity that, to an observer, depends only on the projected area of the surface visible to the observer. The projected area is the area normal to the observer's line of sight:

$$A_\perp = A \cos\theta \qquad (9.19)$$

where θ is the angle from the surface normal to the line of sight. Thus, the reflected energy is distributed exactly as is energy emitted from a black surface; it looks the same to viewers at any angle. Reflected energy so distributed is said to follow Lambert's cosine law; a surface with this property is called a Lambertian surface. A fuzzy object such as a tennis ball or a cloud-covered planet such as Venus represents a good example of a diffuse or Lambertian reflector. Surfaces that are both diffuse and gray may be viewed conceptually as black surfaces for which the emissivity and absorptivity are less than unity.

The energy emitted by a gray surface A_1 is given by Eq. (9.13). The portion of this energy that falls upon a second surface A_2 is given by

$$Q = \varepsilon_1 \sigma A_1 F_{12} T_1^4 \qquad (9.20)$$

This radiation, incident on a nonblack surface, can be absorbed with coefficient α, reflected with coefficient ρ, or transmitted with coefficient τ. From conservation of energy,

$$\alpha + \rho + \tau = 1 \qquad (9.21)$$

If a surface is opaque ($\tau = 0$), then Kirchoff's law states that the surface in thermal equilibrium has the property that, at a given temperature T, $\alpha = \varepsilon$ at all wavelengths. This result like all others is an idealization. Nonetheless, it is useful in reducing the number of parameters necessary in many radiant heat-transfer problems and is frequently incorporated into gray surface calculations without explicit acknowledgment.

A case of practical utility is that of a diffuse gray surface A_1 with temperature T_1 and emissivity ε_1 and which cannot see itself ($F_{11} = 0$, a convex or flat surface), enclosed by another diffuse gray surface A_2 with temperature T_2 and emissivity ε_2. If $A_1 \ll A_2$ or if $\varepsilon_2 = 1$, then the radiant energy transfer between A_1 and A_2 is[2]

$$Q_{12} = \varepsilon_1 \sigma A_1 (T_1^4 - T_2^4) \qquad (9.22)$$

The restrictions on self-viewing and relative size can be relaxed at the cost of introducing the assumption of uniform irradiation. This states that any reflections from a gray surface in an enclosure uniformly irradiate other surfaces in the enclosure. With this approximation,

$$Q_{12} = \sigma A_1 (1 - F_{11})(T_1^4 - T_2^4)/[1/\varepsilon_1 + (1 - F_{11})(1/\varepsilon_2 - 1)A_1/A_2] \qquad (9.23)$$

Equations (9.22) and (9.23) are important practical results in radiant energy transfer, easily specialized to include geometries such as parallel plates with spacing small relative to their size, concentric cylinders, or spheres. Many basic spacecraft energy-balance problems can be treated using the results of this section.

Radiation Surface Coefficient

The foregoing results are obviously more algebraically complex than the corresponding expressions for conductive and convective energy transfer. This should not be taken to imply greater physical complexity; as we have mentioned, the physics of convective transfer is buried in the convective coefficient h_c, which may be difficult or impossible to compute. Nonetheless, there is great engineering utility in an expression such as Eq. (9.12), and for this reason we may usefully define a radiation surface coefficient h_r through the equation

$$Q_{12} = h_r A_1 (T_1 - T_2) \qquad (9.24)$$

It is clear that h_r is highly problem dependent; indeed, even for the simple cases of Eq. (9.22), it must be true that

$$h_r = \varepsilon_1 \sigma (T_1^2 + T_2^2)(T_1 + T_2) \qquad (9.25)$$

Equation (9.25) is more useful than it might at first appear. Though solving for T_1 or T_2 may be part of the problem, thus implying doubtful utility for Eq. (9.25), h_r is often only weakly dependent on the exact values of T_1 and T_2, which in any case may have much less variability than the temperature difference $(T_1 - T_2)$. For example, when T_1 or $T_2 \gg (T_1 - T_2)$, then

$$4T_{\text{avg}}^3 \cong (T_1^2 + T_2^2)(T_1 + T_2) \qquad (9.26)$$

Hence, we may write

$$h_r \cong 4\varepsilon_1 \sigma T_{\text{avg}}^3 \qquad (9.27)$$

which has the advantage of basically decoupling h_r from the details of the problem.

The use of the radiation surface coefficient is most convenient when radiation is present as a heat-transfer mechanism in parallel with conduction or convection. As we shall see, parallel thermal conductances add algebraically, thus allowing straightforward analysis using Eq. (9.25) or (9.27) together with a conductive or convective flux.

9.6 SPACECRAFT THERMAL MODELING AND ANALYSIS

For accurate thermal analysis of a spacecraft, it is necessary to construct an analytical thermal model of the spacecraft. This requires identification of heat sources and sinks, both external and internal, such as electronics packages, heaters, cooling devices, and radiators. Nodes are then defined, usually major items of structure, tanks, and electronic units. Thermal resistance between each pair of nodes that are thermally connected must be determined. This will involve radiation or conduction or both (perhaps even convection but, as observed earlier, not commonly). Thus, determination of the thermal resistance (the inverse of conductance) will involve conductivity of materials and joints as well as emissivity and absorptivity of the surfaces. Once constructed, the model can be used to solve steady-state problems as well as transient problems by the use of explicit or implicit finite-difference codes.

Often the model proceeds in an evolutionary manner, with the nodes initially being relatively few and large and the thermal resistance having broad tolerances. At this stage the model may be amenable to hand calculator analysis and the use of simple codes for quick estimates. As the design of the spacecraft matures, the model will become more complex and detailed, dictating computer analysis. No matter how detailed the analysis becomes, however, a thermal vacuum test of a thermal mock-up or prototype will almost certainly be required since the model requires a host of assumptions unverifiable by any other means, and, as previously observed, the effects of the atmosphere both as a convective agent and a conductor in joints renders thermal testing in atmosphere useless. It is usually desirable to do an abbreviated test on flight units as well as a final verification. The following example demonstrates the basic approach to steady-state thermal modeling.

We consider the insulated wall of a vertically standing launch vehicle LOX tank, illustrated schematically in Fig. 9.6. The LOX is maintained at a temperature of 90 K in the tank by allowing it to boil off as necessary to accommodate the input heat flux; it is replaced until shortly before launch by a propellant feed line at the pad. The tank is composed of an aluminum wall of $\Delta_{al} = 5$ mm thickness and an outer layer of cork with $\Delta_{co} = 3$ mm. The ground and outside air temperatures are both approximately 300 K, and the sky is overcast with high relative humidity. The booster tank diameter of 8 ft is sufficient to render wall curvature effects negligible, and its length is enough to allow end effects to be ignored. It is of interest to determine the steady-state heat flux into the LOX tank so that propellant top-off requirements can be estimated.

The statement of the problem allows us to conclude that radiation from ground and sky at a temperature of 300 K to the wall, as well as free convection from the air to the vehicle tank, will constitute the primary sources of heat input. The LOX acts as an internal sink for energy through

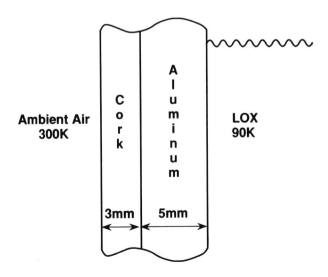

Fig. 9.6   Schematic of LOX tank wall.

the boil-off process; heat transfer to the LOX will be dominated by free convection at the inner wall.

Reference to standard texts yields for the appropriate thermal conductivities,

$$\kappa_{al} = 202 \text{ W/m K}$$

$$\kappa_{co} = 0.0381 \text{ W/m K}$$

and the free convection coefficients outside and inside are approximated as

$$h_{c_o} = 5 \text{ W/m}^2 \text{ K}$$

$$h_{c_i} = 50 \text{ W/m}^2 \text{ K}$$

We assume the cork to have $\varepsilon \cong \alpha \cong 0.95$ and the Earth and sky to have $\varepsilon \cong 1$. Since the outer tank wall is convex, it cannot see itself; thus, $F_{t,t} = 0$. The tank has a view of both sky and ground, in about equal proportions, so $F_{t,s} \cong F_{t,g} \cong 0.5$; however, since we have assumed both to be blackbodies at 300 K, the separate view factors need not be considered. We therefore ignore the ground and take $F_{t,s} = 1$ in this analysis.

For clarity, we at first ignore the radiation contribution, considering only the free convection into and the conduction through the booster wall. The heat flux is unknown, but we know it must, in the steady state, be the same at all interfaces. The problem is essentially one-dimensional; thus, the slab conduction result of Eq. (9.9) is directly applicable. Thus, we may write

$$T_{\text{air}} - T_i = q/h_{c_o}$$

$$T_1 - T_2 = q\Delta_{co}/\kappa_{co}$$

$$T_2 - T_3 = q\Delta_{al}/\kappa_{al}$$

$$T_3 - T_{\text{LOX}} = q/h_{c_i}$$

Adding these results together yields

$$T_{\text{air}} - T_{\text{LOX}} = q[1/h_{c_o} + \Delta_{co}/\kappa_{co} + \Delta_{al}/\kappa_{al} + 1/h_{c_i}]$$

$$\equiv q/U = Q/UA \tag{9.28}$$

The coefficient U defined by Eq. (9.28) is called the overall or universal heat-transfer coefficient between the air and the LOX. As can be seen, the conductive and convective coefficients add reciprocally to form U. This leads to the definition, previously mentioned, of thermal resistance, analogous to electrical resistance. In this problem,

$$R_{al_{\text{conv}}} = 1/h_{c_i}$$

$$R_{al_{\text{cond}}} = \Delta_{al}/\kappa_{al}$$

$$R_{co_{\text{conv}}} = 1/h_{c_o}$$

$$R_{co_{\text{cond}}} = \Delta_{co}/\kappa_{co}$$

and we see that

$$1/U = R_{al_{\text{conv}}} + R_{al_{\text{cond}}} + R_{co_{\text{conv}}} + R_{co_{\text{cond}}}$$

i.e., thermal resistances in series add.

For this problem, we find

$$U = 3.35 \text{ W/m}^2 \text{ K}$$

Hence,

$$q = 703 \text{ W/m}^2$$

Now that the heat flux is known, we can substitute above to find the temperature at any of the interface points if desired. Each interface is a "node" in the terminology used above, connected through appropriate thermal resistances to other nodes. For later use, we note that the outer wall temperature satisfies

$$T_{\text{air}} - T_1 = q/h_{c_o} = 141 \text{ K}$$

Hence,

$$T_1 = 159 \text{ K}$$

Consider now the addition of the radiative flux. From Eq. (9.22), the radiative flux from the tank to the air is

$$q_r = \varepsilon_{\text{tank}} \sigma(T_1^4 - T_{\text{air}}^4) = h_r(T_1 - T_{\text{air}})$$

where, as with Eq. (9.25),

$$h_r = \varepsilon_{\text{tank}} \sigma(T_{\text{air}}^2 + T_1^2)(T_{\text{air}} + T_1) \cong 2.85 \text{ W/m}^2 \text{ K}$$

Of necessity, we take T_1 from the convective solution to use in computing the radiation surface coefficient. If improved accuracy is required, the final result for T_1 obtained with radiation included can be used iteratively to recompute h_r, obtain a new result, etc. This is rarely justified in an analysis such as that done here.

Changing the sign of the radiative flux so as to have it in the same direction (into the tank) as previously, we see that a second, parallel heat flux path has been added to the existing convective flux at the outer wall. This will result in a higher wall temperature than would otherwise be found.

At the wall, the flux is now

$$q_{\text{total}} = q_{\text{conv}} + q_r = (h_r + h_{c_o})(T_{\text{air}} - T_1)$$

which is substituted for the previous result without radiation. Thus, conductances in parallel add, whereas the respective resistances would add reciprocally. When the problem is solved as before, we obtain with the given data

$$q_{\text{total}} = q = 929 \text{ W/m}^2$$

and

$$T_1 = 182 \text{ K}$$

Notice that the radiation surface coefficient method is only useful when the temperature "seen" by the radiating surface is approximately that seen by the convective transfer mechanism.

9.7 THERMAL ENVIRONMENT

The spacecraft thermal environment varies due to a variety of naturally occurring, self-generated effects. Orbital characteristics are a major source of variation. For example, most orbits will involve eclipse at some time during each orbit; however, as the orbit precesses, the time and duration of the eclipse may vary, particularly for a highly elliptic orbit. Obviously, for a spacecraft in interplanetary flight where the orbit is about the sun, the solar intensity will vary as the distance from the sun changes. As discussed in Chapter 4, even the solar intensity experienced in orbit around the Earth will vary seasonally (from an average value of 1388 W/m^2) because of the ellipticity of the Earth's orbit around the sun.

Operational activities alter the thermal environment as well. Orbital altitude may vary, resulting in changing thermal inputs from the planet and possibly frictional heating in the case of low periapsis orbits. Spacecraft attitude may change, resulting in exposure of differing areas and surface treatments to the sun and to space. Onboard equipment may be turned on or off, resulting in changes in the amount of internally generated heat. In the course of thruster firings, local cooling may occur in tanks or lines due to gas expansion at the same time as local heating may occur in the vicinity of hot gas thrusters. Expenditure of propellant reduces the thermal mass of the tanks and the spacecraft as a whole, resulting in differences in transient response.

As the time in space increases, spacecraft surface characteristics change due to ultraviolet exposure, atomic oxygen attack, micrometeoroid/debris impact, etc. This will affect both the absorptivity and emissivity of the surfaces and must be considered in the design of long-life spacecraft.

The final and unpredictable source of change in the thermal environment is anomalies. A thermal failure in wiring may cause loss of part of the solar array power, thus reducing the internally generated heat. A shade or shield may fail to deploy, louvers may stick, etc. Although one cannot predict every possible problem, nor can a spacecraft be designed to tolerate every possible anomaly, it is desirable to provide some margin in the design to allow for operation at off-design conditions.

9.8 SPACECRAFT ENERGY BALANCE

Figure 9.7 shows a generic spacecraft in Earth orbit and defines the sources and sinks of thermal energy relevant to such a spacecraft. Not all features of Fig. 9.7 are appropriate in every case. Obviously, for a spacecraft not near a planet, the planet-related terms are zero. Similarly, in eclipse, the solar and reflected energy terms are absent. In orbit about a hot, dark planet such as Mercury, reflected energy will be small compared to radiated energy from the planet, whereas at Venus the opposite may be true. Solar energy input of course varies inversely with the square of the distance from the sun and can be essentially negligible for outer-planetary missions. These variations in the major input parameters will have significant impact upon the thermal control design of the spacecraft.

The energy balance for the situation depicted in Fig. 9.7 may be written as

$$Q_{\text{sun}} + Q_{er} + Q_i = Q_{ss} + Q_{se} \qquad (9.29)$$

where we have neglected reflected energy contributions other than those from Earth to the spacecraft. This renders the enclosure analysis tractable. In effect, we have a three-surface problem (Earth, sun, and spacecraft) where, by neglecting certain energy transfer paths, a closed form solution

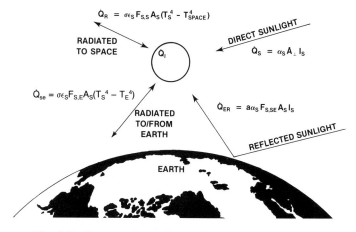

Fig. 9.7 Energy balance for an Earth orbiting spacecraft.

can be achieved. We define

$Q_{\text{sun}} = \alpha_s A_{\perp} I_{\text{sun}} =$ solar input to spacecraft
$Q_{\text{er}} = a\alpha_s F_{s,se} A_s I_{\text{sun}} =$ Earth-reflected solar input
a = Earth albedo (ranges from 0.07 to 0.85)
α_s = spacecraft surface absorptivity
ϵ_s = spacecraft surface emissivity
Q_i = internally generated energy
$Q_{se} = \sigma\epsilon_s A_s F_{s,e}(T_s^4 - T_e^4) =$ energy radiated to Earth
$Q_{ss} = \sigma\epsilon_s A_s F_{s,s}(T_s^4 - T_{\text{space}}^4) =$ energy radiated to space

If certain simplifications are made, such as assuming $T_{\text{space}} \cong 0$, and if we note that the sum of view factors from the spacecraft to Earth and space satisfies

$$F_{s,s} + F_{s,e} = 1 \qquad (9.30)$$

Then, in the equilibrium condition the energy balance equation becomes

$$\varepsilon_s \sigma A_s T_s^4 = \varepsilon_s \sigma A_s F_{s,e} T_e^4 + Q_{\text{sun}} + Q_{er} + Q_i \qquad (9.31)$$

Equation (9.30) may be solved to yield the average satellite external temperature. Note that this result provides no information on local hot or cold spots and does not address internal temperature variations. However, it is useful in preliminary design to determine whether the spacecraft is operating within reasonable thermal bounds.

It is instructive to consider an example solution for Eq. (9.30) for an Earth-orbiting spacecraft. Let us consider a 1-m-diam spherical spacecraft in a 1000-km altitude circular orbit. At the time under consideration the spacecraft is in full sunlight, yet essentially over the dark portion of the Earth. (This might occur in a near-polar sun-synchronous orbit over or near the terminator. Such a dawn-dusk orbit does see some portion of the sunlit Earth, but only near the limb, with consequently little energy input.)

Assuming reasonable values for internally generated heat and for absorptivity and emissivity, we use

$$Q_i = 50 \text{ W}$$

$$\alpha = 0.7$$

$$\varepsilon = 0.9$$

From a 1000-km orbit, Earth's disk subtends 120 deg; hence, a solid angle of π sr, or 25% of the celestial sphere. Thus,

$$F_{s,e} = 0.25$$

With the approximation discussed earlier that $F_{s,se}$, the view factor to the sunlit Earth, is essentially zero; hence,

$$Q_{er} \cong 0$$

The area intercepting sunlight and radiation from the Earth is the projected area of the spherical spacecraft, and is thus a disk; hence,

$$A_\perp = \pi D^2/4 = \pi/4 \text{ m}^2$$

Using a solar intensity value of 1400 W/m² yields for the solar input to the spacecraft

$$Q_{\text{sun}} = (1400 \text{ W/m}^2)\alpha A_\perp = 770 \text{ W} \qquad (9.32)$$

The total surface area of the sphere is

$$A_s T = 4\pi R^2 = 3.14 \text{ m}^2$$

To solve for the temperature of the spacecraft, the thermal equilibrium equation may be rewritten as

$$T_s^4 = F_{s,e} T_e^4 + [Q_{\text{sun}} + Q_{er} + Q_i]/\varepsilon_s \sigma A_s \qquad (9.32)$$

Using the known values from above and $T_e = 290$ K for the Earth, we obtain

$$T_s = 288 \text{ K}$$

The formulation of Eq. (9.32), utilizing Eq. (9.31), makes clear the dependence of spacecraft temperature on the surface α/ε ratio.

References

[1]Wylie, C. R., and Barrett, L. C., *Advanced Engineering Mathematics*, McGraw-Hill, New York, 1982.

[2]Gebhart, B., *Heat Transfer*, 2nd ed., McGraw-Hill, New York, 1971.

[3]Tauber, M. E., and Yang, L., "Performance Comparisons of Maneuvering Vehicles Returning from Orbit," *Journal of Spacecraft and Rockets*, Vol. 25, July–Aug. 1988, pp. 263–270.

[4]Siegel, R., and Howell, J. R., *Thermal Radiation Heat Transfer*, 2nd ed., Hemisphere, New York, 1981.

[5]Hottel, H. C., and Sarofim, A. F., *Radiative Transfer*, McGraw-Hill, New York, 1967.

10
POWER

Spacecraft power has been one of the major limitations on spacecraft design since the beginning of the space age. The earliest orbiting vehicles flown by both the U.S. and the USSR depended on batteries. The limited energy storage capabilities of the batteries available at that time prevented operations of more than a few days. This was not satisfactory for missions of duration suitable for detailed scientific observations or military reconnaissance, and solar power generation arrays quickly appeared on the scene. Although not highly efficient in turning sunlight into electricity, solar arrays (or solar panels) were admirably suited to powering spacecraft. Since no consumables were used in generating electrical power, the life expectancy of the power system was limited only by degradation of the components of which it was composed. Spacecraft operating life of several years became feasible with development of these photoelectric arrays.

Solar panels and batteries in combination powered virtually all unmanned spacecraft launched in the first 15 yr of the space program, with the exception of a few short-lived battery-only systems. Manned spacecraft, with the exception of the solar-powered Skylab, Salyut, and Soyuz, have used batteries or fuel cells.

Solar power systems become unsatisfactory for missions beyond the asteroid belt as the sun's energy becomes more diffuse. With interest developing in outer-planetary missions, a new power source was required. At the same time certain military spacecraft missions required a sturdy compact power source. Both requirements were met by the development of radioisotope thermoelectric generators (RTGs). RTGs convert the heat energy produced by radioisotope decay into electricity via the thermoelectric effect. Power output is independent of solar distance, and lifetime is limited only by component degradation and the half-life of the radioisotope. RTGs are also useful for operations on planetary surfaces where extended dark periods may be encountered. Outer solar system spacecraft such as Pioneer, Voyager, and Galileo as well as the Viking Mars Landers and Apollo Lunar Surface Experiment Packages have all been RTG-powered, as have some Earth Orbiters.

Nuclear reactor-based power systems offer very high power in a compact package for long duration. Such systems tend to be highly independent of environment. After an extensive development program in the 1960s, all space reactor work was terminated as part of the cutbacks in the early 1970s. Only recently has there been a revival of interest in power plants of this type. Although the USSR flies relatively short-lived reactor power systems on an operational basis, the U.S. has flown a single reactor test mission, the SNAP-10A in 1972.

The power subsystem is a major driver in any spacecraft design and is in turn strongly driven by a variety of mission, system, and subsystems considerations. It interfaces directly with every other subsystem and, as a result, requires considerable attention from the systems engineer. Technology is rapidly evolving, especially in the application of automation and development of more efficient conditioning and control circuitry.

10.1 POWER SUBSYSTEM FUNCTIONS

The obvious functions of a spacecraft power subsystem are to generate and store electric power for use by the other spacecraft subsystems. Since these other subsystems may have various requirements for voltage, frequency, cleanliness, or other characteristics, the power subsystem may be called upon to supply them, although a significant system level tradeoff is the decision whether to supply all subsystems the same basic power and leave the meeting of specific local requirements to the subsystem. For example, a specific requirement for a very high voltage in the telecommunications subsystem might be supplied by the power subsystem or by a source within the telecom subsystem that would operate off the basic power bus. Similar tradeoffs exist for special cleanliness requirements (e.g., absence of ripple on a dc line).

Regardless of the conclusion of these tradeoffs, the power subsystem will have the responsibility to control, condition, and process the raw power received from the power source to comply with the needs of the spacecraft system. The subsystem must supply stable and uninterrupted power for the design life of the system.

In order to maintain the long-term reliability of the system, the power subsystem must provide protection to subsystems against reasonably expectable failures in other user subsystems or within the power subsystem itself. For example, a short circuit in a subsystem should not be allowed to drag down the main bus voltage to the point of inducing failure elsewhere in the system. Similarly, failure protection should be implemented in the power subsystem itself to allow for continued functioning of the system (perhaps degraded) following some degree of malfunction.

In the course of normal operation, the power subsystem must accept commands from onboard and external sources and provide telemetry data to allow monitoring of its operation and general health.

Finally, it may be necessary to provide highly specialized power for specific functions such as firing ordinance.

10.2 POWER SUBSYSTEM EVOLUTION

The evolution of spacecraft power subsystems has been characterized by growth from subsystems delivering a few watts to those delivering a few tens of kilowatts (e.g., Skylab). The space station is expected to require about 75 kWe initially with growth to 220 kWe or more postulated relatively rapidly. Line loss and other efficiency and mass factors push toward higher voltages as the power demands increase. Figure 10.1 illustrates the trend and projects broadly what may be anticipated in the near future.

Design lifetime of space systems tends to increase along with power levels as spacecraft become more complex and expensive. As power level and lifetime change, the choice of power source may change as well. Figure 10.2 illustrates the general operating regimes of various types of power sources. There is a substantial overlap between the regimes, and various other considerations may dictate use of some power source at a location in the power vs endurance space which may not appear to be optimum. Figure 10.2 provides a good source for preliminary conceptual thinking in regard to power source choices.

10.3 POWER SUBSYSTEM DESIGN DRIVERS

A variety of considerations may drive the design of the power subsystem. Table 10.1 presents a number of these considerations. Not all will be applicable for every system design, and, conversely, some designs may involve considerations not listed here. However, most cases will be covered. The designer should view Table 10.1 as a checklist to be used as a reminder to cover all points in the initial design and as the design matures to assess the impact of changes.

Discussing the checklist items briefly, the customer or user may have specific requirements such as size, observability, or operational constraints

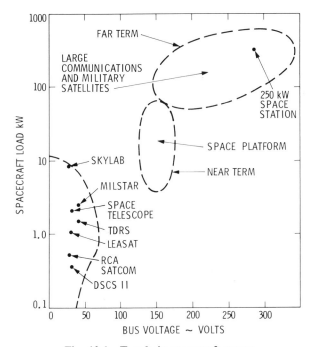

Fig. 10.1 Trends in spacecraft power.

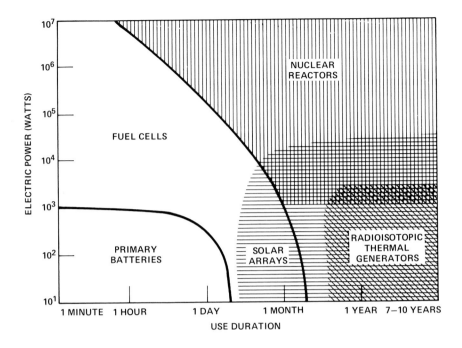

Fig. 10.2 Operating regimes of spacecraft power sources.

that will limit the choices of power source or other subsystem elements. The target planet, Earth or another, and the resultant solar distance will in some cases drive the choice of power source because of limitations of available solar energy per unit area or temperature of exposed surfaces.

Lifetime requirements in given operating environments may also drive the power subsystem. Solar array degradation upon radiation exposure may prevent use of these devices on long-lived spacecraft operating in the van Allen belts, for example. Many spacecraft have a variety of operating modes involving different power levels. The percentage of time in each mode is of great significance and may indicate a hybrid system using more than one power source or particular types of energy storage.

The type of attitude control concepts used will drive the power subsystem in two ways: 1) in terms of configuration constraints involving solar arrays, waste heat radiators, and other elements and 2) in responding to specific power needs of the attitude control devices.

Orbital parameters about a particular body of interest will strongly affect the choice of power source and its configuration as well as the energy storage requirements. The case of a system operating on the surface will often be the most environmentally demanding and the most difficult in terms of deploying large solar arrays and meeting energy storage demands.

Specific mission demands may also impact power subsystem design. For example, a spacecraft that must maneuver rapidly may not be able to tolerate large, flexible solar arrays. A low observable spacecraft may preclude use of concepts involving high temperatures such as RTGs.

Table 10.1   Subsystem design considerations

Customer/user
Target planet, solar distance
Spacecraft configuration
   Mass constraints
   Size
   Launch vehicle constraints
   Thermal dissipation capability
Lifetime
   Total
   Percentage in various modes, power levels
Attitude control
   Spinner
   3-axis stabilized
   Nadir pointer
   Thrusters
   Momentum wheel
   Gravity gradient
   Pointing requirements
Orbital parameters
   Altitude
   Inclination
   Eclipse cycle
Payload requirements
   Power type, voltage, current
   Duty cycle, peak loads
   Fault protection
Mission constraints/requirements
   Maneuver rates
   G-loads

10.4   POWER SUBSYSTEM ELEMENTS

Figure 10.3 presents a typical spacecraft functional block diagram that identifies the major elements involved in such a subsystem. A substantial variety of options exist within each of these elements. Table 10.2 identifies the options most likely to be encountered in normal spacecraft design practice.

10.5   DESIGN PRACTICE

Although details of design practice will vary from one organization to another, some broad rules of general applicability can be set down. These are discussed in this section.

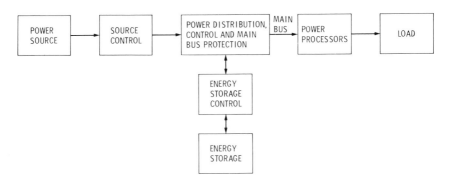

Fig. 10.3   Power subsystem functional block diagram.

Direct Current Switching

As a general goal, switches or relays should be in the positive line to a given element with a direct connection on the negative side. The purpose of course is to allow power to be shut off in the event of a short circuit or other high-current flow failure within the element.

Arc Suppression

In order to maximize effectiveness, arc suppression should be located as close to the source of the arc as possible.

Table 10.2   Power subsystem elements

Power source
Solar photovoltaic
Radio-isotope thermo electric generators
Nuclear reactor—static or dynamic
Radio-isotope dynamic
Solar dynamic
Fuel cells
Primary batteries
Source control
Shunt regulator
Series regulator
Shorting switch array
Energy storage control
Charger
Regulator
Power conditioning
DC-DC convertors
DC-AC inverters
Regulators

Modularity

Modular construction is desirable to simplify testing and to expedite change-out of failed or suspect units during system test or launch preparation.

Grounding

A ground cable is preferable to grounding via structure because it is difficult to maintain high continuity between structural elements, thus increasing the possibility of variations in ground voltage. In addition, any ground circuit current flow is less likely to disturb sensitive components if confined to a properly isolated and connected ground cable.

Continuity

Good continuity should be maintained between structural elements, thermal blankets, etc., to minimize the probability of buildup of static electrical potential or other voltage differences.

Shield Continuity

Shield continuity must be maintained across all connections. A single-point shield ground is desirable to minimize the possibility of shield current flow. Most circuits, especially noise-sensitive or noise-generating circuits, will be shielded.

Complexity

Avoid excessive complexity. A spacecraft power subsystem should be no more complex than is necessary to do the job. Excessive, unnecessary complication will increase design, fabrication, and test costs and increase the probability of failure.

Particular circumstances may force violation of any of the previously mentioned rules in order to meet some overriding requirement. In the absence of such a requirement, however, adherence to these rules is very much recommended and will generally have a most desirable impact on the overall operation.

10.6 BATTERIES

Batteries have been and will continue to be (for the foreseeable future) the primary means of electric energy storage onboard spacecraft. In the following discussion, a variety of terms relating to batteries will be used. These are defined here in order to enhance understanding of what follows:

Ampere-hour =	total capacity of battery (e.g., 40 A for 1 h = 40 A-h
Depth of discharge (DOD) =	percent of battery capacity used in discharge (75% DOD means 25% capacity remaining, DOD usually limited for long cycle life)

Watt-hours =   stored energy of battery, equal to A-h capacity times average discharge voltage

Charge rate =   rate at which battery can accept charge (measured in A)

Average discharge voltage =   number of cells in series times cell discharge voltage (1.25 V for most commonly used cells)

Batteries are divided into two major categories: primary and secondary. The former are distinguished by higher power densities but are generally not rechargeable. They are especially well adapted to one-time events requiring substantial power from minimal mass. Missiles or disposable stages often use primary batteries. In cases where extremely long installed storage is required, e.g., a missile in its silo or a planetary atmosphere probe that is inert during interplanetary transfer, the battery is often dry (i.e., without electrolyte) prior to activation. Upon an activation signal a pyrotechnic valve fires to allow the electrolyte to enter the battery from a separate reservoir. This approach provides a highly reliable quick reaction power source that is nevertheless protected from degradation and requires no maintenance during extended storage.

A major application of this type of battery in long-life space systems is to supply power to activate pyrotechnics and other deployment devices. Such devices typically are operated at the beginning of the mission. For a variety of reasons, such as minimizing power drain or isolating noisy circuits from the main power bus, it may be desirable to operate these circuits from a primary battery completely separate from the main power circuits.

During the early years of the space program the most common type of primary battery was the silver-zinc, usually abbreviated Ag-Zn. This battery has excellent energy density and still is the battery of choice in many cases. In recent years a variety of batteries based on lithium in combination with various other materials have come on the scene. Many of these batteries offer the highest energy density currently available. Some types of lithium batteries have had some major teething problems in early applications, showing a distressing tendency to explode in some situations. Leakage and corrosion problems have also been encountered. However, these problems are yielding to engineering attention and better understanding of battery characteristics, and lithium batteries can be reliably employed in suitable applications.

Table 10.3 provides a list of a variety of primary battery types and the characteristics of each.

The rechargeable battery generally has a much lower energy density (usually defined as watt-hours per unit mass), which is further aggravated by limitations on the depth of discharge. Again silver-zinc batteries were the most commonly used for a number of years and demonstrate good energy density. However, these batteries suffer from limitations in lifetime, especially in applications involving a large number of charge/discharge cycles. As a result, nickel-cadmium (nicad) batteries have become very nearly the standard for spacecraft applications.

In recent years a new type of rechargeable battery has appeared on the scene, the nickel-hydrogen (NiH_2) type. This is a substantial departure from other types of batteries in that its operation involves the generation and recombination of substantial amounts of free hydrogen. As a result, quite high pressures are generated and the case of the battery is in fact a pressure vessel. (Actually, other battery types, e.g., nicads, do generate some internal pressure and require a sufficiently strong case to contain it. NiH_2 battery pressure exceeds that of nicads by a factor of 10.) It has been said that NiH_2 batteries can be as little as half the mass of an equivalent capability nicad, due in part to greater depth of discharge. However, this has not been demonstrated in a flight unit, nor has the anticipated improvement in lifetime been demonstrated. NiH_2 batteries have been successfully flown and, as the technology matures, will probably continue to improve in performance.

As noted earlier, battery average discharge voltage is cell voltage times the number of cells in series. Cell voltage of most of the types discussed here is about 1.25 V. Most spacecraft systems flown to date by the U.S. have been approximately 28-V dc systems; thus, most of the associated battery hardware has also been 28 V dc. This voltage selection reflects the early heritage of spacecraft from aircraft electronics in terms of electronic component design. Generally speaking, this has been satisfactory for the relatively small low-power spacecraft flown to date. However, with larger and more powerful systems becoming more common, higher-voltage sys-

Table 10.3   Battery chemical types

Silver-zinc (Ag-Zn)
    Commonly used early in space program
    Good energy density (175 W-hr/kg primary)
    Limited cycle life (and/or limited depth of discharge)
    Both primary and secondary types
Nickel-cadmium (NICAD or NiCd)
    Most common type today
    Low energy density (15–30 W-hr/kg)
    Long cycle life
    Good deep discharge tolerance
    Can be reconditioned to extend life
Nickel-hydrogen (NiH_2)
    High pressure
    Most figures of merit similar to NiCd in current practice
    Promise of higher energy density and longer life
Lithium batteries
    Several types ($LiSOCl_2$, $Li-V_2O_5$, $Li-SO_2$)
    Very high energy density (650 W-hr/kg, 250 W-hr/kg, 50–80 W-hr/kg secondary)
    Both primary and secondary types
    Higher cell voltage (2.5–3.4 V)

tems are becoming more attractive, and the future will undoubtedly see a continuing trend toward higher voltage (see Fig. 10.1).

Depth of discharge limitations are a tradeoff between battery mass due to the unused capacity and battery lifetime problems and degradation due to repeated deep discharge. Spacecraft in low-altitude, low-inclination orbit around the Earth or another planet typically experience the most severe usage in terms of charge/discharge cycles. The spacecraft will experience eclipse on each orbit. Low orbits having periods of less than 100 min mean that the spacecraft battery will be discharged and charged about 16 times per day. In only 2 yr this amounts to nearly 12,000 cycles. Many modern spacecraft are expected to last substantially more than 2 yr. Eclipse time in a low orbit can be as high as 40% of the orbital period or on the order of 36 min. Spacecraft in synchronous equatorial (geostationary) orbits go for extended periods without encountering eclipse. However, such spacecraft encounter two eclipse seasons each year, with each period being 45 days long. During these periods the spacecraft encounters one eclipse each day ranging in duration from momentary at the beginning and end of the period up to 72 min at the midpoint. Some spacecraft in near-polar sun-synchronous orbits located near the terminator may never be in eclipse. The same may be true for deep space vehicles. This does not necessarily mean batteries are not needed, however. It may be necessary to maneuver the spacecraft off the sun line to obtain proper thruster pointing for course correction. Even if this is not necessary, it may be mass efficient to have a battery on line to handle occasional intermittent high loads rather than oversize the solar arrays to handle infrequent events. For similar reasons batteries may be required even in systems using power sources (such as RTGs) that do not depend on the sun.

Given the power usage of the spacecraft and the maximum allowable depth of discharge (DOD) for the design lifetime of the battery, the battery can be sized by the following simple equation:

$$DOD = \frac{\text{Energy delivered during eclipse}}{\text{Total battery stored energy}}$$

or

$$DOD = \frac{\text{Load power} \times \text{discharge time}}{\text{Capacity} \times \text{battery avg discharge voltage}}$$

where load power is in watts, discharge time is in hours, capacity is in amperes-hours, and battery average discharge voltage is in volts.

The charge rate also drives battery size. A rule of thumb for charge rate is

$$\text{Charge rate} = \frac{\text{Capacity}}{15 \text{ h}}$$

where capacity is in amperes-hours.

In other words, a battery can accept a charge equal to one-fifteenth of its total capacity per hour. In a typical orbiter where 40% of the orbit is spent discharging and 60% charging, the depth of discharge during eclipse is limited by how rapidly the charge can be put back in during the light phase with the rate limited as shown earlier. Thus, the size of the battery in this case is driven not by the maximum allowable DOD but by the charge rate, and the depth of discharge per orbit is limited to 7 or 8%. On the other hand, a geo-stationary spacecraft will encounter only a few hundred discharge cycles in a 10-yr life and have ample recharge time. Much deeper discharge can be tolerated in this case.

The $\frac{1}{15}$ factor is quite conservative. Substantially higher recharge rates may be acceptable for a given battery. (See manufacturers specifications.) The limiting factor is usually temperature.

A simple example is presented to demonstrate preliminary battery sizing.

Problem
Size a nicad battery to support a 1500-W payload in geostationary orbit.

Given
Bus voltage = 28 V dc

Load duration = 1.2 h maximum

Energy density = 15 W-h/lb for 100% discharge

Avg cell voltage = 1.25 V

Maximum DOD = 70%

Computation

$$\text{Number of cells} = \frac{28}{1.25} = 22.4$$

Choose 22 cells for battery voltage of 27.5 V dc.

$$\text{Total capacity} = \frac{1500 \text{ W} \times 1.2 \text{ h}}{0.70 \times 27.5} = 93.5 \text{ A-h}$$

$$\text{Battery capacity} = 93.5 \times 27.5 = 2571 \text{ W-h}$$

$$\text{Battery weight} = \frac{2571}{15} = 171 \text{ lb}$$

Note that

1) It may be desirable to split the battery into two or three individual battery packs for ease in packaging, placement, and balance.

2) Each pack must contain 22 series-connected cells in order to maintain voltage.

3) This computation does not consider redundancy.

In order to obtain maximum life from a nicad battery, a reconditioning process is required. Reconditioning consists of a very deep discharge to the point of voltage reversal followed by recharge under carefully controlled conditions. Figure 10.4 shows the effect of this operation. In the absence of reconditioning, the battery voltage begins to decline and, after four or five eclipse seasons, declines fairly rapidly. On the other hand, with periodic reconditioning the voltage declines only slightly from the "new" level and reaches a steady-state level. (Note that "eclipse season" refers to the two periods per year when a geosynchronous spacecraft is eclipsed once per orbit. Thus, the numbers on the abcissa convert to years when divided by two.)

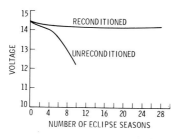

Fig. 10.4   Reconditioning of NiCd batteries.

10.7   PRIMARY POWER SOURCE

Once beyond the relatively small range in which batteries are suitable, choice of a prime power source opens up to include several possibilities. The choice is governed by a variety of factors including required power level, operating location, life expectancy, requirement for orientation flexibility, radiation tolerance, and cost. Figure 10.2 depicts operating ranges that are generally deemed to be suitable for various types of prime power sources based on power level and lifetime. Clearly such curves are broadly indicative only. There is substantial overlap between the various types. Additionally, other factors may drive the choice in a direction that would not be the optimum from the purely technical mass-power-lifetime considerations.

As previously observed, power requirements for spacecraft in general tend to grow larger with time, and the related main bus voltage tends to rise accordingly in an effort to reduce resistive losses and reduce conductor and component mass. Figure 10.1 presents data on past spacecraft and predictions concerning near- and far-term applications, both civil and military. Lifetime required of space assets tends to increase as well due to the very large investment represented by a spacecraft in operation and its associated ground equipment. In the case of science and exploration spacecraft, the lifetime increases because the targets are more distant and/or the missions more complex. In any case, life expectancy becomes an increasingly important factor, especially for spacecraft that cannot readily be serviced or refurbished.

Environmental factors that must be considered include the perhaps obvious one of access to adequate solar illumination. If the spacecraft is too far out from the sun, solar arrays are not a viable choice. Concentrators may extend their usefulness to a limited degree, but eventually the square law wins out. Radiation resistance is another significant consideration. Solar cells tend to be seriously degraded by extensive exposure to radiation. This can be a major consideration for a spacecraft that must operate extensively in the van Allen belts or other high-radiation environments.

In subsequent sections, various prime power sources are discussed, particularly in terms of the capabilities and limitations of each. Detailed technical descriptions of each are beyond the scope of this book, and the interested reader is referred to more specialized literature for such information.

10.8 SOLAR ARRAYS

Regardless of the size of the total array, each array is made up of a very large number of individual cells arranged on a substrate of some type. Although each cell puts out a relatively small current and voltage, proper series and parallel connection can provide any desired current and voltage (within reasonable physical limitations). Cells are made in a variety of shapes and sizes. Probably the most common cells at the time of writing are rectangular cells with dimensions on the order of 2 cm × 4 cm. This shape and size allows for reasonably efficient packing, thus allowing for minimum array size and mass. A well-designed array might have a cell packing density of 90%. Since some minimum spacing and allowance for connections must be provided, it is difficult to improve very much on this number, although innovative techniques may allow some improvement.

Since solar arrays tend to be large for higher-power spacecraft, it quickly becomes impossible to find adequate area on the fixed spacecraft structure. Early low-powered spacecraft did in fact restrict the array area to the spacecraft skin. Most designs were drum-shaped spinning spacecraft, where only about 40% of the array was illuminated by the sun at any time. As power requirements grew, fixed arrays that were not specifically part of the spacecraft structural shell or skin were tried; however, launch vehicle nose fairing dimensions limited the utility of such an approach, and deployable solar arrays made an early appearance. Figure 10.5 depicts a variety of solar array designs.

Deployable solar arrays have typically been rigid paddlelike structures that deploy from the main structure after the spacecraft is injected into orbit. Keeping the array firmly locked to the spacecraft structure during

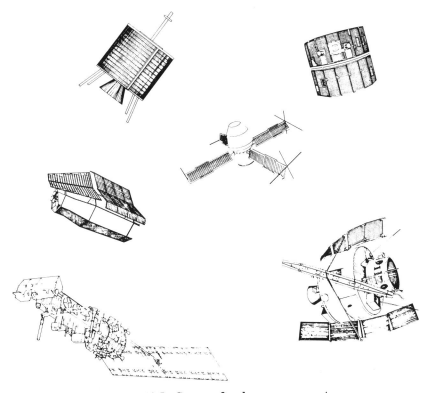

Fig. 10.5   Spacecraft solar array concepts.

launch allows design of extremely lightweight structures. The designs are driven more by the need for rigidity than strength; thus, very thin-section built-up structures are possible. Such structures are highly susceptible to handling damage, and often the primary criterion defining material thickness is the need to handle it during installation. The development of highly rigid composite materials in recent years has greatly enhanced the possibilities in solar array design.

The potential convenience of roll-up solar arrays was recognized early in the history of spacecraft design. However, the technology did not lend itself to such an approach; solar cells were too thick, connections were too stiff, and suitable substrates did not exist. Recently, however, technology has caught up with the concept, and a variety of flexible roll-up and fold-up solar arrays have been demonstrated. Perhaps the most spectacular example was the 12.5-kW array demonstrated during a Space Shuttle mission. This type of design, along with a large variety of deployable rigid array concepts, makes it possible to conveniently package very large arrays for launch. A problem with very large arrays and the attendant high voltage and power levels is that of conductor mass and insulation between circuit elements. This can be particularly trying in flexible arrays and represents one of the practical limits that solar array technology may impose on the

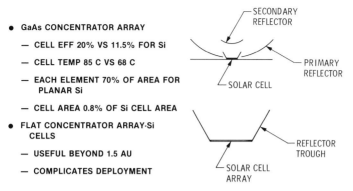

- Gaas CONCENTRATOR ARRAY
  - CELL EFF 20% VS 11.5% FOR Si
  - CELL TEMP 85 C VS 68 C
  - EACH ELEMENT 70% OF AREA FOR PLANAR Si
  - CELL AREA 0.8% OF Si CELL AREA
- FLAT CONCENTRATOR ARRAY-Si CELLS
  - USEFUL BEYOND 1.5 AU
  - COMPLICATES DEPLOYMENT

Fig. 10.6 Solar concentrators.

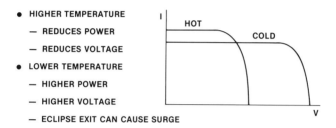

- HIGHER TEMPERATURE
  - REDUCES POWER
  - REDUCES VOLTAGE
- LOWER TEMPERATURE
  - HIGHER POWER
  - HIGHER VOLTAGE
  - ECLIPSE EXIT CAN CAUSE SURGE

Fig. 10.7 Effect of temperature on solar cells.

spacecraft designer. Roll-up arrays offer the lowest mass approach currently available for providing large array areas.

To extend the capabilities of solar arrays to regions farther from the sun, concentrators of various types have been proposed. Figure 10.6 shows two concentrator concepts. The flat concentrator array for use with silicon (Si) cells is basically a trough that increases the collection area relative to the cell area. This concept is useful at solar distances beyond 1.5 A.U. by increasing the energy available for conversion and keeping the cells from becoming excessively cold. Such concentrators can probably extend the useful range of Si cells out to 3–4 A.U. and possibly somewhat farther.

The other concentrator concept is particularly directed toward gallium arsenide cells. These cells are quite expensive, and it is therefore desirable to minimize cell area. Also, these cells prefer a higher temperature than their Si counterparts. The concentrator concept shown concentrates sunlight from a large collection area onto a small cell area. This reduces cell cost and brings the cells to a higher operating temperature than they would encounter otherwise, thus providing a double benefit. An obvious disadvantage is that such concentrators are complex and expensive to manufacture. Any concentrator, even the relatively simple one shown for the Si array, clearly complicates stowage and deployment.

Certain characteristics of solar cells such as temperature dependence and the current-voltage (I-V) curve are of interest to the spacecraft designer. Like all devices of this general type, solar cells have a temperature-dependence characteristic. This may be approximated as about 2 mV/C. A decrease in cell temperature results in cell voltage increase. Figure 10.7 illustrates this dependence. It will be noted that, as voltage increases, the current drops. The effect of this increase in voltage with very cold panels, as in the case of a spacecraft just exiting eclipse, often cannot be ignored in design of the power system, since a major surge may occur.

The shape of the I-V curve shown in Fig. 10.8 is typical of solar cells and important in the design of spacecraft power systems. For minimum mass and maximum efficiency, it is desirable to operate the array at its maximum power point. Since power can be written as I × V, the I-V curve allows determination of the maximum power point. This point can be shown to be at the point where the maximum area rectangle that will fit within the I-V curve intersects the curve. As can be seen in Fig. 10.10, this point thus lies on the knee of the I-V curve. Although some specific applications may dictate operation at some other point, the majority of systems will be designed for maximum power point operation. With the maximum power point for the cells defined, the current and voltage of individual cells is known. This information, in conjunction with the voltage and current requirements of the spacecraft, defines the series-parallel arrangement of the cells in the array.

It is occasionally possible that some transient or other problem could drive the operating point off the knee of the curve and down into the lower

- TYPICAL CURRENT/VOLTAGE CURVE
- MAXIMUM POWER POINT DEFINED
  BY MAXIMUM AREA RECTANGLE
  WHICH FITS WITHIN CURVE
- DESIGN FOR NORMAL OPERATION
  AT MPP

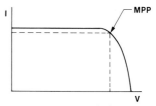

Fig. 10.8   Typical solar cell power characteristic.

- LESS ENERGY AS SOLAR DISTANCE
  INCREASES
- OPEN CIRCUIT VOLTAGE SAME
- INVERSE CASE AS APPROACH SUN
- MUST ALSO CONSIDER TEMPERATURE
  EFFECT

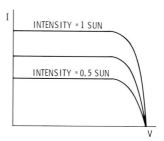

Fig. 10.9   Effect of solar distance.

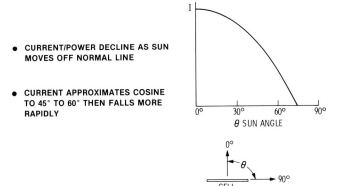

- CURRENT/POWER DECLINE AS SUN MOVES OFF NORMAL LINE

- CURRENT APPROXIMATES COSINE TO 45° TO 60° THEN FALLS MORE RAPIDLY

Fig. 10.10   Effect of sun angle on solar array power.

voltage range. The array then may not be able to pull itself back up to the normal operating point. Having a battery on line helps to stabilize against such an eventuality. Even systems that may not require a battery in normal operation (e.g., never in eclipse) may require a battery for this reason Off-line batteries or other devices that switch in to provide a temporary boost may also be used.

As the solar array moves farther from the sun, the current available drops while open circuit voltage stays the same or may increase due to lower temperatures. This leads to a family of curves similar to that shown in Fig. 10.9. A similar appearing set could be drawn for an array moving in toward the sun with current increasing and voltage dropping slightly as solar distance decreases. Regarding Fig. 10.9, it will be noted that as a result of these changes the maximum power point moves slightly. This will usually need to be considered in spacecraft power system design for planetary vehicles. For example, in the case of a Mars Orbiter one would normally design for the maximum power point corresponding to Mars distance, since that is where the power demand will be greatest due to operation of the science payload and because a large excess of power will be available at the Earth in any case.

It will be obvious that maximum power is available when the sun falls on an array that is normal to the sun line. As the angle between sun line and array normal deviates from 0 deg, one expects that a cosine relationship would obtain. This is indeed the case for angles up to about 60 deg. At larger angles where the sun angle approaches parallel to the array face, the cosine relation begins to break down somewhat due to the finite thickness of the cells and other effects such as specular reflection from the cover glass surface. Figure 10.10 gives an approximate curve shape for current vs sun angle.

As mentioned earlier, radiation has a detrimental effect on solar cells. The general effect of this degradation is shown in Fig. 10.11. Some degradation will take place during any mission of appreciable duration. If the operation takes place in more severe environments, e.g., the van Allen

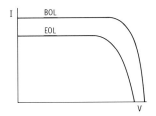

Fig. 10.11   Radiation effect on solar cells.

- RADIATION EXPOSURE REDUCES
  AVAILABLE POWER

- RADIATION REDUCES O/C VOLTAGE
  AND S/C CURRENT

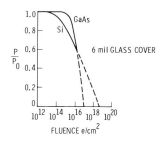

Fig. 10.12   Solar photovoltaic radiation effect.

belts, the degradation per given time will be more severe. Since, presumably, the spacecraft requires as much power late in the mission as in the beginning, the solar array must be sized based on end-of-life or degraded capability rather than on performance of a new array. Radiation environment is discussed in some detail in Chapter 3.

Considerable effort is being expended on development of more efficient and more radiation-resistant solar cells based on gallium arsenide rather than silicon. Figure 10.12 compares the radiation resistance of the two types of cells. Note that at some point the curves cross and silicon may be better again beyond that point. However, if this is real, it occurs at very high fluence and may not be of practical interest.

In terms of efficiency at the cell level, silicon cells deliver a solar-to-electric conversion efficiency of about 11.5%. Gallium arsenide promises about 18%, however, at a cost up to 10 times that of silicon cells. The cost of gallium arsenide cells is dropping rapidly, but, unless the cost comes down substantially, it is probable that gallium arsenide will be limited to applications that demand radiation resistance or where performance needs are so demanding as to justify the additional cost. The concentrator concept discussed earlier may help to reduce the cost of the cells, but the concentrator itself is complex and costly. Careful cost trades that fully assess all the impacts must be conducted before such a decision can be made.

10.9   PRELIMINARY SOLAR ARRAY SIZING

Although detailed design of solar arrays is the function of specialists in the field, it is relatively easy to perform preliminary sizing calculations to provide general characteristics of an array required for a given spacecraft. The problem on p. 419 demonstrates these calculations.

10.10 RADIOISOTOPE THERMOELECTRIC GENERATORS

The radioisotope thermoelectric generator (RTG) is a power source that renders the spacecraft essentially independent of the sun. Although this is an advantage in many cases, it also comes at a price that explains why these units have seen relatively limited use. The RTG functions by converting the heat energy generated by decay of a radioisotope into direct current electricity by means of the thermoelectric effect. In a typical RTG (Fig. 10.13) a central core of radioisotope material is surrounded by an array of thermocouples connected in series parallel. The hot junction of the couples is in contact with the canister containing the radioisotope and the cold junction with the external wall of the RTG from which heat is radiated to space. The efficiency of the RTG is ultimately limited by the conversion capability of the thermoelectric elements.

Modern semiconductor thermoelectrics, such as are used for Galileo, can deliver a conversion efficiency of 10–11%, and research is under way to improve this. Other limitations involve the internal thermal conductivity of the assembly. Considerable effort goes into designing a thermal path with minimum temperature drop from the isotope to the hot junction and from the cold junction to the outer case while at the same time minimizing heat leakage between these points which bypasses the thermoelectrics. Since the conversion efficiency of a given design depends on maximizing the temperature differential across the thermoelectrics, and the upper and lower limits are driven by material limits on the hot side and radiator size (usually) on the low, the importance of minimizing conductive drops or thermal leakage is obvious. These factors plus internal resistance and other losses explain

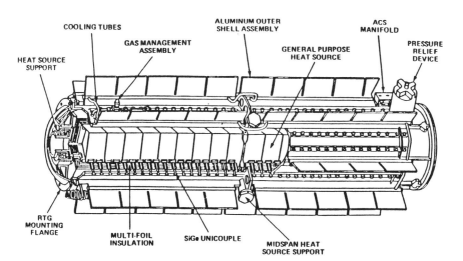

Fig. 10.13   Galileo RTG. (Courtesy of Jet Propulsion Laboratory.)

why overall RTG efficiency is less (e.g. 6 or 7%) rather than equalling the value for the elements.

The RTG used for the Galileo Jupiter Orbiter delivers 298 W $\pm$ 10% at the beginning of life for a mass of about 56 kg. The thermoelectric elements are doped silicon-germanium (Si-Ge). The radioisotope material used is plutonium-238 (^{238}Pu).

Although all radioisotopes exhibit a loss in energy output with time, the 86.7-yr half-life of ^{238}Pu is not the major life-limiting mechanism of current RTGs. Degradation of the thermoelectric elements, caused mostly by dopant migration at the relatively high temperatures involved, is a more significant cause of performance loss. Breakdown of insulators because of temperature and radiation is also a factor.

Although ^{238}Pu has been most commonly used in RTGs, there are other candidates that feature the combination of reasonably long half-life with high energy output which qualify an isotope for RTG use. Table 10.4 lists some candidates. Note that, for long space missions, only strontium-90 (^{90}Sr) has adequate half-life to be a viable candidate in addition to ^{238}Pu.

By their nature, RTGs cannot be turned off in the conventional sense. That is, the radioisotope continues to decay and generate energy regardless of any external action. Similarly, the thermoelectrics will generate electricity whenever a temperature differential exists. Since there is no practical way to control the generation of electricity at an essentially constant rate within the RTG, control must be external. In spacecraft applications, control is accomplished by use of a shunt radiator to dispose of electrical energy in excess of requirements at the time. RTGs are usually stored in a shorted condition. This has the desirable characteristic of reducing the temperature of the RTG during storage because of the Peltier thermoelectric cooling effect.

Aside from cost, particularly of the radioisotope, RTGs have a variety of problems that have tended to limit the use of these devices where other power sources will suffice. The high external temperature and the radiation from the radioisotope are a major problem in ground handling. Special handling equipment is required since the assembly crew cannot directly handle the units without thermal protection (e.g., heavy gloves). This complicates structural assembly and electrical connections and increases installation time. This result is unacceptable since it increases exposure of

Table 10.4   RTG material properties

	Po 210	Pu 238	Ce 144	Sr 90	Cm 242
Half-life (years)	0.378	86.8	0.781	28.0	0.445
Watts/gm (thermal)	141	0.55	25	0.93	120
$/Watt (thermal)	570	3000	15	250	495

the crew to radiation from the radioisotope. This can be of sufficient intensity to mandate the use of oversize crews so that no individual exceeds allowable dose limits. Note that it is important to allow for contingency. If the crew is sized to just reach the exposure limit in the course of a normal installation, then there will be no one available in the event of a problem that requires the RTGs to be removed and reinstalled.

A further problem is the possibility of contamination of the Earth by the radioisotope in the event of a launch failure or decay from orbit. The radioisotopes in RTGs launched to date have been extensively protected against burnup on re-entry. The fuel itself is normally the oxide of the radioisotope and is therefore reasonably strong and resistant to high temperature of itself. The lumps of oxide are then encased in graphite for entry and impact protection. Extensive tests are performed to qualify the fuel elements for this environment. In the case of the Apollo 13 RTG fuel element which entered Earth's atmosphere at lunar return velocity, there was no evidence of any release of radioisotope material during entry or impact.

Similarly, most reasonable launch failure scenarios can be dealt with by the internal protection built into the RTGs. It is possible, however, to postulate a launch failure of such severity that the fuel elements will be shattered and the radioactive material scattered into the atmosphere. This may be acceptable if the material is thoroughly dispersed in the upper atmosphere as very fine particles so that concentrations at any point on the surface are very low when it finally comes down. However, this is very improbable, especially in a launch failure. Extensive analysis is necessary to determine possible environment and to devise protection adequate to ensure that the radioactive material comes down in a condition that minimizes dispersion and allows for recovery. This is especially of concern for plutonium, which, besides being radioactive, is extremely toxic. Analysis indicates that the Challenger accident would not have created a hazard due to radioisotope dispersal had RTGs been aboard.

The radiation from the RTGs is detrimental to spacecraft electronics and instruments, making it necessary to mount the units on booms at some distance from the body of the spacecraft and/or to provide shielding.

10.11 FUEL CELLS

Fuel cells, which have been used to power the later Gemini spacecraft as well as the Apollo and Space Shuttle vehicles, are devices that allow direct conversion of chemical energy into electricity. An oxidizer and a fuel are fed into the cell, which is roughly similar to a battery in internal arrangement. Electricity is generated directly from the oxidation reaction within the cell but without the high temperature and other complications associated with combustion.

Hydrogen and oxygen are the reactants used in currently operational cells. The output of the cells is essentially pure water, which may be used for crew consumption with little or no treatment. Other reactants have been used experimentally, but presently only the hydrogen-oxygen combination has been used operationally.

A frequently suggested possibility is the use of regenerable fuel cells in lieu of batteries for energy storage in large systems, e.g., space station. In this scenario, fuel cells would use stored hydrogen and oxygen to generate electricity during eclipse. During the remainder of the orbit, solar arrays would generate electricity to power the spacecraft and to recharge the fuel cells by electrolysis of the water generated by the cells during operation. The resulting hydrogen and oxygen would then be stored to provide reactant to the cells for the next eclipse period.

The overall mass of a fuel cell system is a function of operating time since the mass of reactant is included. For the fixed mass of the system, however, a value of 1.1 kWh/lb at a power level of 2.6 kW is a good rule of thumb.

10.12 POWER PROCESSOR

The power processor portion of the space power subsystem carries the responsibility for many of the functions listed earlier in this chapter.

The voltage from the power source may vary substantially for a variety of reasons (e.g., temperature, load, external environment). The power processor regulates the voltage supplied to the remainder of the spacecraft system to the specified level within some tolerance, thus protecting the other subsystems from these fluctuations. In some cases, a variety of voltage levels for different functions may be required.

Any electrical noise generated by the power source or control functions must be isolated from the main bus. The main bus must be isolated from any power source faults such as loss of part of a solar array or voltage transients due to exiting eclipse.

Finally, the power subsystem itself and the other spacecraft subsystems must be protected from faults in any one subsystem.

10.13 FUTURE CONCEPTS

Nuclear reactor-based power plants offer considerable promise for the future. For very large power levels, hundreds of kilowatts to megawatts, reactors may be the only viable source in the next several decades. The nuclear reaction supplies heat, which is converted to electricity by a variety of techniques. Candidate techniques include thermionics, thermoelectrics, Stirling engines driving alternators, and various rotating machinery concepts such as Brayton or Rankine cycle.

All conversion concepts require radiators to reject waste heat to space. These radiators become very large for high-power units and present a major design challenge. The dynamic conversion concepts are generally much more efficient than the static and therefore require a smaller reactor and, in some cases, a smaller radiator. The vibration and other disturbances typical of dynamic systems may be a problem in some cases.

Reactors have the advantage that, until they are in operation, they are not highly radioactive and can be handled with relative safety. When in operation, however, the radiation is very intense and much more damaging than that characteristic of RTGs. Heavy shielding is required in even unmanned applications since electronics cannot withstand the radiation from the reactor without attenuation. For manned applications, the shielding and/or separation requirements become much more stringent. Figure 10.14 shows a typical reactor-powered spacecraft design using separation to reduce shield mass. In the configuration shown, a shadow shield is used that only protects a relatively small portion of the volume of space surrounding the reactor, thus saving substantial mass. However, the price paid to save this mass is that close proximity operations outside the shield shadow are precluded. An early concept for the SP-100 nuclear reactor-based space power system was specified at a mass of 3000 kg for a 100 kWe system. The mass-to-power ratio should improve somewhat as size increases. Work is being carried out toward development of much larger power plants in the multimegawatt electrical output range. No details of this work are presently available.

The dynamic isotope system concept is an effort to obtain more electrical power from the same isotope heat source as that used for the RTGs. In this case, the heat from the decaying isotopes is used to heat the working fluid of some type of engine such as the Brayton cycle, organic Rankine cycle, or Stirling, which in turn drives an alternator. Because of the much higher conversion efficiency of these dynamic systems vs the thermoelectric, amounting to a factor of 5–7, much more power can be obtained from a given isotope loading. This has advantages in reducing cost, radiation exposure, and mass. The negative side involves the reduction in reliability

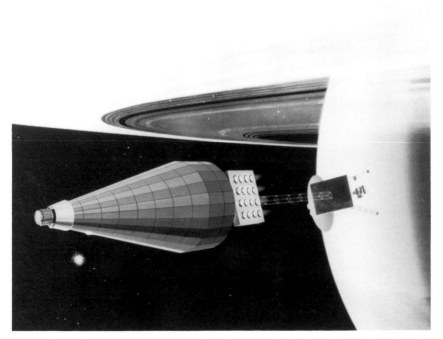

Fig. 10.14   Nuclear-electric spacecraft. (Courtesy of Jet Propulsion Laboratory.)

because of moving parts and the vibration that any dynamic system will tend to generate. Conversion systems of this type have been tested extensively but, to date, have not been flown. A possible disadvantage is the requirement to dump waste heat at a temperature lower than that typical of the thermoelectric approach in some cases, thus requiring larger radiator area. This comment may seem at odds with the earlier one regarding possibly smaller radiators. Either may be true. High efficiency demands a low cold end temperature and thus a larger radiation. However, high efficiency allows a smaller radiator. The result is simply a matter of how the trades work out.

An interesting energy conversion concept for potential future use is alkali metal thermal-to-electric conversion (AMTEC), which is now being tested in the laboratory. This device has no moving parts (if we discount the sodium working fluid circulated by electromagnetic pumps) but offers potential conversion efficiency approaching that of the dynamic systems. AMTEC applies sodium heated by the basic energy source to one side of a ceramic membrane that conducts sodium ions but not electrons. Thus, the positive sodium ions pass through, but electrons tend to accumulate. A conductive film on the membrane collects the electrons, which are then conducted through a load to the downstream side of the membrane to neutralize the sodium ions. A number of problems must be solved including membrane life, sodium condensation management in $0\,g$, and a variety of materials concerns before this concept can be considered for operational use.

Solar dynamic systems are, as the name implies, dynamic machines such as Brayton, Rankine, or Stirling, using the sun as a basic energy source and driving electrical generators. These units can deliver conversion efficiency five to seven times that of photovoltaic arrays. This becomes very attractive when we consider high power levels, say above 100 kW. Photovoltaic arrays become very expensive at high power levels due to the very high cost of solar cells. In addition, the very large array sizes pose attitude control and atmospheric drag problems. The reduction in area of collectors for the dynamic system greatly reduces the drag and stability concerns and becomes attractive costwise as well at high power. However, this approach does carry the usual dynamic system problems of lifetime, possible vibration, and possible attitude control interactions. Also, some of the size advantage may be partially offset by the requirement for waste heat radiators associated with these conversion concepts.

We have referred on several occasions to the need for radiator surfaces to dispose of waste heat. As systems become larger, the significance of the radiator increases until, for very large systems, it may be the largest single item. Present radiator concepts consist of large thin skins, usually metal. The heat to be dissipated may be delivered by conduction, by a pumped fluid loop, or by an array of heat pipes. Any concept that will improve performance is obviously a matter of great interest. Conventional radiators are limited by such factors as allowable material temperature, surface-to-mass ratio, emissivity, and conductivity. A variety of innovative concepts have been proposed to provide higher capability radiators. These concepts include droplet radiators, membrane radiators, rotating band radiators, and other concepts.

The droplet approach offers very high performance because of the large surface area to volume of the droplets and the possibility of allowing a liquid to solid phase change, thus greatly increasing the energy content. However, several practical problems must be solved, including droplet generation and collection (especially while maneuvering) and materials selection.

The membrane concept achieves high efficiency by flowing the fluid down the inside of a contoured rotating membrane. A gas-to-liquid phase change can enhance performance. Small punctures can be tolerated because surface tension in the fluid will prevent leakage. The need to rotate may be a problem, however. Material selection and launch/deployment may also present difficulties.

The rotating band is simply a broad, thin continuous loop of high-temperature metal that moves between heated rollers in the spacecraft, out into space to reject heat, then back to the rollers. Effective transfer of heat to the band is crucial to this concept. A similar approach using a rotating disk has also been suggested.

Problem
Size an array to support a 1500-W load plus battery charge.

Given
Solar cell efficiency = 11.5% at 28°C

Operating temperature = 50°C

Degradation over lifetime = 30% (10 yr)

Sun angle (maximum off normal) = 6.5 deg

Solar intensity (1 A.U.) = 1350 W/m^2

Temperature coefficient = -0.5% per °C

Packing factor = 90% (10% area loss due to cell spacing)

Battery capacity = 90 A-h

Computation
Array voltage must exceed battery voltage for the battery to charge. For these voltage levels a good rule of thumb is 20% above battery voltage. For a 27.5-V battery,

$$\text{Array voltage} = 27.5 \times 1.2 = 33 \text{ V}$$

$$\text{Total power (EOL)} = \text{load} + \text{battery charge}$$

$$= 1500 \text{ W} + \frac{90 \text{ A-h} \times 33 \text{ V}}{15 \text{ h}}$$

$$1698 \text{ W} = 1700 \text{ W}$$

$$\text{Temperature effect} = (50 - 28) \times 0.005 = 0.11$$

$$\text{Array capacity (BOL)} =$$

$$\frac{\text{total power (EOL)}}{\text{degradation} \times \text{cosine sun angle} \times \text{temperature effect}}$$

$$= \frac{1700}{(1 - 0.3)(\cos 6.5 \text{ deg})(1 - 0.11)}$$

$$= \frac{1700}{0.619} = 2746 \text{ W} = 2750 \text{ W}$$

$$\text{Total cell area} = \frac{\text{array capacity (BOL)}}{\text{solar intensity} \times \text{efficiency}}$$

$$= \frac{2750}{1350 \times 0.115} = 17.7 \text{ m}^2$$

For $2 \text{ cm} \times 4 \text{ cm}$ cells $- 8 \times 10^{-4} \text{ m}^2$ per cell,

$$\text{Number of cells} = \frac{\text{Total cell area}}{\text{cell size}}$$

$$= \frac{17.7}{8 \times 10^{-4}} = 22{,}142$$

$$\text{Array size} = \frac{\text{Total cell area}}{\text{Packing factor}}$$

$$= \frac{17.7}{0.9}$$

$$= 19.7 \text{ m}^2$$

11
TELECOMMUNICATIONS

11.1 INTRODUCTION

Telecommunications in space differs from the earthbound version in two major respects: 1) long range, which may be anything from a few hundred to several billion kilometers; and 2) potentially large relative velocity between transmitter and receiver, so that Doppler shift becomes significant (± 50 kHz in S band for low Earth orbit), requiring complex frequency-tracking loops in the receiver. Also, spacecraft in low orbit see very limited communications coverage from any single surface station. A station that can track to within 5 deg of the horizon will view a spacecraft in a 300-km orbit for only 6.5 min, even for a zenith pass. At the opposite extreme, distant spacecraft move very slowly against the background of the fixed stars: thus, the pass time is essentially governed by the rotation of the Earth. Signals from distant spacecraft, because they are very weak, require tracking by large, specialized equipment, such as NASA's Deep Space Network (DSN).

These factors complicate spacecraft design because of the mismatch between the rates of data acquisition and return. In low Earth orbit (LEO) a spacecraft may collect data throughout the orbit period of perhaps 95 min. Given only one downlink station, the spacecraft can dump data only a few times per day. Clearly, the downlink data rate must be many times that of the acquisition rate even with onboard processing and compression of the data. Power limitations and range restrict the rate at which data can be returned from a spacecraft at another planet. Data may be acquired very rapidly during an encounter and then played back at a relatively low rate over a long period.

Moreover, passage through the Earth's troposphere and ionosphere complicates signal propagation, as a result of energy absorption, rotation of polarized signals, etc. We shall examine these effects in more detail later.

Spacecraft telecommunications hardware has power, mass, and volume limitations more extreme than in other applications, even aircraft avionics. Meeting these challenges has led to the technology of low-power, low-mass electronics seen in today's consumer electronics market. As discussed in Chapter 3, spacecraft electronics experiences a variety of environmental stresses, such as mechanical shock and the acoustics and vibration of launch and atmospheric flight. Spacecraft are exposed to radiation that can damage electronics over a period of time. Extremes of thermal environment normally do not unduly affect the electronics of the telecommunications system, which is usually located in the temperature-controlled interior of the spacecraft, but external equipment such as antennas may be strongly driven by thermal design considerations.

Because of its role in accepting ground commands and returning data, the telecommunications system interfaces directly or indirectly with virtually every spacecraft subsystem and experiment. The earthbound end of a link interfaces with tracking stations, and through them with operating agencies around the world.

11.2 COMMAND SUBSYSTEM

The command subsystem allows instructions and data to be sent to the spacecraft. In some cases the command will be acted upon immediately; in others it may be stored to be acted upon when a particular clock time is reached, some event is sensed, or a particular spacecraft state is attained.

The two basic command types may be characterized as relay commands and data commands. The former are functionally equivalent to switch closures, and may provide a simple on/off function or initiate a complex, stepwise operational sequence. Such commands may provide a pulse signal or may latch in a new state until a further command is received (in the switch analogy, a momentary contact vs a toggle).

Data commands, as the name implies, provide information upon which the spacecraft acts, such as the direction and magnitude of a translation maneuver. Later in this chapter we will discuss how such commands are structured.

A complex operation such as a thruster firing to cause a midcourse correction might involve the transmission of a substantial number of data commands involving directions and magnitude of attitude maneuvers, rocket motor burn time or required change in velocity, and maneuvers back to cruise attitude. In addition, a number of relay commands might be required to configure the spacecraft for the maneuver (e.g., science instruments off, telecommunications from high gain to omnidirectional antenna, etc.). A final relay command to enable the sequence of actions would probably be required as well.

This example of midcourse correction illustrates the need for the two types of commands and also the need for delayed commands. The maneuver may need to take place out of sight of ground stations, or the timing may be so critical that it is not acceptable to depend on ground commands, where the communications uplink might be lost at a critical time.

The length and structure of command messages and individual words will depend on the amount of information to be sent and the capability of the equipment. In addition to the actual information, there will be address, identification, and other "formatting" data bits that must be transmitted and can substantially increase the overall data rate.

The component choices for the command subsystem are much the same as those for Earth applications: bipolar transistors, n-type metal-oxide semiconductors (NMOS), and complementary metal-oxide semiconductors (CMOS). Bipolar transistors are slowest and use the most power but are usually the most resistant to radiation. The higher speed and lower power consumption of the conventional metal-oxide semiconductors come at the cost of much greater sensitivity to radiation, unless special radiation-hardening measures are taken in their design.

Radiation causes long-term degradation of components and eventual loss of function. This can occur through a variety of mechanisms, depending on the type of radiation. As a more immediate problem, energetic charged particles passing through the junctions of the components can cause "soft," or temporary, errors. Deposition of sufficient energy in a junction can cause it to "flip," leaving, for example, a logical 1 where a 0 had been. If this particular junction contains a bit that is part of a data command, erroneous data now reside in that register.

In the more common "soft-error" case, the error can be corrected by reloading the command. The damage is not permanent. In the case of CMOS circuitry, the energy deposition can destroy the junction, in what is called a "latchup" condition. Modern CMOS circuits normally have latchup protection for space applications, but availability may be limited.

The smaller the junction—and small size is the means by which high speed and low power consumption are achieved—the lower the energy required to cause the phenomena just discussed. Thus, the improvements in electronic component technology which have allowed us to design more capability into given power and volume constraints have simultaneously increased the susceptibility to radiation damage. Since the problem primarily concerns space operations, most research into radiation-hardened electronics has been done by NASA and the Departments of Defense and Energy. Production of such components is limited, which makes them expensive. One obvious solution to the radiation problem is shielding. Unfortunately, this is of limited practicality, as discussed in Chapter 3.

11.3 HARDWARE REDUNDANCY

Redundancy is an important factor in achieving the reliability required for long life. A common and straightforward approach simply uses two completely separate parallel systems. Although this probably (but not certainly) improves reliability, a single failure in each string will still cause loss of the command function. A more sophisticated approach employs redundancy at the subsystem level, with cross strapping such that a given subassembly can be used in either string.

In this arrangement one or more failures can occur in each string, but as long as there are no duplicate failures in each string (i.e., at least one of each type of subassembly is working), a working command system can be assembled by selective cross strapping between the strings.

Control and management of redundancy is a complex issue, which, when improperly done, can lead to serious pitfalls. The redundancy scheme must be examined with care to avoid inadvertent and irreversible switchovers, possible untested modes, etc. Also, care must be taken so that the system is truly redundant; for example, two fully duplicated strings operating off a single fused power cable are not redundant.

11.4 AUTONOMY

With the development of computer capability and the undertaking of more complex missions at more distant targets, spacecraft have become more autonomous. The trend is away from very detailed command sequences as described earlier and toward high-level commands. As an example, a spacecraft might simply be commanded to apply a specified delta-V in a given direction. It would then autonomously compute and execute the required attitude maneuvers and the rocket motor burn. Still more advanced (and not yet practical) spacecraft would perform navigation onboard and autonomously decide that a course correction is required and perform it.

The advantages of autonomy are obvious. The size of the ground operations crew and support team is reduced—a major savings since for long missions the cost of flight operations can easily exceed that of development and launch. Reliability is enhanced since success is no longer as dependent on the link to Earth.

However, an autonomous spacecraft for a given mission will always be more complex than a ground-controlled machine. (This overlooks very simple spacecraft such as the early Explorers and Pioneers, which had no uplink, used a single operating state, and were always on, collecting data and sending it to the Earth whether anyone was listening or not.) This increased complexity implies a greater variety of spacecraft states and operating modes to be considered in system design and testing. Since it is probably impossible to test every conceivable mode, a great deal of consideration must go into the design of a system free of traps and testable with reasonable time and effort.

11.5 COMMAND SUBSYSTEM ELEMENTS

Figure 11.1 presents a functional block diagram of a typical spacecraft command subsystem. The major elements that make up the subsystem are defined in the diagram and are discussed in the following paragraphs.

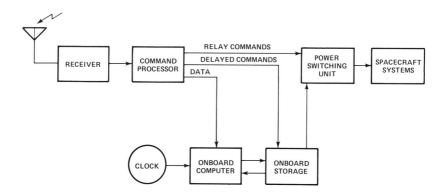

Fig. 11.1 Spacecraft command system block diagram.

Antennas

For LEO missions the uplink, or command antenna, will usually be omnidirectional to facilitate communication from ground stations while the aspect angle is changing during the pass. Deep space missions, in contrast, require directional high-gain antennas and thus attitude and articulation control. But for use near the Earth such missions also carry omnidirectional antennas to aid operations and to avoid overdriving the receiver with excessive gain at short range. Capability to send uplink commands through a low-gain omnidirectional antenna should be retained for emergency use even at long range. Anomalous behavior resulting in loss of high-gain antenna pointing may cause loss of the mission if no other way exists to get commands into the spacecraft. The capability to broadcast commands from the Earth at very high power to the omnidirectional antenna might save the spacecraft.

Receivers

The receiver may be either of two types: tuned radio frequency (TRF) or, more commonly, superheterodyne. Details of the two types go beyond the scope of this text. Briefly, the TRF is a radio-frequency (RF) amplifier, tuned for a narrow bandwidth around the transmitted frequency, followed by a detector or demodulator stage and several stages of low-frequency amplification. In a superheterodyne receiver the received signal is shifted in frequency (heterodyned or mixed) to a frequency lower than the transmitted one. Two or more shifts are common, and the resulting signal is amplified and filtered at each stage to provide greater sensitivity to weak signals and better selectivity for rejection of unwanted signals.

Modulation

The receiver and uplink may use amplitude modulation (AM), phase modulation (PM), or frequency modulation (FM). The choice depends on several factors, including required signal-to-noise ratio (SNR), desire for graceful degradation, available RF bandwidth, required data rate, hardware complexity, and compatibility with existing ground tracking systems.

FM and PM systems can operate with low RF signal-to-noise ratios than AM systems because FM and PM provide improved performance at the cost of greater bandwidth. FM and PM systems suffer the penalty of a threshold effect. As the SNR on the RF link decreases, the performance of the link degrades very slowly until the threshold SNR is reached. As the SNR progresses below the threshold, the performance of the link drops precipitously. AM systems do not exhibit this behavior. They require more SNR to achieve a given performance, but degrade gracefully as the SNR is reduced. (The reader can experience this with an automobile radio when driving away from a station. FM will remain reasonably clear up to some distance from the station, when the signal abruptly deteriorates and is lost. AM becomes progressively weaker, probably with distortion increasing, but will remain audible through the noise for a long period.)

A high degree of frequency selectivity is essential in spacecraft receivers to enhance SNR and to reduce sensitivity to electromagnetic interference (EMI). During ground testing and launch, spacecraft operate in a very "signal-rich" environment and even in orbit will be illuminated by unwanted signals. Frequency selectivity is essential in such situations.

In addition to amplifying the signal and filtering out noise and EMI, the receiver demodulates the signal and provides the information-bearing portion of the received signal to the command decoder and processor.

It is a cardinal rule that the command receiver—at least in U.S. spacecraft—is always on. If a command exists that can turn off the receiver, and such a command is inadvertently sent (erroneous commands do occur), there would be no way to undo the damage. Of course, mechanisms could be devised to turn it back on after a time, or some other recovery approach could be employed, but the straightforward and therefore preferred approach simply has the receiver permanently on.

The command decoder (not shown separately in Fig. 11.1) may be viewed as the first stage in command processing. The decoder first inspects the identifier bits that make up a part of each command word. These bits identify the word as a command and are used to synchronize the bit stream along command word boundaries. They also contain the address of the intended spacecraft. The decoder verifies that the word is a command and is intended for this spacecraft. In the case of an encrypted command string, appropriate decryption algorithms must be applied in the process. The decoder then passes the bit string to the command processor.

Command Processor

The command processor in a modern spacecraft usually employs a microprocessor. In early spacecraft the processor was usually a simple hardwired (and therefore inflexible) circuit. The processor interprets the command for proper destination and required action. As a precaution against erroneous action, commands are checked for validity using parity bits or more sophisticated error detection and correction schemes. The processor then sends the signal to the appropriate destination: the computer, an onboard storage memory unit, or directly to a power-switching unit.

The central (command) computer executes the more complex operations involving data commands and delayed commands. In early spacecraft the "computer" was little more than a programmable sequencer. With increasing sophistication, spacecraft computers have become quite capable data processors. The rise of microprocessor technology allows a distributed architecture in which most of the computational power resides in the subsystems, and the central computer functions primarily as a coordinator. Unless an extremely large amount of data must be stored, solid-state memory usually proves sufficient.

The power-switching elements are the interface circuits between the command subsystem and the remaining spacecraft subsystems. These elements may consist of pulsed or latching relays or solid-state switches. More complex operations may require sequences of steps to achieve the desired

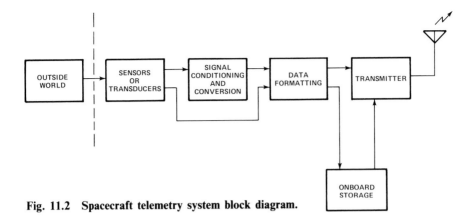

Fig. 11.2 Spacecraft telemetry system block diagram.

goal. The switching elements may be operated immediately by relay command or by the computer, based on sequences previously loaded.

Telemetry Subsystem

The telemetry subsystem takes engineering or scientific data and prepares it for transmission to the ground. Figure 11.2 presents the functional block diagram of a typical spacecraft telemetry subsystem.

The data are generated by sensors or transducers responding to events in the "outside world." In this context the outside world may be other subsystems in the spacecraft that generate data concerning their status and condition, or events in the surrounding environment as sensed by the science subsystem. Parameters that are measured often include acceleration, angular rate, angular position, pressure, temperature, density, resistance, voltage, current, intensity, electric field, magnetic field, and radiant energy.

It may be said that the sensors and transducers are not truly part of the telemetry system, but rather of the subsystem in which they reside. This argument carries most weight in regard to the science instruments. In any case, these elements provide the signals upon which the telemetry system operates.

Signals from the sensors or transducers are rarely in a form immediately suitable to the data-formatting element. Generally, the signal must be "conditioned" and converted from analog to digital. Signal conditioning converts data to a form that is acceptable to the telemetry system. Weak signals may be amplified or excessively strong and attenuated. High- or low-pass filters are used to remove bias or noise; notch filters are used to remove high-intensity signals in a specific frequency band. A signal's dynamic range may be compressed, perhaps by the use of a logarithmic amplifier. Every effort will be made to achieve isolation between signals to ensure accuracy of the data.

Analog-to-digital conversion (ADC), one of the major functions of signal conditioning, will usually be required because, generally, the real world is analog; it varies more or less continuously across the range of the

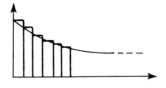

Fig. 11.3   Analog to digital conversion.

phenomenon being observed. Spacecraft data, on the other hand, can be handled more efficiently in digital form, as a series of discrete steps.

Figure 11.3 shows the process schematically. The smooth curve represents the phenomenon being measured. The digital approximation is represented by the "stair steps," where the quantized level is proportional to the average value during the sampling period or the value at the sampling instant. The sampled value is restricted to one of a finite number of allowable values so that the digitized data can be represented by a finite number of bits. The analog input could be a voltage curve, between prescribed limits, which a transducer produces in response to a phenomenon. The curve may be directly proportional to the phenomenon or may have been modified, perhaps by a log amplifier, to make the range compatible with the telemetry system. The output of the digitizer is a binary word for each digitized data point. The data are quantized into one of 2^n levels, where n is the number of bits in the data word. Several types of analog-to-digital converters are available to do this. The flash ADC can operate at rates in the tens of megahertz range, but is noisy. The successive-approximation register can handle data rates of hundreds of kilohertz. The integrating ADC can handle only tens of hertz, but produces very clean data. The details of the logic circuitry associated with different types of ADCs are beyond the scope of this text.

Several errors are inherent in the ADC process and the subsequent reconversion to analog that most data undergo on the ground. As with any telemetry system, the measured signal will be embedded in noise. A noisy analog input may fool the ADC into digitizing at a level higher than the actual level of the signal. An experienced human analyst looking at the noisy curve in raw analog form may be able to reject the noise and infer the true data. In the digitized and reconstructed data the analyst will not see the original data but rather a quantized and reconstructed version of the signal. As a result, it may be harder to estimate the correct data, or even to recognize that the data are noisy.

The quantization process can only represent a finite number of levels, and data at any point may fall between the levels. If eight bits are used to transmit each quantized level, then 256 levels can be represented; thus, the potential error is less than 0.4%. This may or may not be sufficient; additional bits, requiring a larger ADC, may be required.

Handling more than one data type requires some form of multiplexing by the telecommunication system. This usually takes one of three forms.

Frequency-division multiplexing (FDM) subdivides the frequency bandwidth of the telemetry downlinks (Fig. 11.4) and allocates the various data streams to separate portions of the available bandwidth. In the temporal

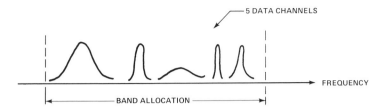

Fig. 11.4 Frequency-division multiplexing.

sense the data may be viewed as going out in parallel. The subdivision of the bandwidth is not necessarily equal; higher-rate data streams must be allocated wider bandwidths. This approach is common when one or more channels of high-rate data (e.g., video) are to be returned. It is possible to apply the other schemes discussed later to the individual channels within the FDM structure, and these channels may be analog or digital.

Time-division multiplexing (TDM) employs temporal separation to assign different sets of bits within a data frame to different users, as depicted in Fig. 11.5. The frame repeats continuously, with each user occupying the assigned bits in a cyclic fashion. Note that the cycle may be subcommutated, with more than one user sharing a particular set of bits in a subpattern within the overall sequence. By convention, frames are limited to 2048 bits with word lengths of 6–64 bits. Within these limits virtually any level of subframe definition is possible.

The final type of multiplexing, code-division multiplexing (CDM), sends the data in parallel over the same bandwidth during the same time period, but encoded using spread-spectrum techniques so that the individual streams can be separated at the receiver.[1]

The problem of "aliasing" is a major concern in any digital communication. If a band-limited, sampled signal is to be reconstructed accurately, it must be sampled at a rate at least twice the maximum frequency contained in the signal. This minimum sampling rate is called the Nyquist rate. A simple example will illustrate the problem. Assume that a sinusoidally varying signal at 1000 Hz is being sampled at 1000 Hz (half the required Nyquist rate). In this example the digitizer would always sample the same point on the waveform, and the data appear constant—a straight line with no indication of the true signal.

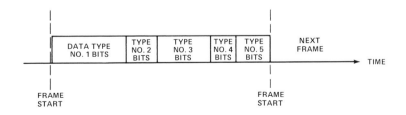

Fig. 11.5 Time-division multiplexing.

Table 11.1 Sample rates based on Nyquist criterion[a]

Signal type	Analog bandwidth, kHz	Typical resolution, Bits	Nyquist digital rate, kbps
Voice	4	7	56
Music	20	10	400
BW TV	4600	4	36,800
Color TV	4600	10	92,000

[a]Binary keying assumed. More complex coding schemes can be used to achieve lower bit rates.

As stated, the Nyquist criterion is

$$f_s \geq 2f_{\max}$$

The factor of two is, in practice, too low to achieve accurate representation of the sampled data. A more realistic sampling rate will be five or more times the maximum frequency.

It may not be necessary to use the full bandwidth of a measured signal to obtain useful mission information. An example is the use of the Earth's magnetic field for attitude control. A 20-Hz variation in the field may be of substantial scientific interest, but is of no interest for attitude control, where 1 Hz would normally suffice.

Sampling rates based on the Nyquist criterion are shown in Table 11.1. As noted earlier, real sample rates must be substantially higher, as shown in the final column. The very high data rates indicated usually are reduced by a variety of coding and data-compression techniques.

With TDM the bit stream or data stream comprises sequentially sampled data types combined in a specified frame pattern. This process of sequential data sampling is referred to as "commutation." Figure 11.6 schematically represents a major data frame from a continuing stream of bits. The major frame is repeated continuously in the telemetry stream and is subdivided

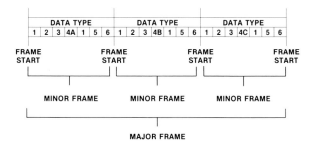

Fig. 11.6 Subcommutation and supercommutation.

into three minor frames. In this example each minor frame contains seven data elements.

In the (somewhat improbable) event that the commutation rate perfectly matches the desired sample rate for all data elements, each data element would appear once per subframe, and each subframe would be identical. In fact, however, there are always parameters that vary so slowly that sampling at a lower rate is acceptable. The same data-element position can thus carry many different low-rate parameters in each subframe, each sampled at less than the basic frame rate. This "subcommutation" is represented by data elements 4A, 4B, and 4C in Fig. 11.6. The majority of data elements, represented by 2, 3, 5, and 6 in the figure, will be sampled at the basic rate. Note, however, that Element 1 appears twice in each minor frame, and is thus sampled at twice the basic rate. This is called "supercommutation."

The engineering telemetry requirements of a spacecraft propulsion unit can illustrate this practice. The temperature of the propellant line, tank, and pressurant bottle will change slowly and might be subcommutated as Element 4. Tank and bottle pressures will change more rapidly and would be sampled at the basic rate. Thrust chamber pressure—the most dynamic and crucial to performance—might be supercommutated as Element 1, or at an even faster rate.

For proper telemetry interpretation various additional information must be included along with the measurement data. The major frames and often the minor frames will include additional bits for frame synchronization and identification. Extra bits will often be included for error detection and correction. The minimum would be a single bit for parity check. Incorrect parity would indicate that an odd number of bits had been erroneously received, but there would be no indication as to where the error had occurred. More complex schemes for error correction require additional bits and greater encoding and decoding complexity.

To allow the decoding system to synchronize with the downlinked data, frame synchronization bits are sent at the beginning of major or minor frames. The number of bits, by convection, will be 33 or less, with 24 being common for a major frame. Sending synchronization bits with each frame ensures proper synchronization, regardless of phase or frequency errors since the previous frame. Without this concern synchronization bits would be needed only at the beginning of the stream.

A final type of information required is a time tag. In most cases, for the data to be useful, it must be accurately annotated (commonly within 1 ms) as to the time of acquisition, and the time data included in the data frame. The time base is usually supplied by a stable crystal oscillator, often the same oscillator that comprises the clock in the onboard computer.

Telemetry formats vary from system to system, and the ratio of "overhead" bits to actual data will vary to some degree. In all cases the overhead is substantial, and the required telemetry bit rate will always be higher than the information bit rate. For many reasons, including power and antenna size, it is desirable to reduce the number of bits that must be transmitted. A variety of data-compression and data-encoding schemes have been devised and are in current use. This rather specialized field will not be covered here.

Onboard Processors

The onboard computer is an essential subsystem in most modern spacecraft. Early onboard "computers" were little more than timers enhanced with some modest logic circuitry. Processing capacity has grown with the maturation of computer and electronics technology until today most spacecraft are highly capable electronic devices.

Spacecraft computers have reflected the trends in ground computers. During the 1960s a central "mainframe" architecture was the norm, and this approach is still used. However, the development of microprocessor technology and distributed networks has led to their adoption in spacecraft architecture. In this approach the central computer coordinates and directs traffic between equally powerful processors located in various subsystems. Redundancy and cross strapping can provide remarkable flexibility of operation and increase system reliability. Selection of a system architecture should be based on the type of operation anticipated, cost, and complexity.

As with any spacecraft subsystem, the onboard processor should take minimum power, weight, and volume consistent with the required performance. The hardware must perform well under the usual environmental stresses, including temperature extremes, thermal cycling, hard vacuum, shock and vibration, and radiation. In addition, the hardware must be resistant to electromagnetic interference.

These special requirements on space hardware may cause a 5- to 10-yr lag in the technology being flown in spacecraft as compared to nonflight applications. In addition, special testing and flight qualification increase cost, as does limited production of space-qualified components and subsystems.

Some information regarding current processors (as of 1990) and future trends may help to put the foregoing discussion in perspective.

Throughput of an onboard processor is a complex function of architecture, word length, instruction set, and cycle time, and must usually be determined by benchmark runs of the system. The basic capability will be degraded 10% or more by error detection and correction codes. Cycle times currently are in the 80-ns to 1-μs range, with word lengths normally of 16–32 bits (although 18-, 24-, and 64-bit words sometimes appear). Early 1990s radiation-hardened processors operating within this range of parameters can supply on the order of 5 MIPS for the typical mix of instructions encountered in spacecraft operations. For comparison, commercial reduced instruction set computer (RISC) processors of this era yield a throughput on the order of 20 MIPS.

Directly addressable memory of a few megabytes is readily available. This can be expected to increase rapidly with improving storage technology. Off-line storage up to 10^9 bits is available. Redundant memory, especially for critical command storage, is important for reliability.

Input/output (I/O) channel capacity typically can have a range of 1–15 μs per word for serial ports. Second-generation machines offer direct memory access (DMA) capability in the 100-kbps to 20-Mbps range.

The typical processor has a mass of 5–50 kg, depending on the capacity of the machine, the technology in use, and other factors. Similarly, power can vary from 5 W to as much as 150 W, depending on speed, memory size and type, and architecture. Small microprocessors represent the low end of the size and power range; complex mainframes with extensive memory represent the high end.

The issue of parts qualification has been discussed previously in this and other chapters. Space-qualified ("class S") parts are expensive and typically have long delivery times. In fact, they may be impossible to obtain unless the project will pay for flight qualification and maintenance of a special production and test operation. This situation occurs because manufacturers show more interest in the commercial market, with production of millions of units, than the much more limited aerospace market. One option is the use of class B parts, which are functionally equivalent to class S but lack the pedigree and the screening and testing that define class S. An in-house screening and burn-in program to provide the effective equivalent of class S parts may be a cost-effective answer. There are many subtleties in the parts selection and screening business, far more than we can treat here. Consultation with specialists in reliability, safety, and quality assurance (RS & QA) is to be recommended.

A requirement for radiation hardness presents a problem, especially for spacecraft operating above low orbit and incorporating modern electronics. The soft-error rate experienced on the first Tracking and Data Relay Satellite (TDRS) was on the order of one per day. Extra shielding usually is not effective in reducing this rate because of the high energy of the offending particles. Special components and failure-tolerant architectures may be required. Current spacecraft processor technology offers total dose hardening to over 1 Mrad, prompt dose upset tolerance of 10^9 rad/s and survival tolerance of about 10^{12} rad/s, and a single-event upset rate of less than 10^{-10} errors/bit/day, with latchup protection.

Onboard Storage

Mass storage is required when data cannot be sent over the downlink at the time it is taken or at the rate of acquisition. Storage may also be required for computers engaged in operations requiring reference data.

Table 11.2 compares types of mass storage commonly in use today: tape recorders, bubble memory, and solid-state memory. Although easily the slowest due to the inherently serial access and less reliable because of moving parts, tape recorders retain the lead in terms of cost, power, and data density per bit. Tape recorders are chosen for storage of large amounts of observational data. Bubble memories are good for moderate amounts of data in difficult environments. Their relatively high power requirement presents a problem in some applications. Additional problems may result from the requirement for serial access and the relatively bulky nature of bubble memory units. Limited availability may also cause difficulties. Solid-state memory offers moderate but constantly improving capacity, and is used in conjunction with computers as well as for storage of observed data. Solid-state memory has the greatest promise in the long

Table 11.2 Onboard storage technologies

Parameter	Tape recorders	Bubble memory	Solid state
Power consumption	Medium	High	Low
Write speed	$\sim 10^7$ BPS	$\sim 10^5$ BPS	$\sim 2 \times 10^8$ BPS
Read speed	Slow, serial access	$\sim 5\ \mu\text{sec}$, random access	$\sim 30\ \mu\text{sec}$, random access
Storage density	$\sim 10^7$ bits/in^3	$\sim 10^6$ bits/in^3	$\sim 10^6$ bits/in^3
Storage capacity	$\sim 10^{10}$ bits	$\sim 10^7$ bits	$\sim 5 \times 10^8$ bits
Operating temperature	0–50°C	-55°C to $+85^\circ$C	-55°C to $+85^\circ$C

term because of continuing progress in reducing the size of electronic components.

Newer concepts may come into play in the future, such as high-density storage on optical disks. Although write/erase capability has only recently become available, optical disks have tremendous potential in the near-term for nonvolatile storage. This could be used to make very large amounts of reference information available to the onboard computer, thus allowing much greater autonomy in spacecraft operations.

Modulation Methods

The modulation scheme is the method by which command and telemetry systems encode a "baseband" information-bearing signal upon an RF carrier. The carrier, $S(t)$, is an RF signal characterized as

$$S(t) = A(t) \cos[\omega(t)t + \phi(t)] \tag{11.1}$$

where A is the amplitude, ω the frequency, and ϕ the phase angle of the signal. All of these quantities can vary as a function of time. Three modulation schemes are commonly used. AM varies the amplitude of the RF carrier wave signal according to the baseband signal. FM varies the ω of the carrier according to the baseband signal. Both of these schemes are in common use in terrestrial radio systems. The third scheme, PM, varies the ϕ with the baseband signal.

All three modulation schemes offer particular advantages, some touched on briefly in the section on command. The selection of the best method will depend on the application, available bandwidth, required signal-to-noise ratio, and the capability of the ground stations. However, it is true that few modern communications systems employ AM modulation.

Both analog and digital input signals may be used with any of the three methods. Discrete amplitude changes, frequency shifts, or phase shifts can represent the 1's and 0's of digital data. Similarly, continuous changes in amplitude, frequency, or phase can represent an analog signal. Modern systems use digital modulation almost exclusively. However, there may be cases where analog modulation offers advantages (e.g., with high-bandwidth video data).

Pulse-code modulation (PCM), a technique for converting analog signals into digital form rather than a modulation method as AM, FM, or PM, samples the signal and quantizes it into one of 2^n levels at a sample rate appropriate to the application. Since the allowed 2^n levels are finite, the unique level of each sample can be represented by a digital word n bits long. These data words are then formed into a serial bit stream arranged in minor frames that in turn make up major frames, as discussed earlier. Figure 11.7 diagrams the process of converting from the original analog waveform to the final digital bit stream.

This digital bit stream varies, or "keys," the carrier signal by one of the methods discussed earlier. If each bit is encoded onto the carrier independent of the other bits in the bit stream, the digital modulation is called binary. In binary amplitude-shift keying (BASK), the amplitude of the

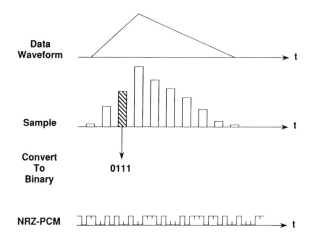

Data
Waveform

Sample

Convert
To
Binary   0111

NRZ-PCM

Fig. 11.7   Pulse code modulation.

basic signal is varied to represent a 1 or a 0. (Morse code is a simple example of BASK.) In binary frequency-shift keying (BFSK), two frequencies, close to but distinct from the carrier, represent the 1 and 0. This may be accomplished, for example, by switching between the outputs of two crystal oscillators, one just above and the other just below the basic frequency. Binary phase-shift keying (BPSK) transmits a sinusoidal carrier with one of two allowable phases that represent a 1 or a 0. Of these three techniques PSK and FSK are the most common in space communications.

If often seems to the casual observer that there are as many digital coding schemes as there are practitioners of the art. Figure 11.8 shows eight common approaches. The system engineer, unless specifically involved in coding schemes, need not be able to interpret them. The principal differences between the schemes are the bandwidth efficiencies and the ease of determining the clock rate and bit boundaries. It is important to recognize that a variety of schemes exist and to ensure consistency across subsystem interfaces.

Figure 11.9 qualitatively compares performance of the various modulation schemes as a function of signal power and SNR. A point of interest is the very sharp threshold of the FM and PCM schemes as power and SNR decrease. This graphically displays the phenomena discussed earlier.

Antennas and Gain

Antennas are generally categorized as omnidirectional ("omni") or directional. The latter type come in a variety of types, with beamwidths ranging from a few tenths to several tens of degrees. The beamwidth of a directional antenna is defined as the angle between the $-3\,dB$ (half-power) points relative to the power on the boresight axis. Figure 11.10 shows this and other beam characteristics.

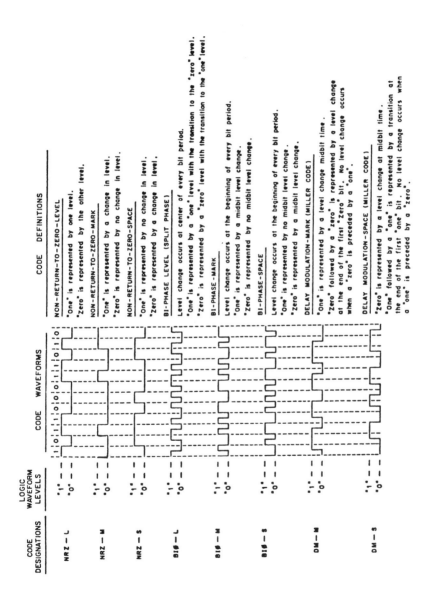

Fig. 11.8 Digital encoding schemes.

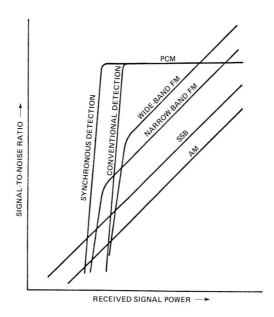

Fig. 11.9 Signal-to-noise behavior of various modulation techniques.

The gain of a directional antenna relates the power on the boresight axis to that of an ideal isotropic radiator, a point source that radiates energy equally in all directions. The gain of an isotropic radiator is 0 dB. An antenna with gain gathers up the radiation that is distributed evenly to the celestial sphere by an isotropic radiator and concentrates it into a smaller area. Thus, the gain of an ideal parabolic dish illuminating 1 deg^2 of sky is 41,253. Put another way, the product of gain G and the subtended angle ϕ (or field of view) is always 41,253 deg^2 for an idealized antenna:

$$G\phi = 4\pi \text{ sr} = 41,253 \text{ deg}^2 \qquad (11.2)$$

However, real antennas are not this efficient. For example, the gain-beamwidth product of a parabolic dish is approximately

$$G\phi = 2.6 \text{ sr} = 27,000 \text{ deg}^2 \qquad (11.3)$$

This is a good rule of thumb for parabolic antennas, although coefficient values as low as 20,000 and as high as 30,000 are common in communications literature.

Since the goal is to provide a certain minimum signal strength at the receiver, a trade-off always exists among transmitter power, antenna beamwidth, and the derived pointing accuracy. Figure 11.11 gives field-of-view (FOV) definitions and provides the equations for computing them.

Beamwidth is also related to the frequency of transmission f, transmitted wavelength λ, and the speed of light c. The gain of an antenna as a function

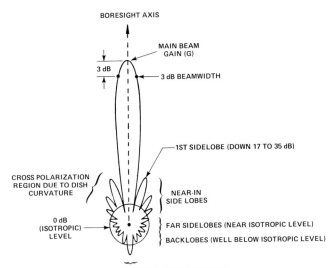

Fig. 11.10 Antenna pattern.

$$\phi_{\text{STERADIANS}} \triangleq \frac{\text{SUBTENDED SPHERICAL SURFACE AREA}}{r^2}$$

∴   4π STERADIANS IN THE ENTIRE SPHERE

CONICAL FOV:

$$\phi = 2\pi \left(1 - \cos\frac{\theta}{2}\right) = 4\pi \sin^2\frac{\theta}{4}$$

RECTANGULAR FOV:

$$\phi = \alpha\beta$$

Fig. 11.11 Solid angle.

of the area of the aperture A, or diameter D for a circular aperture, is given by

$$G = \eta \frac{4\pi f^2 A}{c^2} = \eta \frac{\pi^2 f^2 D^2}{c^2} \tag{11.4}$$

where η is the antenna efficiency. Common satellite directional antennas typically have η values of 0.50–0.80.

If we assume that the antenna has a rectangular FOV with beamwidth in one plane of $\theta = \sqrt{\phi}$, then

$$G\theta^2 = 2.6 \text{ sr} = 27{,}000 \text{ deg}^2 \tag{11.5}$$

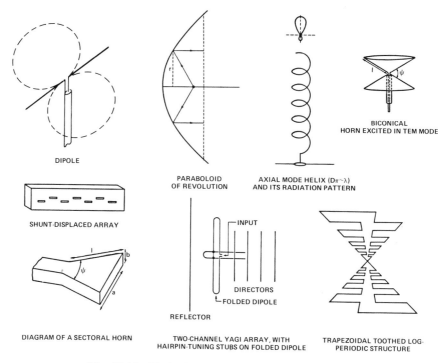

DIPOLE

PARABOLOID
OF REVOLUTION

AXIAL MODE HELIX (Dπ~λ)
AND ITS RADIATION PATTERN

BICONICAL
HORN EXCITED IN TEM MODE

SHUNT-DISPLACED ARRAY

DIAGRAM OF A SECTORAL HORN

INPUT

DIRECTORS
FOLDED DIPOLE

REFLECTOR

TWO-CHANNEL YAGI ARRAY, WITH
HAIRPIN-TUNING STUBS ON FOLDED DIPOLE

TRAPEZOIDAL TOOTHED LOG-
PERIODIC STRUCTURE

Fig. 11.12   Typical space communications antennas.

The beamwidth θ for a parabolic dish can be approximated by

$$\theta = \frac{164\lambda}{\pi D} \tag{11.6}$$

Figure 11.12 shows a variety of antenna types, particular characteristics of which go beyond the scope of this book. Table 11.3 presents gain and effective area for several types.

Antenna polarizations, mentioned earlier, can either be linear with vertical or horizontal orientation or circular with left or right orientation. Improper matching of transmit and receive antenna polarization will result in losses ranging from moderate to severe. Linear to circular mismatch will result in a loss on the order of 3 dB. A left/right mismatch of circular polarization will result in a loss of 25 dB or greater. Worst of all is a vertical/horizontal mismatch in linear polarization, where the losses theoretically become infinite; losses in the range of 25–35 dB are common.

Figure control—maintaining a very smooth and accurate surface shape—must be exact for parabolic and phased-array antennas. It is desirable to maintain the surface contour within 1/20 of a wavelength. For high frequencies or large antennas this becomes very demanding on structural design and fabrication. An empirical factor for loss in antenna gain

Table 11.3 Gain and effective area of several antennas

Type of antenna	Gain	Effective area
Isotropic	1	$\lambda^2/4\pi$
Elementary dipole	1.5	$1.5\,(\lambda^2/4\pi)$
Halfwave dipole	1.64	$1.64\lambda^2/4\pi$
Horn (optimum)	$10A/\lambda^2$	$0.81A$
Parabolic reflector (or lens)	$6.2-7.5(A/\lambda^2)$	$0.5A-0.6A$
Broadside array (ideal)	$4\pi A^2/\lambda^2$	A

due to surface roughness is

$$g = \exp\left(-4\pi^2 E_{\text{rms}}^2 A\right) \tag{11.7}$$

where g is the gain/loss factor, E_{rms} the average surface error in fractions of a wavelength, and

$$A = \frac{1}{1 + (D/4F)^2} \tag{11.8}$$

where D is the diameter of the antenna, and F is the focal length.

Radio-Frequency Link

The received signal power over a communication link, when combined with the total noise power, provides the fundamental measure of the quality of service available for communications.

Consider a transmission of $P_t\,W$ to a receiver system R m away. If the $P_t\,W$ are transmitted through an isotropic radiator, then the spherically expanding wavefront will have a flux density of [2]

$$F = \frac{P_t}{4\pi R^2}\ \text{W/m}^2 \tag{11.9}$$

when the front arrives at the receiver. If the same power is transmitted through an antenna of gain G_t, then the flux density at the receiver will be

$$F = \frac{G_t P_t}{4\pi R^2} = \frac{\text{EIRP}}{4\pi R^2}\ \text{W/m}^2 \tag{11.10}$$

where $\text{EIRP} = P_t G_t$ is the effective isotropic radiated power. This is simply the power that would have to be transmitted through an isotropic radiator to achieve the same flux density obtained from an antenna of gain G_t.

At the receiver, an antenna with physical area A_r and effective area $A_e = \eta A_r$ intercepts a portion of the flux density, its total received power being

$$P_r = FA_e = \frac{P_t G_t A_e}{4\pi R^2} \; \text{W} \qquad (11.11)$$

The gain of the receiving antenna can be expressed in terms of its area as

$$G_r = \frac{4\pi A_e}{\lambda^2} \qquad (11.12)$$

The received power, therefore, will be

$$P_r = P_t G_t G_r \left(\frac{\lambda}{4\pi R} \right)^2 \text{W} \qquad (11.13)$$

The term $[\lambda/(4\pi R)]^2$ is known as the path loss. This is not an absorption loss, but rather a "dilution" of the transmitted energy as the wavefront expands in traveling toward the receiver. In terms of the path loss the received power will be

$$\text{Power received} = \frac{\text{EIRP} \times \text{Received antenna gain}}{\text{Path loss}} \text{W} \qquad (11.14)$$

Path loss of some 200 dB characterizes links between Earth stations and geosynchronous satellites. As a result, high-gain antennas, low-noise receivers, and careful selection of modulation method and bandwidth are required to achieve acceptable signal-to-noise ratios over that long link. Large path loss is a distinctive feature of space communication systems.

Several other loss mechanisms may be present as well. These additional losses are included with the path loss in the previous equation to arrive at the overall received signal power.

Multipath loss refers to the signal being reflected to the receiver off other objects. Since these signals have traveled a greater distance, they arrive after the main signal and may interfere destructively or appear as noise. The most common example of this is the "ghost" on a television screen caused by reflection of the signal from nearby mountains, overflying aircraft, or other large objects.

Faraday rotation is a problem for linearly polarized signals. As the linearly polarized electromagnetic field of the RF signal penetrates the Earth's magnetic field, it is rotated as shown in Figs. 11–13. Any rotation between the polarization direction of the antenna and that of the incoming signal causes a loss that can become very large as the rotation approaches 90 deg, as shown in Fig. 11.13. This problem can be avoided in space applications by using circular polarization. Thus, one often sees helical antennas for HF/VHF/UHF links and cruciform antennas for X band in low- and midgain applications.

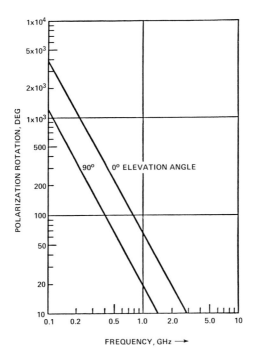

Fig. 11.13   Faraday rotation of polarized waves.

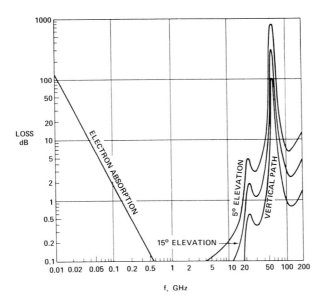

Fig. 11.14   Total absorption loss (excluding weather).

Losses related to atmospheric absorption afflict both high- and low-frequency ends of the scale (Fig. 11.14). Signal loss due to absorption by ionospheric electrons first becomes noticeable at 500–600 MHz and begins to be significant at about 100 MHz as frequency is decreased. Atmospheric absorption causes essentially no loss between about 600 MHz and 4 GHz. This convenient window explains the extensive use of S band in the 2-GHz range for spacecraft communication.

The desire for higher data rate and narrower beamwidth pushes spacecraft designers toward higher frequencies. However, as shown in Fig. 11.13, atmospheric losses rise rapidly with increasing frequency in the higher ranges due to absorption by oxygen and by water vapor. The figure shows data for an average atmosphere. Severe clouds, fog, or rain will greatly increase the loss. The absorption region begins at approximately X band (8 GHz), and systems in this frequency range perform quite satisfactorily except in case of significant rain at the receiving site. Systems operating at Ku band (12–14 GHz) are further into the absorption range but still satisfactory in good weather. Note the significance of elevation angle. This simply reflects that, at lower elevation, the signal must traverse a longer path through the atmosphere. The higher frequency ranges may thus be more useful for geostationary communications satellites, which are usually well above the horizon for typical Earth station locations. Above Ku band, absorption rises rapidly. Dips do occur, however, since the offending molecules absorb at discrete frequencies. The two windows of potential future interest are in the vicinity of 35 and 94 GHz—regions referred to as Ka band and "millimeter wave," respectively. For high elevation angles and good weather these bands may have application, but there is always the risk of data loss due to inconveniently timed bad weather.

Note that most of these concerns do not apply for spacecraft-to-spacecraft communication. This, plus unstable politics on Earth, has generated interest in placing large relay spacecraft in Earth orbit to communicate with planetary spacecraft using K band or higher frequency. This would reduce power requirements and antenna size on the deep space vehicle. Data from the Earth orbiter would then be relayed to the ground using frequency bands better suited to traversing the atmosphere. Cost has prevented pursuit of the idea, but it has much to recommend it.

Noise

The signal power from a communications link is always received in the presence of noise. Given this noise, the problem of receiving and correctly decoding the transmitted signal becomes a problem in estimation theory, a branch of mathematical statistics. The noise degrades the ability of the receiver to estimate the transmitted signal precisely, leading to incorrect bits in a digital signal or, more obviously, overt corruption of an analog signal.

Strictly speaking, noise is any received power that interferes with the desired signal, a definition that includes many things that could be "signals" to other users. Spacecraft communications engineers and radio astronomers have fundamentally differing views about which electromagnetic waves constitute "signal" and which constitute "noise." The principal figure of merit used to specify the quality of a communications link is the ratio of signal power to noise power, the SNR. Interfering signals aside, most noise is of thermal origin, either from blackbody emission as discussed in Chapter 9, or from thermally induced motion of electrons in the receiver circuitry, referred to as Johnson noise. Noise that is not thermally generated will usually be approximated in its effects by assuming that it is thermally induced, as we shall discuss later.

As an aside, it should be noted that the discussion in this section is relevant only for information transmission at conventional radio frequencies. Optical communications links are becoming quite common in space applications, both internally to the spacecraft and between vehicles. Thermal noise is not typically important in optical and electro-optical communications systems, which are dominated by shot noise, quantum noise, or the optical interference background, depending on the wavelength and system characteristics.

Noise sources. Electrons in receiver circuits are thermally agitated when the circuits operate above 0 K. This motion causes a random voltage or current to be induced in all portions of the circuit. Subject to the assumption that we are operating at radio frequencies with circuits of moderate temperature (at least a few tens of Kelvins), the noise power from a given device in a given bandwidth interval may be approximated by

$$P_N = kTB \tag{11.15}$$

where P_N is the noise power (in W), k is the Boltzmann constant (1.38×10^{-23} W-s/K), T is the device temperature, and B is the bandwidth interval.

A few comments are in order. T is referred to as the effective noise temperature of the device and is commonly used as a figure of merit in characterizing noise performance in a communications system. With the agitated electron conceptual model discussed earlier, the total noise power will be due to the collective output of a multitude of individual oscillators. The central limit theorem of statistics thus guarantees the noise to be Gaussian; i.e., the noise power per unit frequency interval at any given instant will have a level drawn from a normal distribution, characterized by a mean N_0 and a standard deviation σ. If N_0 is constant across all frequencies (always an idealization, since this implies infinite total energy content in the noise), then the noise is called "white," by analogy to white light as an equal mixture of all colors. White Gaussian noise (WGN) is the standard assumption in communications link analysis.

As indicated earlier, the WGN assumption cannot ever be strictly correct. Nonetheless, receiving system bandwidths are normally quite small with respect to the frequency scale over which N_0 varies significantly. Thus, the WGN assumption with constant noise power spectral density N_0 given by

$$N_0 = kT = P_N/B \tag{11.16}$$

is normally quite good. The utility of this concept is so great that communications engineers commonly characterize even highly colored (e.g., nonthermal) noise by an equivalent "noise temperature." This would be selected to yield a white noise power level comparable in its effects to that of the actual colored noise.

Noise in the communications link is not limited to that generated in the receiver circuitry. As we discussed in Chapter 9, any object with a temperature above 0 K emits electromagnetic radiation distributed across all wavelengths according to Planck's law, Eq. (9.15) in Chapter 9 (modified,

Table 11.4   Effective noise temperature of various sources

Source	Temperature
Sun	6000 K
Earth	290 K
Galaxy	Negligible above 1 GHz
Sky	30–150 K
Atmosphere	Noise due to absorption and reradiation

of course, by the fact that no surface is ideally "black"). Though blackbody noise is highly colored, it is again modeled in its effect on communications systems by assuming WGN at an equivalent temperature, usually that of the blackbody spectral peak. Thus, the sun typically acts as a noise source at approximately 6000 K. Table 11.4 shows effective noise temperatures for several potential sources. Figure 11.14 depicts the combined effect of common noise sources as a function of frequency and for several elevation angles.

Noise Figure

Another common measure of merit for the noise performance of a device is the noise figure, defined for a "two-port" device as

$$F_n = \frac{\text{Noise output of a real device}}{\text{Noise output of an ideal device with input at temperature } T_0}$$

(11.17)

where $T_0 = 290$ K is the standard reference temperature.

To develop a mathematical expression for noise figure, note that an ideal device adds no additional noise to the signal passing through it. The noise at its output will be the noise at the input, amplified by gain. Thus, the noise figure is the actual noise power output divided by the noise power output due only to the input noise generated by the input termination at temperature T_0. The input noise power P_{N_i} to the device due to the input at temperature T_0 will be kT_0B_n. If P_{N_o} is the actual noise output power from the real device, then the noise figure is

$$F_n = \frac{P_{N_o}}{GP_{N_i}} = \frac{P_{N_o}}{GkT_0B_n} = \frac{P_{N_i}G + \Delta P_N}{GkT_0B_n} = 1 + \frac{\Delta P_N}{GkT_oB_n} \qquad (11.18)$$

where ΔP_N is the noise at the output of the device due to the device itself, and G is the two-port device gain. Note that $F_n \geq 1$; the best two-port noise figure a device can achieve is a value of one.

If we represent the noise added by the device, ΔP_N, as an additive noise source at an effective noise temperature of T_e at the input of the device, then the noise power added by this source has a value of $\Delta P_N = kT_eB_n$, and the noise figure is

$$F_n = 1 + \frac{T_e}{T_0} \qquad (11.19)$$

This gives the relationship between noise temperature and noise figure for a two-port device.

If a two-port device is purely passive, such as a transmission line or attenuator, the effective input noise temperature T_e will be a function of the device's thermodynamic temperature T_t and the available loss L,

$$T_e = T_t(L - 1) \qquad (11.20)$$

The available loss is defined as $L = 1/G$. This equation gives the effective noise temperature of the device referred to the input of the passive device. The effective noise temperature of the passive device referred to the output of the device T_l is

$$T_l = T_t(1 - 1/L) \qquad (11.21)$$

This effective noise source at the output can be referred to the input by multiplying the effective output noise temperature T_l by the available loss of the device:

$$T_e = LT_l = T_l/G \qquad (11.22)$$

The noise figure of a cascade of devices can be calculated from the noise figures and gains of the individual stages. Let F_1, F_2, and F_3 be the two-port noise figures for the first, second, and third stages in a cascade of devices; G_1, G_2, and G_3 be the power gains of the three stages; and T_1, T_2, and T_3 be the effective input noise temperatures of the three stages. It can be shown that

$$F_{123} = F_1 + \frac{F_2 - 1}{G_1} + \frac{F_3 - 1}{G_1 G_2} \qquad (11.23)$$

where F_{123} is the equivalent two-port noise figure of the cascade of the three stages. A similar result can be obtained for the equivalent input noise temperature of the cascade T_{123}:

$$T_{123} = T_1 + \frac{T_2}{G_1} + \frac{T_3}{G_1 G_2}$$ (11.24)

These results can be extended to larger numbers of stages. The technique can be used to calculate the overall effective noise temperature of an entire receiving system.

Noise Figure for a Receiver System

Consider the noise performance of an entire receiver system, including the antenna, with a system noise figure, as diagrammed in Fig. 11.16. The "electronics" portion of the receiver is modeled as a two-port device with noise figure F_n, or, equivalently, noise temperature T_e. Figure 11.17 shows a block diagram of the equivalent model of the system. The receiver model implicitly includes any passive loss components between the antenna and the receiver.

All noise, other than that generated internally by the receiver, is modified by an additive noise source of value $kT_a B_n$. The temperature T_a is called the effective antenna temperature. Note that no additional noise enters the

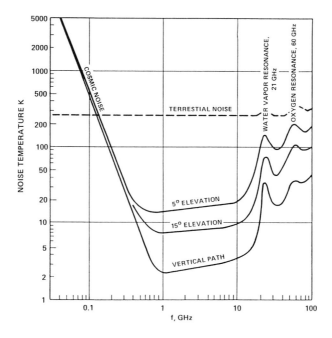

Fig. 11.15   Composite link noise plot (excluding weather).

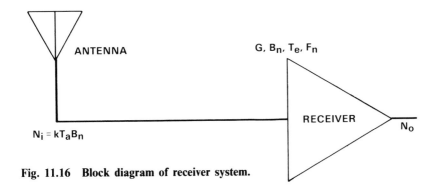

Fig. 11.16 Block diagram of receiver system.

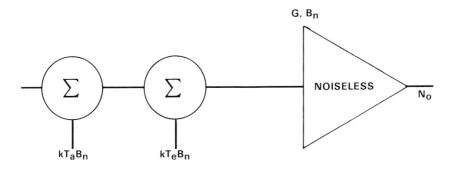

Fig. 11.17 Equivalent model of receiver system.

system through the antenna in this model. All external noise sources, such as galactic noise, atmospheric noise, and warm-body emission, are included in the appropriate selection of the value of T_a. If $T_s = T_a + T_e$ is the effective system noise temperature, then the noise power output from the system will be

$$P_{N_o} = kT_s B_n G \tag{11.25}$$

For the system modeled in Fig. 11.15, the noise figure can be written as

$$F_s = \frac{kB_n G(T_a + T_e)}{kB_n GT_0} \tag{11.26}$$

from which we find

$$F_s T_0 = T_a + T_e = T_s \tag{11.27}$$

$$F_s = \frac{T_a}{T_0} + \frac{T_e}{T_0}$$

$$= \frac{T_a}{T_0} + F_n - 1 \tag{11.28}$$

Note that $F_s \geq 0$. The noise performance of the entire system can be modeled in terms of either T_s, or T_e and T_a, or F_s. Because they provide equivalent results, the choice is primarily one of convenience.

Noise figures and effective noise temperatures for various types of amplifiers appear in common references. However, the reader should note that authors and designers do not always use these same (or even a consistent) set of definitions for these figures of merit. One should always determine the fundamental definitions being used before basing system performance on stated numerical noise-performance figures.

The total noise power, derived from the specified system noise temperature (or noise figure), can be combined with the received signal power to yield a communication system's SNR. From Eqs. (11.13) and (11.15).

$$\text{SNR} = (1/4\pi)^2 (1/k)(\lambda^2/BR^2)G_t P_t (G_r/T) \tag{11.29}$$

From the point of view of a receiving ground station, all parameters in Eq. (11.29) are fixed by nature or by the spacecraft except for the term (G_r/T). This term, the ratio of the receiving antenna gain and system noise temperature, essentially defines the quality of the communications link with

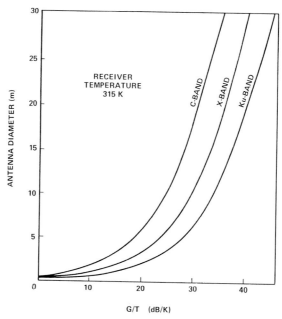

Fig. 11.18 System G/T vs antenna aperture.

a given spacecraft. Figure 11.18 gives system G/T vs antenna aperture for several frequency bands under the assumption of a typical system noise temperature of 315 K.

Communications in noise. As we have indicated, the SNR is the fundamental quantity characterizing the quality of a communications link. This is because the data rate, or *channel capacity*, of the link is directly related to the SNR. The fundamental theory of communication in the presence of noise was developed by Shannon;[3] all subsequent work in the field is an extension of his efforts. According to Shannon's theorem, the error-free channel capacity of a link is given by

$$C = B \log_2(1 + \text{SNR}) \tag{11.30}$$

where C is the channel capacity in bits/second, and B is the link bandwidth in hertz. Thus, if we consider a standard C-band communications satellite transponder with a nominal bandwidth of 36 MHz and assume SNR = 15, the theoretical channel capacity is $C = 144$ Mbps. In actual practice, such a transponder could handle one color television signal or 1200 voice links.[4] Each voice channel is assigned a nominal 56 kbps data rate, adequate for normal speech, and so the link carries a total data rate of 67 Mbps. This is less than 50% of Shannon's error-free limit, and the satellite link will contain some bit errors, typically on the order of one in a million.

This simple example illustrates some important points with respect to Shannon's theorem. The theorem tells us what the limit is, but not how to reach it. Today's satellite communications links employ very sophisticated coding techniques to achieve their capacity. Presumably even more complex techniques can be found to increase channel capacity, but Shannon's theorem gives no clue. Equation (11.30) gives the limit for error-free transmission, but says nothing about how system performance, i.e., bit error rate, degrades as the limit is exceeded. Real systems degrade as shown in Fig. 11.19, which plots bit error rate as a function of, essentially, SNR.

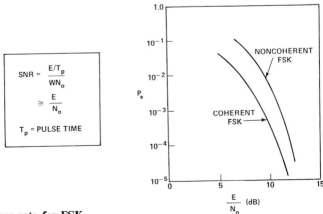

Fig. 11.19   Bit error rate for FSK.

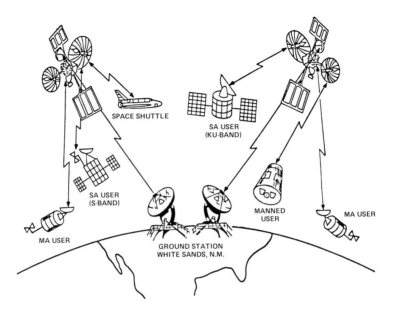

Fig. 11.20   TDRSS concept.

The analysis leading to this result is beyond the scope of this text, but is found in standard references on digital communication theory.[5]

11.6 TRACKING

Tracking stations fulfill three major functions: receive spacecraft telemetry and route it to the control center for processing and distribution; uplink commands from the control center to the spacecraft; and obtain Doppler range-rate data and spacecraft azimuth and elevation for orbit determination. Not all stations necessarily have all these functions. Some may perform tracking or data reception only.

Uplinks and downlinks are supported by various stations in the following frequency bands: S band, 2–4 GHz; C band, 4–6 GHz; X band, 7–9 GHz; Ku band, 12–14 GHz; and Ka band, 20–30 GHz.

United States government owned and operated tracking systems include the following:

1) GSFC STDN, the NASA Goddard Space Flight Center Space Tracking and Data Network;

2) JPL DSN, the NASA Jet Propulsion Laboratory Deep Space Network;

3) TDRSS, the NASA Tracking and Data Relay Satellite System;

4) DMA TRANET, the Defense Mapping Agency Transit Network;

5) USAF SGLS, the USAF Space Ground Link System;

6) USN NAG, the USN Naval Astronautics Group; and

7) NORAD SPADATS, the North American Air Defense Command Space Data Acquisition and Tracking System.

The TDRSS consists of two large satellites in geostationary orbit that relay between low-orbit satellites and the ground, thus eliminating the problem of long communication gaps. This system will eventually replace GSFC STDN. Unfortunately, TDRSS is not well suited to support small low-cost satellites that cannot carry the rather cumbersome equipment necessary. However, it is admirably suited to support the Space Shuttle and very large satellites such as the Hubble Space Telescope. Figure 11.20 shows the TDRSS concept and Fig. 11.21 the coverage patterns of the TDRSS satellites. For comparison Figs. 11.22 and 11.23 show the coverage by the worldwide STDN network in S band and C band, respectively. The JPL DSN specializes in missions to the moon and beyond. The others listed are operated by branches of the military for their programs. The choice of a given system will usually be driven by the sponsor and the mission. The various systems are generally not compatible with one another.

Tracking stations are used to supply position and velocity data needed for orbit determination algorithms discussed in Chapter 4 and in common references such as Bate et al.[6] or Battin.[7] The accuracy of the resulting orbit and the ability to propagate the estimated orbit forward in time are ultimately limited by the tracking precision. For this reason it is instructive to consider state-of-the-art tracking system precision with respect to position and velocity determination and prediction.

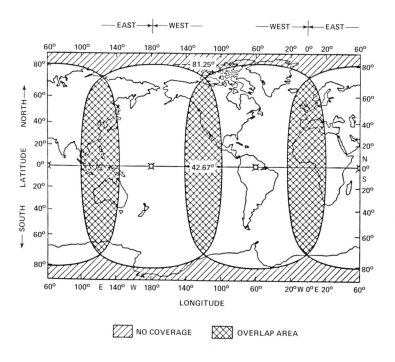

Fig. 11.21   Geosynchronous ground coverage, 0° elevation.

28.5° INCLINATION
400 km CIRCULAR ORBIT
5° ELEVATION

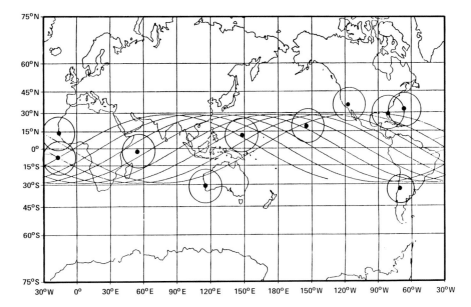

Fig. 11.22   STDN S-band coverage.

28.5° INCLINATION
400 km CIRCULAR ORBIT
5° ELEVATION

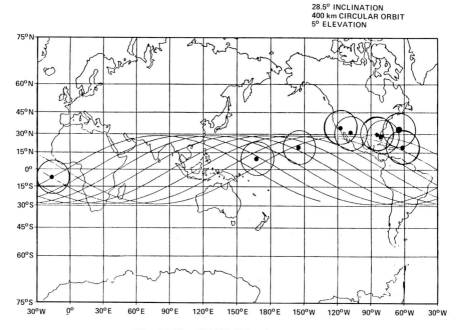

Fig. 11.23   STDN C-band coverage.

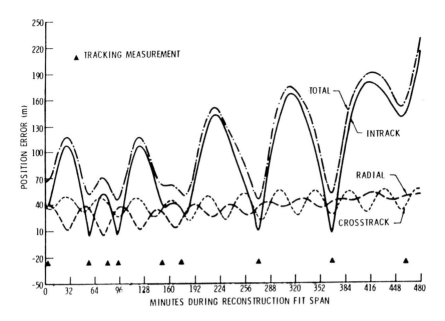

Fig. 11.24   Position reconstruction error estimate.

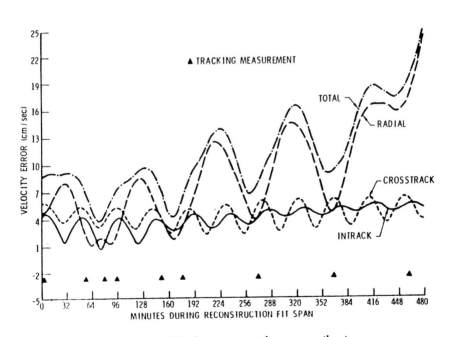

Fig. 11.25   Velocity reconstruction error estimate.

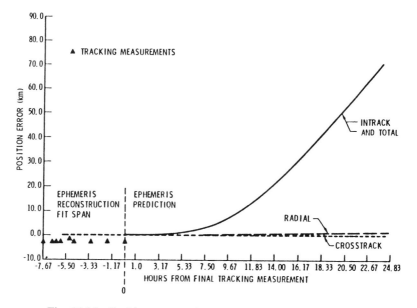

Fig. 11.26   Position error estimates reconstructed and predicted.

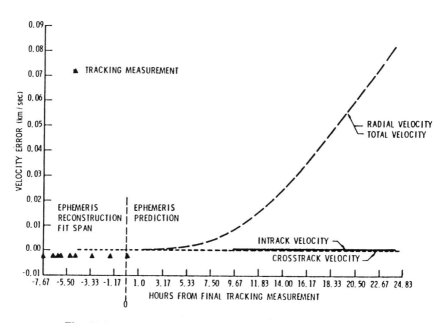

Fig. 11.27   Velocity error estimates reconstructed and predicted.

Table 11.5   Tracking accuracy at Kwajalein Missile Range[a]

| Sensor | Accuracy (bias ± RMS noise) | |
	Range (m)	Angle (μrad)
ALCOR (BCM)	1.5 ± 0.3	55 ± 50
(SKIN)	0.7 ± 0.003	55 ± 165
MMW	0.3	70
TRADEX	1.7 ± 0.004	75 ± 75
ALTAIR	2.0 ± 0.5	175 ± 175

[a]These numbers are for a "good target" (conical, symmetrical). These are range and angle errors at pierce point. Position errors can be derived from these numbers but are trajectory dependent.

Figure 11.24 shows a typical position estimate for a low Earth orbiting satellite as a function of time as *reconstructed* from a tracking data file. The emphasis on reconstruction is deliberate; note that the actual measurements are made at sporadically indicated times. These offer the greatest accuracy, on the order of 10 m for all three velocity components (selected here to be radial, in track, and cross track). In between the actual measurements, the ability of interpolation algorithms to specify precisely the spacecraft location grows poorer, reaching a maximum halfway between data points. Figure 11.25 shows similar behavior for reconstructed velocity.

Figures 11.26 and 11.27 depict the ephemeris error growth in the hours following cessation of measurements. Note the rapid growth of intrack position error in Fig. 11.26. This is due primarily to the effect of atmospheric drag, which, as discussed in Chapter 4, is extremely difficult to quantify.

Probably the best single radar tracking facility available to the United States is at the Kwajalein Missile Range (KMR). Table 11.5 describes the capability of the various KMR radars for targets with and without beacons. Although these particular systems may not be available or of interest for a given mission, they are presented here to give the reader an idea of the accuracy to be expected from a good tracking radar.

References

[1]Fthenakis, E., *Manual of Satellite Communications*, McGraw-Hill, New York, 1984.

[2]Skolnik, M. I. (ed.), *Radar Handbook*, McGraw-Hill, New York, 1970.

[3]Shannon, C. E., "A Mathematical Theory of Communications," *Bell System Technical Journal*, Vol. 27, 1948, pp. 379–423 and 623–651.

[4]Martin, J., *Communications Satellite Systems*, McGraw-Hill, New York, 1978.

[5]Viterbi, A. J., and Omura, J. K., *Digital Communications and Coding*, McGraw-Hill, New York, 1979.

[6]Bate, R. R., Mueller, D. D., and White, J. E., *Fundamentals of Astrodynamics*, Dover, New York, 1971.

[7]Battin, R. H., *An Introduction to the Mathematics and Methods of Astrodynamics*, AIAA Education Series, AIAA, Washington, DC, 1987.

[8]Meer, D. E., "Noise Figures," *IEEE Transactions on Education*, Vol. 32, May 1989, pp. 66–72.

INDEX

Ablative cooling, 257
Absorption, 444
Absorptivity, 381, 382
Accommodation coefficients, 128
Acoustic loads, 46, 357
Active control, 301
Active thermal control, 372
Aerobraking, 249, 271
Aerocapture, 249, 270
Aerodynamic drag, 33, 64
Aerodynamic load, 357
Aerodynamic noise, 45
Aerodynamic torque, 292
Aliasing, 429
Alkali metal thermal-to-electric conversion
    (AMTEC), 418
All-up test, 16
Aluminum, 37, 365
Aluminum-lithium, 366
Analog-to-digital conversion (ADC), 427
Angle factor, 385
Angular momentum, 90, 106, 286
Antennas, 436
Antisatellite (ASAT), 81
Apogee, 107
Apollo, 16, 27, 157, 164
Apollo program, 65
Argument of periapsis, 101
Ariane, 163
Articulating platforms, 349
Arc suppression, 400
Ascending node Ω, 107, 109
Ascent trajectories, 192
Asteroid materials, 37, 38
Asteroid missions, 33
Asteroids, 30
Astrodynamics, 85
Astronomical, 328
Atlas, 163, 190
Atlas II, 216
Atlas-Centaur, 212
Atmosphere models, 122, 233
Atmospheric absorption, 444
Atmospheric density, 124
Atmospheric drag, 120, 124
Atmospheric entry, 147
Atmospheric Explorer, 19
Atmospheric losses, 444
Attitude, 276

Attitude control, 277
Attitude determination, 277
Attitude matrix, 312

B-plane, 96, 248
Ballistic, 237
Ballistic coefficient, 127, 235, 236
Ballistic entry, 237, 239, 240
Bandwidth, 278
Bank angle, 251
Barges, 45
Batteries, 11, 395
Beamwidth, 23, 436, 438
Beryllium, 366
Betacloth, 368
Binary amplitude-shift keying (BASK),
    435
Binary frequency-shift keying (BFSK), 436
Binary phase-shift keying (BPSK), 436
Bit error rate, 451
Blackbody, 129
Blackbody noise, 446
Block diagram, 8
Blunt body, 260
Boundary-layer cooling, 175
Boundary-layer theory, 263
Broken plane maneuvers, 135
Brown dwarf, 29
Bubble memory, 433
Bunnysuits, 43

Calendar time, 111
Carbonaceous chondrite, 38
Carbon-carbon, 367
Carbon monoxide, 39
Catalytic wall, 266
Celestial mechanics, 85
Celestial sphere, 282
Centaur, 29, 178, 216, 223
Central force, 88
Central limit theorem, 280, 445
Challenger accident, 29
Chamber pressure, 181
Channel capacity, 451
Characteristic equation, 303, 310
Circular velocity, 91
Clean room, 42
Clohessy-Wiltshire equations, 156

459

CMOS, 68
Code-division multiplexing (CDM), 429
Collisions, 23
Colored noise, 446
Combustion cycles, 176
Comets, 30
Command subsystem, 422
Communications satellites, 23, 24, 69, 325
Commutation, 430
Compensator, 302
Concentrator array, 409
Concentric flight plan, 157
Conduction, 82, 371, 394
Configuration, 385
Configuration design, 325
Continuity, 401
Control law, 302
Control moment gyros (CMG), 307
Convection, 66, 371, 374, 377
Convective heating, 378
Convective heat flux, 378
Conversion efficiency, 412
Coplanar transfers, 135
Corona discharge, 329
Corrosion, 42
Cosmic ray, 68
Cosmos, 36, 954
COSPAR International Reference
    Atmosphere, 233
CO_2 band, 314
Crew morale, 36
Critical charge, 68
Cross-range maneuvers, 249
Cryostat, 373

Damping ratio, 310
Data rate, 443, 451
Debris cloud, 80
Deep Space Network (DSN), 26
Defense Meteorological Satellite Program
    (DMSP), 342
Delrin, 74
Delta, 78
Deployable booms, 348
Depth of discharge, 401
Diffuse surface, 385
Direct current switching, 400
Discoverer, 240
Disturbance Compensation System
    (DISCOS), 120
Docking, 152
Doppler shift, 421
Drag, 124
Drag coefficient, 127
Drag loss, 192
DSN, 452, 453
Dual spin, 337
Dual-spin configuration, 306

Dual-spin satellite, 291, 340
Dual-spin spacecraft, 306
Dump cooling, 175
Dust, 42

Earth observation, 18
Earth observation satellites, 326
Earth resources satellites, 327
Eccentric anomaly, 92
Eccentricity, 88, 100
Effective isotropic radiated power, 441
Electric propulsion, 10, 35
Elliptic orbits, 91
Emissivity, 381, 382
Encke, 30
Energy balance, 81, 392
Energy sink, 291
Engine cooling, 174
Entry corridor, 248
Entry heating, 255, 378
Entry heating rate, 261
Entry heat load, 260
Entry vehicle testing, 17
Ephemeris time, 112
Epoch, 105
Equilibrium glide, 243
Equilibrium gliding entry, 261
Error signal, 301
Escape velocity, 91
Euler angles, 282
Euler equations, 288
Euler parameter, 284
Exhaust velocity, 165
Expansion ratio, 166
Expansion-deflection (ED) nozzles, 167
Explorer, 1, 290
Extendable exit cones (EECs), 170
External tank (ET), 189
Extraterrestrial materials, 37
Extravehicular activity (EVA), 67
Extravehicular operations, 67

F-1 engine, 176, 177
Faraday rotation, 442
Fiberglass, 366
Fiber-optic gyros, 318
Fields and particles, 328
Figure control, 440
Film coefficient, 378, 380
Fins, 372
Fixed-head star trackers, 316
Flat plate, 259, 263
Flat spin, 290
Flight acceptance (FA) criteria, 332
Flight tests, 15
Flight-path angle, 107
Fltsatcom, 339

Fly-by-wire, 24
Fourier's law, 375
Fracture mechanics, 359
Free convection, 82
Frequency-division multiplexing (FDM), 428
Fuel cells, 11, 395, 415
Functional requirements, 7, 8

G/T, 451
Galileo, 29, 240
Galileo mission, 31, 32
Gallium arsenide, 412
GCI, 106
Gemini, 157, 219
Geocentric inertial (GCI) system, 102
Geopotential altitude, 186
Geostationary, 22
Geostationary orbital slots, 23
Geosynchronous Earth Orbit (GEO), 22
Geosynchronous orbit, 80
Geosynchronous satellites, 442
Geosynchronous spacecraft, 406
Giotto, 77
Glaser, Peter, 33
Glide, 237
Gliding entry, 242
Graphite-epoxy, 82, 367
Gravitational potential, 117
Gravity-assist trajectories, 148
Gravity-gradient attitude control, 300
Gravity-gradient force, 64
Gravity-gradient stabilization, 299
Gravity-gradient torque, 293
Gravity assist, 31
Gravity loss, 193
Gray surface, 385
Greenwich Mean Time (GMT), 112
Grounding, 401
Ground handling, 41
Ground strap, 43
Ground transportation, 44
Ground-Based Electro-Optical Deep Space Surveillance (GEODSS), 109
Gyroscope, 318

Halley, 30
Halley's Comet, 30
HCI, 106
Heaters, 372
Heating rate, 255
Heat flux to the wall, 257
Heat load, 255
Heat pipes, 372, 418
Heat sinking, 256

Heliocentric inertial (HCI) system, 103
Hemispherical-resonator gyroscopes, 318
Hermetically sealing, 82
High Energy Astronomical Observations Spacecraft (HEAO-2), 20
Hill equations, 156
Hohmann orbit, 31
Hohmann transfer, 31, 139
Horizon scanners, 313
Horizon sensors, 314
HOTOL, 245
HS, 340, 376
HST, 308
Hubble Space Telescope, 20, 45
Hybrid propulsion, 198
Hyperbolic orbits, 94
Hyperbolic passage, 94
Hyperbolic velocity, 96

Impulsive transfers, 130
Inclination, 100
Inconel, 366
Inertial upper stage (IUS), 225
Inertia matrix, 286
Infrared astronomy satellite (IRAS), 373
Insulation blankets, 372
International atomic time, 111
International Cometary Explorer, 150
International Infrared Astronomical Telescope (IRAS) spacecraft, 20
International Sun-Earth Explorer (ISEE), 85, 327
International Ultraviolet Explorer (IUE) observatory satellite, 25
Interplanetary transfer, 143
Invar, 82
Isotropic radiator, 438
IUS, 29

J-2 engine, 176
Jitter, 278
Johnson noise, 445
Julian dates, 112
Jupiter, 29

Kapton, 54, 368
Kepler's equation, 92, 94, 98, 99
Kepler's law, 85
Keplerian orbital elements, 100
Kevlar, 74
Kick motors, 198
Kinetic energy, 90, 287
Kirchoff's law, 386
Kwajalein Missile Range (KMR), 457

L/D, 235, 236
Lagrangian points, 85, 116
Lambert problem, 138
Lambert's cosine law, 385
Lambertian reflector, 386
Laminar flow clean room, 42
Laplace, 114
Large space structures, 33
Large structures, 363
Latchup, 423
Lateral vibration, 49
Launch environment, 45
Launch site, 108
Launch vehicle tests, 18
Law of universal gravitation, 86
Librations, 300
Line of apsides, 119
Lithium batteries, 402
Load alleviation, 359
Loads analysis, 359
Loh's solution, 252, 254
Longitudinal vibration, 49
Low Earth Orbit (LEO), 15
Low-thrust planetary trajectories, 32
Luna 3, 9, 27
Lunar crust, 37
Lunar module descent engine, 199
Lunar Orbiter, 27
Lunar roving vehicle, 65
Lunar transfer, 150
Lunokhod, 27

Magnesium, 365
Magnetic field, 69
Magnetic torque, 295, 307
Magnetometer, 315
MAGSAT, 300
Major axis rule, 290
Major axis spin rule, 307
Mariner 10, 27, 31, 300
Mars Orbiter, 411
Mars Sample Return, 32
Mascons, 120
Mass fraction, 188
Mass properties, 351
Mass ratio, 187
Material selection, 50
Mean anomaly, 92
Mercury Orbiter, 129
Mercury spacecraft, 237, 240, 256
Metal matrix composites, 366
Meteor bumper, 76
Methane, 39
Method of patched conics, 142
Microgravity, 61
Micrometeoroid flux, 74
Micrometeoroids, 74
Mir, 21

Modularity, 401
Modulation, 435
Molniya, 22
Molniya communications spacecraft, 326
Molybdenum, 366
Moment-of-inertia ratio, 355
Momentum-bias system, 306
Momentum wheels, 306
Moment of inertia, 354
Multidiscipline team, 4
Multipath loss, 442
Mylar, 367

NASA's Deep Space Network (DSN), 421
Natural frequency, 310
Navigation satellites, 18
Neptune, 29
Newton's law of cooling, 378
Newtonian flow, 127
Nickel-cadmium (nicad) batteries, 402
Nickel-hydrogen (NiH$_2$), 403
Niobium, 366
NMOS, 68
Noise, 445
Noise figure, 446, 448
Noise power, 445
Noise temperature, 445, 446
Non-Keplerian motion, 114
NORAD, 452
Nozzle contour, 173
Nozzle expansion, 166
Nuclear power, 12, 35
Nuclear reactors, 30
Nuclear waste, 40
Nutation, 289
Nutation dampers, 297
Nyquist criterion, 430
Nyquist rate, 429

Oblateness, 118
Onboard computer, 432
Onboard storage, 433
Optical disks, 435
Orbit design, 30
Orbit determination, 109, 138
Orbital angular momentum, 91, 97
Orbital debris, 77
Orbital elements, 99, 105
Orbital inclination, 18
Orbital lifetime, 127
Orbital maneuvers, 130
Orbital precession, 18
Orbiting Solar Observatory, 19
Orbiting Deep Space Relay Satellite
  (ODSRS), 26
Outgas, 52
Outgassing, 52
Oxygen, 37

Parabolic antennas, 438
Parabolic dish, 438, 440
Parabolic orbits, 98
Parallax, 276
Passband, 278
Passive thermal control, 371
Path loss, 442
Payload Assist Modules (PAM-A and PAM-D), 226
Payload ratio, 188
Payload specialist, 21
Periapsis, 88
Periapsis passage, 101
Perigee, 107
Perturbation methods, 151
Perturbation theory, 151
Perturbations, 101
Pioneer, 27
Pioneers 10 and 11, 28, 82, 298
Pioneer Venus, 240
Pitch program, 192
Planck's law, 383, 446
Plane changes, 131
Planetary entry, 18
Planetary observation, 328
Plug nozzles, 167
PMOS, 68
Pogo, 49
Pointing, 279
Polarizations, 440
Potential energy, 90
Power spectral density, 445
Primary, 402
Primary battery, 402
Principal axis, 287
Progress, 21
Propellant manufacturing, 38
Pulse-code modulation (PCM), 435

Radiation, 67, 371
Radiation cooling, 174
Radiation effects, 67
Radiation pulses, 25
Radiation surface coefficient, 387
Radiative cooling, 256
Radioisotope thermoelectric generators (RTGs), 12, 29, 331
Random vibration, 47
Ranger, 27
Reaction jets, 308
Reaction wheels, 303
Reconditioning, 406
Redundancy, 423
Refractory metals, 366
Regenerative cooling, 174
Regeneratively cooled rocket engines, 378
Rendezvous, 152
Retrograde orbit, 18

Right ascension of ascending node, 101
Rigid body dynamics, 288
Ring-laser gyros, 318
RL-10, 178
Robotics, 21
Rocket performance, 186
Roll-up solar arrays, 408, 409
RTG efficiency, 414

Safety factor, 360
Salyut, 21, 34
Satellite geodesy, 119, 120
Satellite lifetimes, 122
Saturn, 29
Saturn 1, 17
Saturn 5, 16, 45, 177
Scale height, 234
Science working group, 6
Seasat, 19
Secondary, 402
Secondary propulsion, 67
Semi-latus rectum, 100
Semimajor axis, 100, 107
SGLS, 452
Shannon's theorem, 451
Shielding, 71
Shield continuity, 401
Shock, 46
Shock spectra, 46
Shock transients, 45
Shuttle external tank, 45
Shuttle Orbital Maneuvering System (OMS), 206
Sidereal time, 113
Signal conditioning, 427
Silicon cells, 412
Silver-zinc batteries, 402
Similar-stage approximation, 190
Single-axis attitude determination, 311
Single-event upsets, 67, 68
Single-impulse transfer, 135
Skin friction coefficient, 259, 263, 266
Skip entry, 237, 245, 246, 247, 261
Skylab, 19, 21, 34
SNAP-10A, 396
SNR, 445, 450, 451
Soft-error, 423
Solar-electric propulsion, 33
Solar arrays, 348, 407
Solar cells, 407
Solar dynamic systems, 331
Solar flares, 72
Solar flux, 124
Solar Max, 81
Solar maximum mission, 54
Solar Mesosphere Explorer, 46, 362
Solar observation, 327
Solar photovoltaic systems, 330

Solar power satellites, 33
Solar pressure, 33
Solar radiation, 129
Solar radiation pressure, 64, 128, 129
Solar radiation pressure torque, 294
Solar sails, 33, 129
Solar wind, 130
Solid-state memory, 433
Solid Rocket Boosters (SRBs), 190
Soyuz, 395
SP-100, 417
Space colonies, 36
Space environment, 49
Space observation, 19
Space Shuttle, 16, 152, 163
Space Shuttle entry profile, 242
Space Shuttle main engine (SSME), 165, 167, 176
Space Shuttle payload accommodations, 201
Space stations, 34
Space systems engineering, 2
Space-processing, 21
Spacecraft configuration, 72
Spacecraft environment, 41
Spacecraft power, 395
Spacecraft propulsion, 9
Specific impulse, 165, 166
Sphere of influence, 95, 114
Spike nozzles, 167
Spin stabilization, 297
Spot shielding, 72
Spray cooling, 175
Staging, 186
Stagnation point, 256, 266
Stagnation point heating, 264, 267
Stationkeeping, 158
Static electricity, 43
Star trackers, 308, 316
STDN, 452, 453
Steel, 365
Steering loss, 193
Stefan-Boltzmann law, 381
Sticking friction, 305
Strapdown platform, 318
Structural coefficient, 188
Structural materials, 365
Sun-synchronous orbits, 119
Supersonic nozzle, 164
Surface contact, 231
Surveyor, 27, 164
Sun sensors, 312
Sun shields, 372
Synchronous orbit, 22
Systems engineering, 2, 337
Systems engineering requirements, 3
S4 upper stages, 178

Tantalum, 366
Tape recorders, 433

TDRSS, 453
Technology tradeoffs, 12
Teflon, 74, 368
Telemetry subsystem, 427
Television and Infrared Observation Satellite (Tiros), 19
Temperature, 81
Temperature gradient, 81, 255
Thermal control, 81, 371
Thermal environment, 391
Thermal protection, 256
Thermal resistance, 389
Thermal stress, 358
Thermal vacuum, 54
Thermionics, 416
Thermoelectric cooling, 373
Thermoelectrics, 416
Three-axis attitude determination, 311, 315
Three-body problem, 116
Throttling, 197, 198
Throughput, 432
Thrust equation, 164
Thrust loss, 193
Thrust vector control, 184, 197
Time-division multiplexing (TDM), 429
Timekeeping systems, 110
Time-of-flight problem, 138
Titanium, 37, 365
Titan 3, 190, 219
Torque, 287
Total dose, 67
Tracking, 279
Tracking and Data Relay Satellite (TDRS), 433
Tradeoff analysis, 9
Tradeoffs, 6
Trajectory design, 138, 142
Trajectory optimization, 192
Transfer function, 303
Transfer orbit stage, 227
Transformation matrix, 283
Transpiration cooling, 175
True anomaly, 88
TTL, 68
Tungsten, 366
Two-body motion, 86
Two-impulse transfer, 137
Type acceptance (TA) criteria, 332

U. S. Standard Atmosphere, 59, 233
Universal formulas, 108
Universal time (UT), 112
Uranus, 29
USAF Space Command, 77

Vacuum, 50
Vanguard, 119

Van Allen radiation belts, 69, 72, 398
Venera, 27
Vernal equinox, 102
Vibration, 45, 357
View factor, 385
Viking, 27, 32
Viking Orbiter, 14, 74
Vis-viva, 90
Voshkod, 240
Vostok, 240
Voyager, 334
Voyager 1, 29
Voyagers 1 and 2, 28, 82
Voyager spacecraft, 14

Wall heat flux, 265, 266
Water vapor, 54
Weather satellites, 23, 25, 327
Weightlessness, 61, 65
White Gaussian Noise (WGN), 281, 445
Wien's displacement law, 383
Wind gust, 45
World Administrative Radio Conference
  (WARC-79), 23

Zonal harmonic, 118
Zond, 246